Renewable Energy Sources towards a Zero-Emission Economy

Renewable Energy Sources towards a Zero-Emission Economy

Guest Editor

Piotr F. Borowski

MDPI

Basel • Beijing • Wuhan • Barcelona • Belgrade • Novi Sad • Cluj • Manchester

Guest Editor
Piotr F. Borowski
Faculty of Business and
International Relations
Vistula University
Warsaw
Poland

Editorial Office
MDPI AG
Grosspeteranlage 5
4052 Basel, Switzerland

This is a reprint of the Special Issue, published open access by the journal *Energies* (ISSN 1996-1073), freely accessible at: https://www.mdpi.com/journal/energies/special_issues/ESWU6524AP.

For citation purposes, cite each article independently as indicated on the article page online and as indicated below:

Lastname, A.A.; Lastname, B.B. Article Title. *Journal Name* **Year**, *Volume Number*, Page Range.

ISBN 978-3-7258-6234-4 (Hbk)
ISBN 978-3-7258-6235-1 (PDF)
https://doi.org/10.3390/books978-3-7258-6235-1

Contents

About the Editor

Piotr F. Borowski

Piotr F. Borowski, DSc, Eng., Professor at the Faculty of Business and Management, Chairman of the Scientific Council for the Management and Quality Science Discipline, and former Vice-Rector for Science, has been included in the prestigious Top 2% list for the fourth time. This distinction confirms his outstanding achievements in business management, particularly in the energy sector. In the context of the zero-emission economy, his research contributes to shaping the global approach to energy transformation, highlighting adaptability as a key development strategy for modern energy enterprises. Professor Borowski serves as a full-time professor at Vistula University (Faculty of Business and International Relations) and at the Nicoulaus Copernicus Superior School. He earned his engineering degree from the Faculty of Power and Aeronautical Engineering at the Warsaw University of Technology (1994) and completed his PhD at the Warsaw School of Economics (2005). His research focuses on management in the energy sector, the zero-emission economy, green innovations, and organizational development. He is the author of more than 150 scientific publications, including over 38 indexed in the Web of Science and Scopus databases. He has extensive international research experience in countries such as Guinea, Ethiopia, and Egypt, and has completed research fellowships at St. John's University (New York) and the University of St. Thomas (Addis Ababa). He has also lectured at universities in Viterbo (Italy), Faro and the Azores (Portugal), and Rouen (France). His scientific achievements have earned him numerous awards, including the Bronze Cross of Merit, the Medal for Long Service, and the Gold Medal Sapientia et Veritas for contributions to Polish science. Professor Borowski is also the author of numerous publications on ethical and servant leadership and on the integration of moral and spiritual values in contemporary organizations.

energies

MDPI

Editorial

Renewable Energy Sources Towards a Zero-Emission Economy

Piotr F. Borowski [1,2]

[1] Faculty of Business and International Relation, Vistula University, 3 Stokłosy Str., 02-787 Warsaw, Poland; pborowski@autograf.pl or p.borowski@vistula.edu.pl
[2] Department of Economics and Management, Khazar University, 41 Mahsati Str., AZ1096 Baku, Azerbaijan

1. Introduction

Transforming the economy towards climate neutrality is one of the greatest challenges of our time [1]. Achieving zero emissions requires profound changes not only in the energy sector but also in industry, agriculture, and entire socioeconomic systems [2]. Renewable sources—such as wind, solar, hydropower, and biomass—are the foundation of this objective, and their growing importance is reflected in both scientific research and public policy.

Renewable energy sources (RESs) are a key pillar of the contemporary energy transformation, essential to achieving global sustainable development goals and reducing greenhouse gas emissions [3]. Faced with intensifying climate change and growing energy demand, the global economy increasingly requires stable and sustainable energy resources. This, in turn, necessitates the search for new, efficient, and environmentally friendly green energy solutions [4].

Transitioning to zero-emission energy systems is not only a technological challenge but also a social, economic, and organizational one. Therefore, research in this area must cover not only the issues of technological efficiency and their impact on the environment but also management models and potential integration with existing energy infrastructure [5].

This Special Issue focuses on these challenges, bringing together studies that explore innovative pathways towards a climate-neutral economy.

2. The Importance of Renewable Energy Sources

The development of renewable energy sources, such as photovoltaics, wind power, biomass, and hydropower, is the foundation for building a climate-neutral economy [6]. In the context of global energy transformation, regional aspects cannot be overlooked. These are effectively addressed, for example, by research on the progress of renewable energy deployment in different countries and continents. These studies show that even in developing countries, the implementation of renewables contributes to improved access to energy, job creation, and the stimulation of technological innovation. At the same time, they highlight barriers related to financing and the need for appropriate energy policy design.

Photovoltaics is currently one of the fastest-growing segments of renewable energy. Technological advances make it possible to reduce solar panels' production costs, increase their efficiency, and integrate them with modern energy storage systems [7]. Similarly, wind energy, both onshore and offshore, is becoming increasingly competitive compared to traditional sources, and its development enhances energy security.

However, it should be emphasized that the rapid expansion of photovoltaics also entails environmental challenges [8]. The production of panels requires significant energy input and the use of raw materials such as silicon, silver, and rare earth metals, the extraction and processing of which can create additional environmental burdens. Another critical issue is the disposal of used panels. Although panels' lifespans are usually 20–30 years, in the coming decades, the problem of growing volumes of photovoltaic waste will emerge. The lack of effective recycling systems may negatively affect the environment through the emission of toxic substances and increased pressure on landfills. Therefore, research on renewable energy sources increasingly focuses on developing technologies for material recovery from panels and creating closed-loop product life cycles in line with the principles of a circular economy [9].

Similar environmental challenges apply to wind energy. Although it is considered one of the cleanest energy sources, the construction of wind farms requires the installation of vast foundations that interfere with soil and aquatic ecosystems. In addition, the production of turbine components, such as blades and towers, requires large amounts of steel, concrete, and composites. The extraction and processing of these materials also generate emissions and may negatively impact the environment. Particularly problematic is the recycling of turbine blades made of fiberglass or carbon fiber, whose durability is an advantage during operation but becomes a challenge at the end of their life cycle [10]. For this reason, studies on wind energy increasingly cover not only improvements in turbine efficiency but also the development of more sustainable construction materials and methods for disposal and recycling [11].

Biomass and hydropower, in turn, stabilize the energy system, enabling flexible responses to fluctuations in wind and solar power production [12]. However, their responsible use, with respect for ecosystems and local environments, is crucial [13].

3. Carbon Capture and Storage as a Component of Transformation

Although renewable energy sources are the foundation of a zero-emission economy, they alone are insufficient to completely neutralize greenhouse gas emissions. In many sectors—such as heavy industry, cement production, and steelmaking—technological processes generate emissions that are difficult to avoid. For this reason, carbon capture and storage (CCS) technologies are playing an increasingly important role [14].

CCS systems involve separating CO_2 from industrial processes and fuel combustion, followed by its safe transport and storage in geological formations, including under the seabed, which allows natural geological structures to be used as long-term gas reservoirs [15,16]. Combined with renewables, these technologies can significantly accelerate the achievement of climate neutrality. It is also worth highlighting the development of solutions based on direct air capture (DAC), which may become an effective complement to the energy–climate relationship in the future, enabling emission reductions in hard-to-decarbonize sectors [17].

4. Management Perspectives and Innovative Business Models

The energy transition is not only a matter of technology but also of new business models and corporate management strategies. Companies operating in the energy sector, as well as in related industries, must adapt to dynamically changing market, regulatory, and social conditions. Innovative business models based on the concepts of the circular economy, resource sharing, and the creation of energy ecosystems provide a response to these challenges [18].

In practice, this involves the development of services such as energy prosumerism, where consumers simultaneously become producers of energy, and "energy-as-a-service" models, which enable clients to benefit from renewable energy without the need to invest in infrastructure. Another important direction is the expansion of public–private partnerships, which allow for more effective financing and the implementation of innovative technologies.

From a corporate management perspective, it is also crucial to implement tools for assessing climate and environmental risks, enabling decision-making aligned with long-term sustainable development goals. Companies must incorporate ESG (environmental, social, and governance) criteria into their strategies; this will not only support emission reductions but also build competitive advantages in the global market [19].

5. Research Gaps and the Need for Technological Integration

Research on renewable energy sources has traditionally focused on improving energy efficiency and optimizing technologies. However, a key research gap concerns the integration of various renewable sources with CCS systems and with energy and industrial infrastructure. Developing comprehensive models that incorporate not only technical aspects but also environmental, social, and economic dimensions is essential.

Another critical issue is the development of smart grids, which allow for dynamic energy flow management, the integration of distributed sources, and the adjustment of supply for demand [20,21]. In this context, research should concentrate on synergies between different technologies, the creation of hybrid systems, and the use of artificial intelligence and big data to optimize grid performance.

6. Conclusions

Renewable energy sources are an indispensable component of advancement toward a zero-emission economy, yet their full potential can only be realized when combined with other technologies, such as carbon capture and storage (CCS). Equally important is the development of innovative business models and modern corporate management strategies, which enable the effective practical implementation of energy transition.

Research objectives that need to be addressed include the integration of various technologies, the analysis of their long-term impacts on the economy and the environment, and the development of mechanisms that will facilitate the broad adoption of innovative solutions. In this context, it is also essential to investigate the strategies of energy enterprises, which increasingly rely not only on the deployment of advanced technologies—such as renewable energy and CCS systems—but also on innovative market structures. Examples include the development of "energy-as-a-service" models, prosumer ecosystems, public–private partnerships, and flexible financing frameworks for energy projects, all of which accelerate the adoption of green solutions.

The analysis and implementation of such strategies will fill an important research gap, as they will help to understand how innovative technologies and new business models can interact to create a coherent system that will accelerate the energy transition. Progress in these areas is a prerequisite for achieving climate neutrality, enhancing energy efficiency, and fostering sustainable development, thereby contributing to the creation of a resilient global economy capable of withstanding future challenges.

7. Future Actions and Recommendations

Future research and practical initiatives should focus on the integrated implementation of renewable energy sources, along with complementary technologies such as carbon capture and storage (CCS) and energy storage, to maximize decarbonization potential across sectors. It is recommended to develop a comprehensive framework that assesses not only the technical efficiency of these systems but also their long-term economic, environmental, and social impacts.

Furthermore, the development of innovative business models and business management practices should remain a priority. Actions should include energy-as-a-service solutions, expanding prosumer networks, supporting public–private partnerships, and designing flexible financing mechanisms for green projects. These initiatives can accelerate technology implementation, reduce market entry barriers, and support the transformation of energy markets towards sustainable development.

Fostering knowledge exchange between academics, industry professionals, and policymakers is also essential. Collaborative platforms, international partnerships, and open data initiatives can facilitate the dissemination of best practices and innovative solutions, supporting a resilient global energy transition that can address future environmental, social, and economic challenges.

Funding: This research received no external funding.

Conflicts of Interest: The author declares no conflicts of interest.

References

1. Hassan, Q.; Viktor, P.; Al-Musawi, T.J.; Ali, B.M.; Algburi, S.; Alzoubi, H.M.; Al-Jiboory, A.K.; Sameen, A.Z.; Salman, H.M.; Jaszczur, M. The renewable energy role in the global energy Transformations. *Renew. Energy Focus* **2024**, *48*, 100545. [CrossRef]
2. Nyangchak, N. Emerging green industry toward net-zero economy: A systematic review. *J. Clean. Prod.* **2022**, *378*, 134622. [CrossRef]
3. Elkhatat, A.; Al-Muhtaseb, S. Climate change and energy security: A comparative analysis of the role of energy policies in advancing environmental sustainability. *Energies* **2024**, *17*, 3179. [CrossRef]
4. Chen, W.; Alharthi, M.; Zhang, J.; Khan, I. The need for energy efficiency and economic prosperity in a sustainable environment. *Gondwana Res.* **2024**, *127*, 22–35. [CrossRef]
5. Borowski, P.F. Business evolution: Emerging challenges for companies in achieving zero-carbon performance. In *Organizational Development, Innovation, and Economy 5.0*; Routledge: Abingdon, UK, 2024; pp. 184–196.
6. Filippov, S.P. Transition to a carbon neutral economy: Opportunities and limitations, current challenges. *Therm. Eng.* **2024**, *71*, 18–35. [CrossRef]
7. Xiang, Y.; Tang, Q.; Xu, W.; Hu, S.; Zhao, P.; Guo, J.; Liu, J. A multi-factor spatio-temporal correlation analysis method for PV development potential estimation. *Renew. Energy* **2024**, *223*, 119962. [CrossRef]
8. Guo, J.; Liu, X.; Yu, J.; Xu, C.; Wu, Y.; Pan, D.A.; Senthil, R.A. An overview of the comprehensive utilization of silicon-based solid waste related to PV industry. *Resour. Conserv. Recycl.* **2021**, *169*, 105450. [CrossRef]
9. Maghraby, Y.R.; Ibrahim, A.H.; Tayel, A.; Azzazy, H.M.E.S.; Shoeib, T. Towards sustainability via recycling solar photovoltaic Panels, A review. *Sol. Energy* **2025**, *285*, 113085. [CrossRef]
10. Majewski, P.; Florin, N.; Jit, J.; Stewart, R.A. End-of-life policy considerations for wind turbine blades. *Renew. Sustain. Energy Rev.* **2022**, *164*, 112538. [CrossRef]
11. Yang, M.; Chen, L.; Wang, J.; Msigwa, G.; Osman, A.I.; Fawzy, S.; Rooney, D.W.; Yap, P.S. Circular economy strategies for combating climate change and other environmental issues. *Environ. Chem. Lett.* **2023**, *21*, 55–80. [CrossRef]
12. Wieruszewski, M.; Mydlarz, K. The potential of the bioenergy market in the European union—An overview of energy biomass resources. *Energies* **2022**, *15*, 9601. [CrossRef]
13. Borowski, P.F. Water and hydropower—Challenges for the economy and enterprises in times of climate change in Africa and Europe. *Water* **2022**, *14*, 3631. [CrossRef]
14. English, J.M.; English, K.L. An overview of carbon capture and storage and its potential role in the energy transition. *First Break* **2022**, *40*, 35–40. [CrossRef]
15. Rasool, M.H.; Ahmad, M.; Ayoub, M. Selecting geological formations for CO_2 storage: A comparative rating system. *Sustainability* **2023**, *15*, 6599. [CrossRef]
16. Borowski, P.F. Hydrogen as an energy source and as a form of energy storage. *Econ. Environ.* **2025**, *93*, 941. [CrossRef]
17. Sodiq, A.; Abdullatif, Y.; Aissa, B.; Ostovar, A.; Nassar, N.; El-Naas, M.; Amhamed, A. A review on progress made in direct air capture of CO_2. *Environ. Technol. Innov.* **2023**, *29*, 102991. [CrossRef]
18. Siuta-Tokarska, B.; Kruk, S.; Krzemiński, P.; Thier, A.; Żmija, K. Digitalisation of enterprises in the energy sector: Drivers—Business models—Prospective directions of changes. *Energies* **2022**, *15*, 8962. [CrossRef]
19. Borowski, P.F. Economic and technological challenges in zero-emission strategies for energy companies. *Energies* **2025**, *18*, 898. [CrossRef]

20. Avwioroko, A.; Ibegbulam, C.; Afriyie, I.; Fesomade, A.T. Smart grid integration of solar and biomass energy sources. *Eur. J. Comput. Sci. Inf. Technol.* **2024**, *12*, 1–14. [CrossRef]
21. Khalid, M. Smart grids and renewable energy systems: Perspectives and grid integration challenges. *Energy Strategy Rev.* **2024**, *51*, 101299. [CrossRef]

energies

MDPI

Article

Increasing Energy Efficiency Based on the Kaizen Approach

Armenia Androniceanu [1], Ioana-Catalina Enache [2,*], Elena-Narcisa Valter [2] and Florin-Felix Raduica [2]

1 Administration and Public Management Department, Faculty of Administration and Public Management, Bucharest University of Economic Studies, 010374 Bucharest, Romania
2 Department of Engineering Graphics and Industrial Design, Faculty of Aerospace Engineering, University Politehnica of Bucharest, 060042 Bucharest, Romania
* Correspondence: catalina.enache@upb.ro

Abstract: The energy crisis, an increasing concern of the industrial sector, is characterized by increased energy costs and low energy availability. As part of an effort to increase energy efficiency, the present study assessed the impact of applying lean tools on the energy consumption of a manufacturing organization. Using the Kaizen approach, the organization's processes were analyzed to identify the main vulnerabilities and make the required changes to increase consumption performance. Several technical improvements were performed to reach increased energy efficiency. Results showed an energy efficiency increase; consumption was reduced by up to 7.5% in the production line, 3.5% in the extruder stage, and up to 20% for the injection stage of the manufacturing process. The results suggest that the organization's energy reduction targets have been achieved. Standardization of procedures was useful in the development of Kaizen projects which could be applied with certain limitations. The implications of these results for understanding energy consumption reduction are discussed.

Keywords: energy efficiency; manufacturing; Kaizen; continuous improvement

1. Introduction

Technological advance and the development of society are based on the use of energy, from which derives the importance of increasing energy efficiency in organizations. It is obvious that productivity is influenced by the increase in energy costs [1], so there is an incentive to find solutions that lead to efficiency of consumption [2,3]. The decrease in energy consumption and the reduction of energy costs for a manufacturing process can be determined by a detailed analysis of the energy flows [4,5]. The manufacturing process is a complex one and requires the involvement of the human factor through the workforce efforts, of the equipment and of the technologies necessary for manufacturing the final product [6]. In the current conditions, increasing energy efficiency is a priority for all organizations and a strategic objective by which an organization becomes more competitive in the sector and can win the competition with other organizations. The common framework for promoting energy efficiency in the European Union is established according to Energy Efficiency Directive 2012/27/EU (EED) and Directives 2018/2002/EU and aims to reduce energy consumption by 32.5% by 2030 [7,8]. According to the data published by EUROSTAT for 2020 in Europe, energy consumption in the industrial sector represents 26.1% of the total consumption [9,10]. The analysis of the data provided by EUROSTAT [9] for the 27 Member States of the European Union shows that between 2015 and 2020 the energy consumption in the industrial sector had a slight increase, as can be seen in Table 1.

The increase in prices for energy and raw materials influences inflation, so low energy consumption will generate positive macroeconomic effects [11]. In other words, a reduction in the cost of consuming energy for a given process can lead to the redirection of saved revenues to other services, goods, and processes that use energy. Energy efficiency by reducing energy consumption can have notable effects on production costs but also on the environment, as part of the transition to a carbon-neutral economy [12]. In 2021, CO_2 emissions from energy combustion and industrial processes accounted for almost 89% of greenhouse gas emissions in the energy sector [13]. The objectives of an energy efficiency organization can be achieved by using new technologies or upgrading existing ones. Kaizen is an approach centered on process, product, people, and the environment. According to this approach, each component in the chain of conception, production, and sale can be improved [14]. Kaizen is a tool used by organizations around the world, which aims at continuous improvement achieved through small steps and by eliminating losses and which can adapt to the specific needs of each organization. The main reasons for which we used Kaizen are the following: identifying losses, reducing energy consumption, replacing and improving equipment with high energy consumption, and increasing the organization's profit and investments. The case study is conducted for an organization that is a market leader in home appliances. They seek to reduce waste in their production processes by applying lean tools and investing in improving the technology used. Different models of refrigerators are assembled on the five production lines of the organization. Some of the parts used in the assembly process are made internally. The work is structured as follows. Section 2 includes aspects regarding energy efficiency and the Kaizen approach developed in the literature. Section 3 contains the research methodology by which the implementation of the Kaizen approach is achieved, and Sections 4 and 5 describe the steps for the implementation of the Kaizen approach to increase energy efficiency within an organization. Section 6 suggests future research that derives from the limitations of the case studies presented, whilst the last chapter includes the conclusions of the current work.

Table 1. Energy consumption in the last 5 years in Europe [9].

Europe Energy Consumption	2015	2016	2017	2018	2019	2020
Final consumption—total energy used (terajoules)	909,447	927,747	940,587	942,623	937,857	885,764
Industry sector (terajoules)	233,689	237,994	240,289	242,693	239,416	231,211
Percentage (%)	25.7	25.7	25.5	25.7	25.5	26.1

2. Literature Review

2.1. Energy Efficiency

Organizations need to implement energy management strategies for better efficiency to minimize different costs [15–17]. To identify the potential for energy efficiency, a reference point must be established. It is necessary for choosing a method by which to determine energy efficiency. Boyd, G.A. in the paper "Comparing the statistical distributions of energy efficiency in manufacturing: a meta-analysis of 24 Case studies to develop industry-specific energy performance indicators (EPI)" presents the measures and methods by which one can identify the reference point [18]. In order to calculate the reference point, the parameters influencing the energy consumption must be kept constant. The first way to determine the reference point is by measuring the energy intensity, which represents the ratio between energy consumption and activity or can be identified as the energy inputs and outputs from an organization over a certain period of time [19]. It is a useful physical value to measure energy consumption or energy performance. The energy efficiency of an industrial process will increase if the energy consumption is reduced or if for the same amount of energy consumed the level of production or the productivity of the labor force increases [12,19]. The values of energy intensity can be compared across time or between several organizations or at the sector level to which they belong, based on specific parameters. The issue with establishing measures regarding energy consumption is the lack of data within organizations on energy management techniques [15]. The monitoring and analysis of energy consumption for manufacturing systems are essential for the application of consumption reduction methods [20]. Energy efficiency measures (EE) can refer to a physical approach by optimizing process parameters or to a theoretical optimization of process parameters that may have practical limitations due to high calculation costs and the accuracy of results [21]. Another improvement measure consists in investments, which is based on identifying economic factors and barriers [22]. The factors that influence energy efficiency from the point of view of the production process are related to the equipment runtime [23,24]. Avoiding idling, using the equipment optimally, the existence of an operational plan but also the use of new technologies can reduce energy consumption [25–27]. It was found that for the systems operated by electric motors (EMDS), the highest consumption is recorded for the electric motor [28]. The consumption of electric motors and the systems they engage within is estimated to be between 43–46% of global energy consumption [29]. Globally there are regulations on the efficiency of electric motors and the replacement of low-efficiency motors with motors of a higher efficiency class and with low costs [30]. IEC 60034-30 and NEMA MG 1-2014 standards present the classes of "international efficiency"—IE of electric motors [30,31]. Table 2 contains an extract from the IEC and NEMEA standards showing the level of efficiency depending on the energy class.

Table 2. Electrical motors efficiency classes [30,31].

Energy Classes		**Efficiency Levels**
IEC (International)	**NEMEA (USA)**	
IE 1	-	Standard
IE2	Energy Efficient EPACT	High
IE3	Premium	Premium
IE4	Super-Premium	Super-Premium
IE5	Ultra-Premium	Ultra-Premium

The European Commission through its regulation 2019/1781 of 1 October 2019 presents the calendar of energy efficiency requirements for motors and electricity consumption up to 2030 [32].

2.2. Kaizen

In a competitive market environment, the application of modern process management approaches can contribute to the improvement of performance on the productivity of an organization [33]. The improved productivity of an organization will also be positively reflected in the energy efficiency of the entire organization. This efficiency can be achieved with minimal costs if the management of the respective organization applies the principles and tools offered by quality engineering. For example, lean tools are applied to increase the economic productivity of organizations and eliminate waste [34–36]. Several methods can be applied to reduce or eliminate losses and to achieve a certain level of productivity [37,38], which can be applied depending on the size of the organization and its needs. Some of the commonly used methods are the following:

- Kaizen—5S (Sort, Set in order, Shine, Standardize, Sustain) [39]
- Benchmarking [40]
- Kaizen—Continuous Improvement [41]
- Kaizen—Just in Time [42]
- Pull-principle and Kanban [43]
- Visual Management in Production [44]
- Zero waste [45]
- Value Stream Mapping [46]

- Efficient and ergonomic workstations [47]
- Poka Yoke [48]
- Six Sigma [49]

The Kaizen method is part of the plethora of lean tools offered in the literature and can be used by organizations to achieve the organization's objectives, to increase profit, to reduce losses called Muda, and to reduce expenses through the involvement of each employee [50].

The Kaizen method was developed by Iami in 1986 and was first applied in the automotive company Toyota to improve the company's efficiency, productivity, and competitiveness [51]. It was then applied to boost the productivity of other organizations by reducing costs [52–54]. Currently, this method is applied in various fields both in large enterprises and also in small and medium-sized enterprises [55].

The Kaizen method is used to eliminate and reduce losses or Muda. Muda in production is associated with any activity or process that consumes resources without creating value. In the literature [14] there are 7 losses—Muda that once detected and treated properly allow the organization to be more competitive and to have greater efficiency in the processes specific to that organization. These Muda are [14]: waste from overproducing, waste of time (waiting), waste from transporting, waste from over-processing, waste of inventory, waste of motion, and waste caused by defects and reworking.

There are many empirical studies [41,56–59] related to the implementation of Kaizen in different fields of activity, but only some of these focus on improving the process by making energy consumption more efficient. In the pharmaceutical industry, the Kaizen method was used to show how the involvement of operators and managers in the operations of the production process created a circle of positive ecological improvement [56]. In the custom furniture processing industry, the Kaizen method has been used to improve quality and productivity [57]. The Kaizen method has been successfully used both in the automotive industry [58] and in the food service industry [59]. Niranjana Sundararajan in her paper presents a case study through which the elimination of losses and the improvement of productivity in the production of fasteners were obtained through small and systematic improvements and the application of Lean-Kaizen Manufacturing Techniques [41]. There are some limitations to Kaizen implementation:

- reduced employee motivation [60];
- poor cooperation between employees and management (management) [61];
- limited ability to manage continuous improvement [62];
- poor definition of purpose [63];
- the structure of the organization must align with Kaizen principles [64];
- lack of organizational culture with values necessary for this approach [65].

In order to achieve a certain level of performance and the organization's objectives, several managerial actions are applied. In a company, there are many options [56] but the strategic actions applied should lead to benefits in the simplest way with the lowest cost [66]. Applying the Kaizen methodology involves nine steps that can be adapted depending on the study conducted in a particular organization.

1. Select the target process
2. Create a team
3. Set project goals and plans
4. Observe the process
5. Analyze the process
6. Create an improvement plan
7. Implementation
8. Presentation
9. Standardize and monitor

Kaizen methodology is used in various fields. A study based on Online Radio Frequency Identification-based Facility Performance Monitoring System (ORFPM) [67] by a company that remanufactures agricultural clutch discs, water pumps, torque amplifiers, and other industrial equipment has collected data to improve process efficiency. In the disc department, the challenge was that long transportation time led to few good results of parts flow between departments. The proposed framework used RFID, lean strategy, and Kaizen. As a result, the team improved the value stream map (VSM) and overall material flow.

The authors of [68] describe the improvement of operational and sustainability performance using lean management specifically Gemba Kaizen. The team found that the cutting process was incorrectly done and led to waste. New routines were put in place to reduce loss. This resulted in the reduction of shrinkage from 4.4% to 2.4% which led to reduced food waste by 50% and as a secondary outcome the stocks index was reduced by 33% in a cold meat section of a hypermarket store.

A case study [69] that took place in a metalworking company that designs and develops fluid flow valves shows that through lean leadership (LL) and lean tools (LTs) quick wins can be achieved. Daily Kaizen meetings involving the workforce to generate ideas cause rapid improvement over a small period after implementation. A six-dimensional survey revealed that by applying 70% of proposed ideas the quality awareness increased from average to good.

In [70], the authors present the results of a data rich systematic review focusing on improving performance in public hospital emergency departments. Among challenges that can be overcome using Kaizen and lean concepts are patient flow, high waiting time, high length of hospital stays, health safety and cost reduction. For example, waiting time reduction is the most important benefit because it impacts the mortality rate. The high patient flow is reduced with Kaizen events using health information systems (HIS). Implementing tools such as Kaizen events and lean healthcare improve care capacity by an average of 30% and reduce patient waiting time by 50% on average.

In [71], the authors describe a framework developed in the ambulatory healthcare clinic setting. The goal was to improve patient throughput. Identifying problems in this context was done through Kaizen and lean healthcare methodology. Maps were created to improve patient flow. Kaizen events were held to increase access to ambulatory clinics. The Heijunka approach was used to establish a better patient scheduling strategy. Flow-time efficiency was increased. Inventory was reduced using Just in Time (JIT) principles. The results highlighted an improvement of the overall quality of service.

In [72], the authors describe applying leadership and Kaizen principles in education. The results conclude that transformational leadership and regular Kaizen meetings improve clinical research performance in the laboratory and encourage volunteers to contribute to the research effort.

In this paper, we present two case studies for increasing energy efficiency by studying and treating losses (Muda) in the energy system of an organization. We have chosen to apply the Kaizen methodology because it is useful in identifying losses by using the internal resources of the organization and can be adapted to its needs.

The Kaizen approach used here followed the standard applicability steps adapted to the analyzed areas, as presented in Section 3.

3. Research Methodology and Kaizen

The implementation of the Kaizen approach depends on the setting of objectives at the organizational level. The strategic objective of the organization where the case study presented in this paper was performed was to reduce energy consumption without changing the technological process. To achieve this objective, we proposed the use of the Kaizen method.

The steps we follow to increase energy efficiency based on the Kaizen approach are the following:

Step 1: Determining the energy consumption in the organization and develop maps/graphs that reflect the total consumption on each production line.

Step 2: Analysis of the area of the organization with the highest consumption and determination of losses (Muda).

Step 3: Analysis of a process for which energy distribution solutions are investigated.

Step 4: Identifying and implementing solutions that can be applied to reduce energy consumption—eliminating Muda previously detected.

Step 5: Checking the implemented solutions.

To have a necessary reference point in identifying the economic potential, the amount of energy consumed at the level of the organization is measured and a complete map of the energy consumption is made.

From an economic or technological point of view, it is necessary to formulate certain scenarios to improve energy consumption.

As a strategy in the Kaizen approach, we propose to identify the process and, respectively, the machine tools or the area in the organization where the case study is carried out, which have the highest energy consumption [73]. We aim to detect the Muda that are responsible for reducing energy efficiency.

We establish a set of specific indicators that measure energy consumption, then we perform a quick and reliable analysis to determine the factors that influence energy consumption. Such an analysis is oriented toward identifying the largest consumers of energy. The identification of the existing potential must also be analyzed from the point of view of the improvement barriers [74]. In order to achieve the objectives assumed in the case study, we have gone through the next six steps of the Kaizen approach [56].

1. Process selection and detection of «Muda» losses.
2. Setting goals by prioritizing the most important Muda.
3. Analysis of the initial situation.
4. Establishing an action plan and finding solutions to eliminate or mitigate Muda losses detected.
5. Application of the solutions proposed by our team.
6. Checking solutions and standardizing them at the organizational level.

The aim is to eliminate or diminish areas of inefficiency and turn them into opportunities for the organization through the Kaizen approach. After the first step of the Kaizen methodology, in which the highest consumers in the organization are identified, it will be determined which other areas to focus on to reduce energy consumption. The organization establishes strategic actions according to the assumed objectives. By defining the strategic actions, we identify the competitive advantages of the organization having as limitations the associated risks and the existing or potential obstacles. In the medium and long term, the energy efficiency strategy establishes the general objectives of the organization and defines the way in which the organization can act to achieve them.

4. Case Study Based on the Kaizen Approach

In this case study, we use the Kaizen approach to reduce electricity consumption within an organization that produces refrigerating appliances. The strategic objectives of the organization are increasing profit, producing competitive products, reducing losses and reducing electricity expenses. The objective we pursue using the Kaizen approach is to reduce electricity consumption by 5%. We recommend the Kaizen approach because this methodology allows the identification of losses and the increase of the organization's profit by reducing electricity expenses, but also the production of competitive products. The case study focuses on identifying Muda which is responsible for wasting electricity without a precise purpose. It analyzes from an engineering and application point of view the energy consumption necessary for carrying out the activities in the organization. The production process is analyzed in smaller specialized groups to outline the ideas that can lead to the reduction of electricity consumption and to the identification of solutions that can be adopted.

Creating and understanding a hierarchical structure on strategic energy-saving goals is important in the Kaizen process for establishing an action plan and engaging employees at all levels of the organization.

Monitoring energy consumption is necessary to analyze the use of energy and to identify the largest consumers.

Analysis of the use of energy–loss identification (Muda) is carried out by:

- Engineering analysis that requires:
  - Determination of cumulated electricity consumption in different areas of the organization
  - Comparing electricity consumption for different areas of the organization
- The actual analysis of the process that requires:
  - Identifying the area and the process with the highest electricity consumption
  - Determining critical places
  - Analysis of energy consumption
  - Value analysis by formulating solutions that can be applied to reduce electricity consumption.

The operational objectives of the organization for the electricity-saving activities are related to the energy efficiency of the processes and equipment and the energy efficiency of the products. In this case study, we have focused on increasing the energy efficiency of processes and equipment. Figure 1 shows a graphic that highlights the operational objectives of the organization related to their energy efficiency.

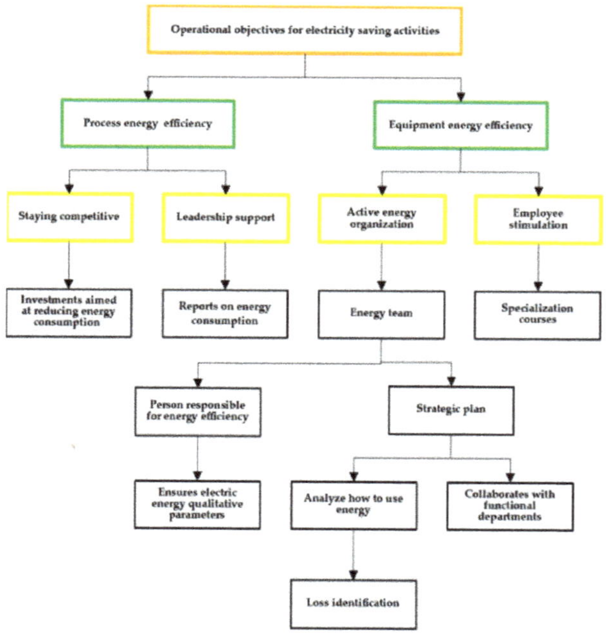

Figure 1. Operational objectives for energy-saving activities.

Operational objectives related to the energy efficiency of products will be the objective of another case study.

The organization wants to remain competitive in the market and this objective can be achieved by supporting investments in new technologies and equipment aimed at reducing energy consumption. Investments regarding energy consumption can be approved by the organization's management following reports on the energy efficiency of the organization.

The reports to the management of the organization are elaborated by a team that monitors the energy consumption and identifies the losses that in the case of our study represent Muda. To identify the losses, it is necessary to involve the employees who can be stimulated by attending specialization courses.

4.1. Step 1: Determining the Energy Consumption in the Organization

To obtain the necessary data for the analysis, the electricity consumption was monitored over a period of one year for 44 points in the organization. In Table 3 we find the energy consumption per day and the arithmetic average of the measurements taken over one week specific to Production Line 1.

Table 3. Consumer monitoring on Production Line 1.

	Day Consumption (MWh)	Weekly Average Consumption (MWh)
Production Line 1	5.701	5.114
Consumer 1	0.12	0.15
Consumer 2	1.578	1.532
Consumer 3	1.785	1.122
Consumer 4	0.109	0.161
Consumer 5	0.622	0.653
Consumer 6	0.381	0.308
Consumer 7	0.006	0.005
Consumer 8	0.023	0.016
Consumer 9	0.118	0.095
Consumer 10	0.081	0.063
Consumer 11	0.102	0.087
Consumer 12	0.084	0.086
Consumer 13	0.09	0.1
Consumer 14	0.109	0.203
Consumer 15	0.493	0.533

As a result of the monitoring carried out, a map (Figure 2) resulted in which the main consumers of electricity are marked. This map gives us a clear picture of each work area in the enterprise. Figures 2 and 3 show that there are five production lines and nine areas serving the production lines in the organization.

The situation of the electricity consumers is highlighted in Figure 3 from which it appears that for the production lines the highest consumption is recorded on Line 1 and for the areas serving the production lines the highest consumption is recorded in the maintenance area.

Our case study will focus on reducing energy consumption in Production Line 1.

The electricity consumption on Production Line 1 is influenced by the number and energy class of the existing electric motors.

In Table 4 for each area of the organization is indicated the number and energy class to which the existing motors larger than 5.5 kW belong. In the organization, there are 405 motors divided into 4 different classes of efficiency: IE3, IE2, IE1, and smaller than IE1. Table 4 shows that 20% of the engines in operation are smaller than IE1, 60% of the engines have the efficiency class IE1, 16% have the efficiency class IE2 and only 2% of the flat motors in operation have the energy class IE3.

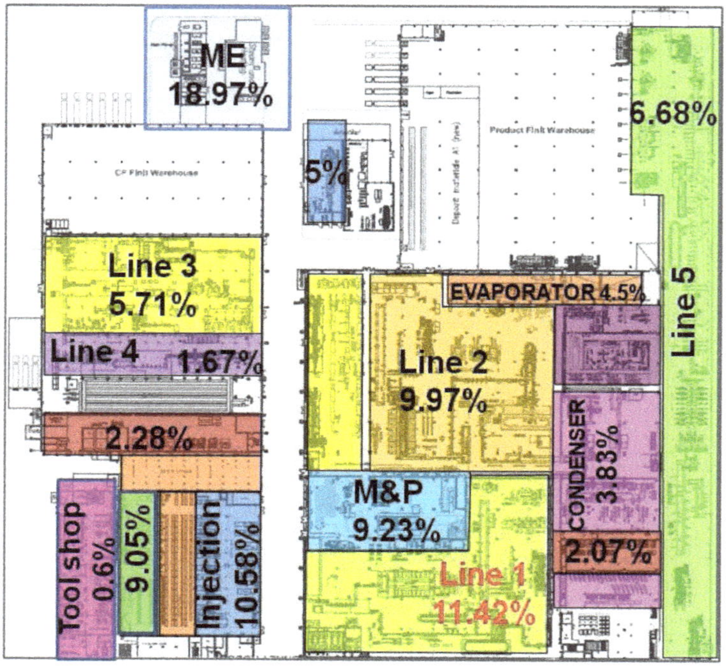

Figure 2. Electricity consumers map. ME—Maintenance Equipment. EPS—Expanded Polystyrene area. M&P—Mechanical and Painting area.

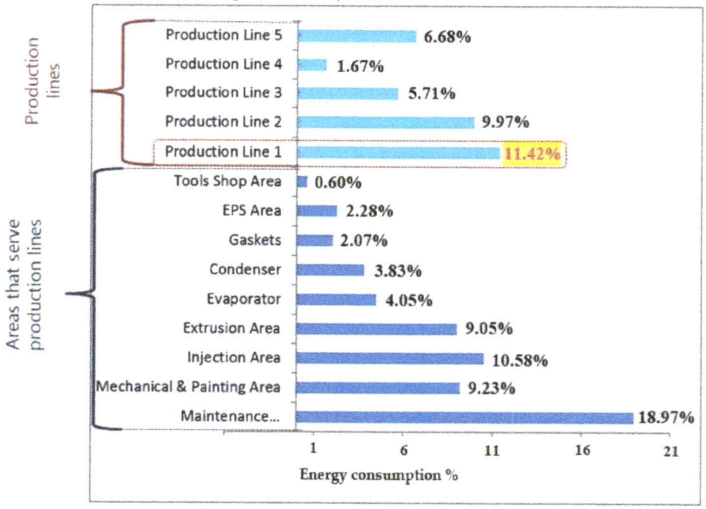

Figure 3. Energy consumption for different areas of the organization.

Table 4. Existing engines.

	Motor Energy Efficiency Class				
	<IE1	**IE1**	**IE2**	**IE3**	**Total Engines**
Efficiency Levels		Standard	High	Premium	
Line 1	3	25	14	2	
Line 2	17	33	-	-	
Line 3	10	19	3	-	
Line 4	-	-	4	-	
Line 5	-	-	28	1	
Total production lines	30	77	43	3	153
Tool shop area	9	5	-	-	
EPS area	14	24	-	-	
Condenser	-	8	-	-	
Evaporator	-	18	-	-	
Injection and Extrusion area	18	49	10	2	
Mechanical and Painting area	12	42	9	-	
Maintenance	-	20	7	5	
Total components by areas	53	166	26	7	252
TOTAL	83	243	69	10	405

Having a clear situation of the electricity consumers presented in Figure 3, together with the data from Table 4 where the types and number of motors existing in the factory are highlighted, will lead to the identification of the process that needs to be improved from an energy point of view. The results show that the highest energy consumption is achieved by Production Line 1.

4.2. Step 2: Analysis of the Area of the Organization with the Highest Consumption

On Production Line 1, two foaming processes are performed for different components of the refrigeration unit and a thermoforming process. Motors of energy efficiency classes IE1, IE2 and IE3 can be found on this production line. Table 5 contains the existing motors on Production Line 1 for the identified processes.

The monitoring of energy consumption for Production Line 1 resulted in a different consumption for each process identified in Figure 4. The consumption for foaming component 1 represents 57% of the total consumption recorded on Production Line 1 so this process must be analyzed thoroughly to identify solutions that can lead to the reduction of energy consumption.

Table 5. Main consumers of energy of choice for Line 1.

	Process	Equipment	Motor	Power [kW]	rpm	No.	Efficiency Class
Production LINE 1	Foaming component 1	2 Composite foaming installations 1	Confirmatory spindle training	5.5	725	16	IE1
				7.5	960	4	IE1
			Polyol	30	1475	2	IE2
			Isocyanate	22	732	2	IE2
			Hydraulic	11	1470	2	IE3
			Ventilator	11	1440	4	IE2
	Foaming component 2	2 Foaming installations component 2	Polyol	5.5	1440	1	IE2
			Isocyanate	5.5	1440	1	IE2
			Hydraulic	5.5	1440	2	IE2
			Ventilator	5.5	1440	1	IE2
			Shaft drive motors	22	3000	1	<IE1
	Thermoforming	Equipment 1	Hydraulic	22	1465	1	IE1
			Vacuum pump	7.5	1440	1	IE1
		Equipment 2	Hydraulic	22	1465	1	IE1
			Vacuum pump	7.5	1440	1	IE1
		Thermoforming equipment 1	Hydraulic	5.5	1450	1	IE1
			Vacuum pump	11	1455	1	IE2
		Thermoforming equipment 2	Vacuum pump	5.5	1440	1	<IE1
		Thermoforming equipment 3	Vacuum pump	5.5	1450	1	<IE1

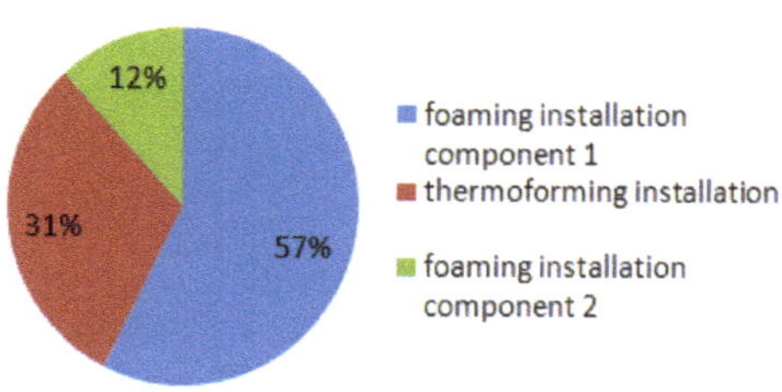

Figure 4. Energy consumption on Production Line 1.

4.3. Step 3: Analysis of a Process for Which Energy Reduction Solutions Are Investigated

Setting a target is essential in the development of the Kaizen project. On component 1 foaming installations, there are 5 categories of monitored motors: isocyanate pumps, polyol pumps, hydraulic motors, motors used for ventilation, and motors used to drive shafts (Table 6). To identify the losses, it is necessary to analyze the consumption for each category of motors separately. Using the data from the energy consumption monitoring program from Table 6, it appears that the largest losses are due to isocyanate pumps and polyol pumps for which consumption of 10.42 MWh was recorded for each category, which means that for Production Line 1, the electricity consumption for the two categories of motors represents 74% of the amount of energy recorded as being consumed.

Table 6. Motors consumption for foaming installation component 1.

No.	Motors	Real Consumption MWh	Consumption Percentage
1	Isocyanate pumps	10.42	37%
2	Polyol pumps	10.42	37%
3	Hydraulic motors	3.71	13%
4	Fan motors	1.7	6%
5	Shaft drive motors	2.1	7%

The objective of the organization is to reduce energy consumption by 5%. To achieve this objective, we proposed to reduce consumption for foaming installation component 1 by 5%. A decrease of 5% means a decrease in electricity consumption from 28.34 MWh to 26.93 MWh. A 5% reduction in consumption for foaming installation component 1 can be achieved if we reduce the consumption for each motor by 5%, Figure 5.

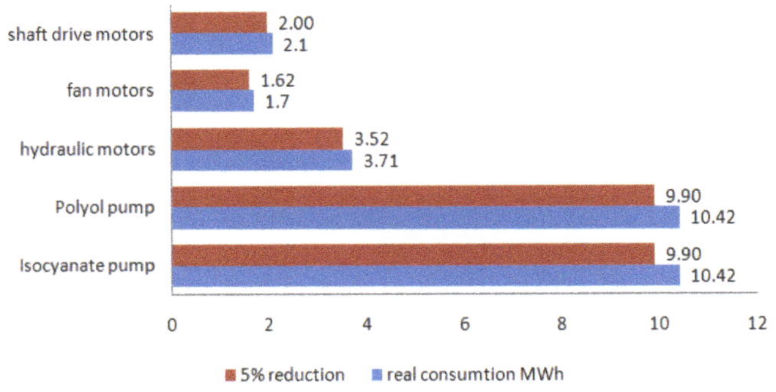

Figure 5. Proposed target for reducing energy consumption at foaming installation component 1.

Identifying solutions to reduce energy consumption requires detailed knowledge of the foaming process for component 1.

The component 1 foaming process is carried out with the help of 20 foaming plants positioned in parallel on two branches A and B that work simultaneously (Figure 6). Each foaming installation has mounted a motor that has the role of lowering and raising the conformers (Figure 7). In total, we have 20 spindle conformer drive motors of IE1 energy class. For the same process, 2 isocyanate motors and 2 polyol motors were identified, both having the IE2 energy class. The area formed by the 20 conformers will be analyzed to improve the foaming process.

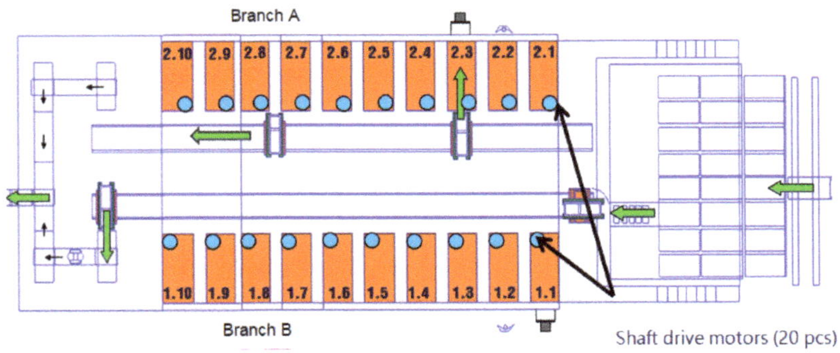

Figure 6. Conformer layout.

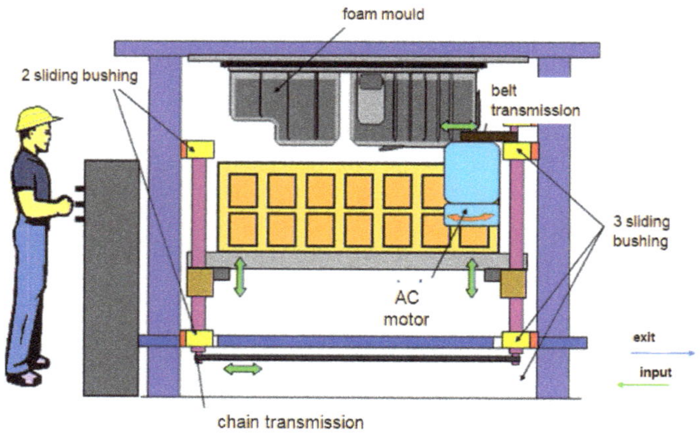

Figure 7. The operation mode of the conformer.

In the foaming process, the motor engages 2 shafts located in parallel (Figure 6). A shaft is engaged directly by the motor through a belt drive. The rotational movement of the shaft is also transmitted to the secondary shaft using a chain transmission. An operator drives the motor that by the transmitted movement raises or lowers the conformers to the foam mold Figure 7.

The motors that drive the shafts have 2 revs and run-in alternating current.

We can calculate the motor speed using Equation (1)

$$n = \frac{60 \cdot f}{P} \quad (1)$$

where:

n = synchronous motor speed [rpm]
f = frequency of the supply voltage [Hz]
P = number of motor pole pairs

For the motor, the frequency of the supply voltage is 50 Hz and uses at low speed 4 pairs of poles, and at high speed 42 pairs of poles. To calculate the speed, we have replaced in Equation (1) the given values.

$$\text{Slow speed}: \quad n_{lenta} = \frac{60 \cdot 50}{4} = 750 \text{ rpm} \quad (2)$$

$$\text{Fast speed}: \quad n_{rapida} = \frac{60 \cdot 50}{42} = 1500 \text{ rpm} \quad (3)$$

From Equations (2) and (3) it follows that the motor can run at a slow speed of 750 rpm (rotations per minute) and a fast speed of 1500 rpm. The motor helps to raise and lower the conformer which can be in the closed or open position. The open position is where the product is fed and foamed and the open position is where the foamed product is discharged. The conformer ascent is done in 3 steps.

Step 1—the conformer ascends slowly (the necessary settings are made to prepare for the fast ascent of the conformer).

Step 2—the conformer ascends quickly until near the closing area of the conformer.

Step 3—the conformer ascends slowly to the closing position.

As shown in Figure 8, the slow to fast speed receding from slow to fast and vice versa is done at an interval of about 0.5 s, suddenly by changing the engine winding. Removing the conformer from the stand for lifting or lowering is done by direct start with contactors. Monitoring of the motor's operation shows that the transition from one step to another is achieved with large jumps in current intensity. It can also be observed that the current intensity does not keep the same value during a step, which can lead to certain operating errors. The 20 motors used to drive the shafts recorded a consumption of 1704 kWh.

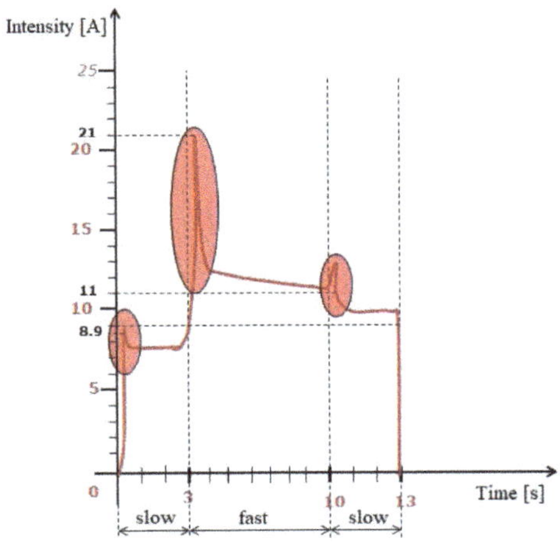

Figure 8. Engine intensity measured over time.

4.4. Step 4: Identifying Solutions That Can Be Applied to Reduce Energy Consumption

The high energy consumption for the component 1 foaming installation is influenced by the consumption of the motor entraining conformer shafts, polyol pumps, and isocyanate pumps. Each case was analyzed to find a suitable solution. The motors that operate the polyol and isocyanate pumps are of the IE2 energy class (Table 4). A solution to this problem is to invest in more advanced motors of the IE3 energy class in which speed cab be varied. Conformation shaft drive motors are of energy class IE1 (Table 4) and a viable solution would be to replace them with motors of energy class IE3/IE4. Another problem related to the energy consumption for these motors refers to the fact that the removal of the conformer from rest is done by starting directly with capacitors. For this problem, there are two solutions. The first solution would be to use hydraulic pumps for operating the conformer, but this solution leads to very high costs. A second solution consists in installing a speed adjustment system that from an economic point of view is preferable compared to the first solution proposed. The scheme in Figure 9 shows an overview of the proposed solutions.

Following the analysis of the solutions in Figure 9, we proposed and piloted three major changes to reduce the electricity consumption of Production Line 1, which had the highest energy consumption.

The first change we proposed to reduce energy consumption was to install Altivar 71 [75] speed dimmers that were designed to generate significant energy savings in industrial processes. The speed adjustment system was replaced by installing speed dimmers resulting in the reduction of energy losses at variable loads. This system offers the possibility of increasing the frequency from 50 to 80 Hz, and implicitly the progressive increase of the speed from 0 to speeds higher than 1500 rpm as well as the leveling of the absorbed current.

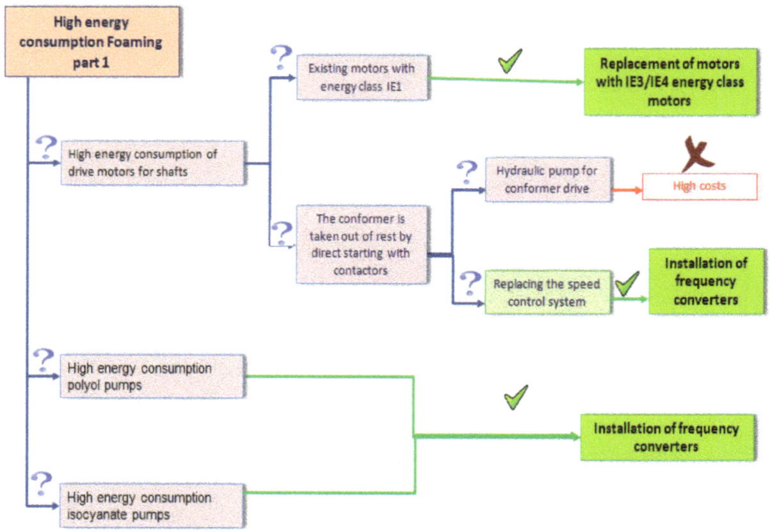

Figure 9. Solution analysis.

After the implementation of the system, a new set of measurements was made to make a comparison between the initial and the current mode of operation (Figure 10). Such before and after comparisons are frequently used in Kaizen projects [41,76].

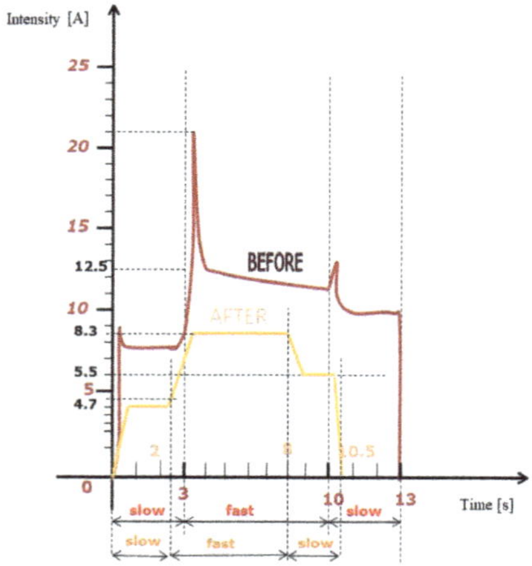

Figure 10. Engine intensity measured over time after improvement.

The improvements obtained from the installation of the speed variators are the following:

1. Removal from rest for ascent or descent of the conformer occurs progressively from 0 to 30 Hz, with an acceleration of 0.4 s.
2. The transition from slow to fast speed occurs progressively with acceleration from 30 Hz to 60 Hz in 0.4 s.
3. The transition from fast to slow speed occurs in 0.4 s from 60 Hz to 30 Hz.
4. The conformer shutdown takes place in 0.4 s from 30 Hz to 0 Hz.
5. Speed increase in a shorter time from 13 s to 10.5 s.

To these improvements we can add:

1. It has a safety function to ensure that the motor stops and prevents accidental restarts.
2. Protects bearings, bearings, and shafts.
3. Contactors are no longer replaced.
4. A single-speed motor can be used.

The second change proposed was to replace the old motor of the efficiency class IE1 (Table 4) with a new motor from the energy efficiency class IE3. A before/after comparison reveals that there are no problems with the installation of the new motor (Figure 11). The motors have the following characteristics: they can develop a power of 5.5 KW, at a frequency of 60 Hz resulting in a speed of 1500 rpm if we replace them in Equation (1).

Figure 11. Engine positioning before and after applying the Kaizen approach.

The advantages of using IE3 motors are:

- Saving energy and reducing CO_2 emissions
- Extra performance
- Low maintenance costs
- Comfort of use

The third change proposed was to replace the polyol and isocyanate pumps for branch A and branch B, with 2 IE3 motors with frequency converters with lower energy consumption.

4.5. Verification of Implemented Solutions

A new set of measurements regarding energy consumption was made after the implementation of the solutions. The solutions adopted are focused on replacing motors that have low efficiency with motors from higher efficiency classes. The acquisition costs are substantial but can be compensated by the decrease in electricity consumption. Instability in electricity prices [77] is a factor influencing production.

The new set of measurements regarding the energy consumption compared to the initial situation leads to a result that exceeds the proposed target. The graph in Figure 12 is a histogram that presents a comparison of the initial situation, the proposed target, and the current situation for each type of motor and pump.

With the new improvements, a saving of 7.5% was reached, from an initial consumption of 28.34 MWh to 26.22 MWh, a decrease that led to a saving of 20,630 € annually for the organization. Reducing energy consumption has a positive impact on the organization's energy costs.

In the first stage, the research was carried out only for Production Line 1. It was then extended to other production lines or areas that serve the production lines. By replacing the motors that had low energy efficiency, the organization managed with a total investment of 15,480 € to reach an annual profit of 194,853 €. This profit is due to annual energy savings that reached −1.139 MWh.

The organization's profits will be used for other projects mainly aimed at reducing energy consumption. In the next section, we present the application of Kaizen methodologies in other areas.

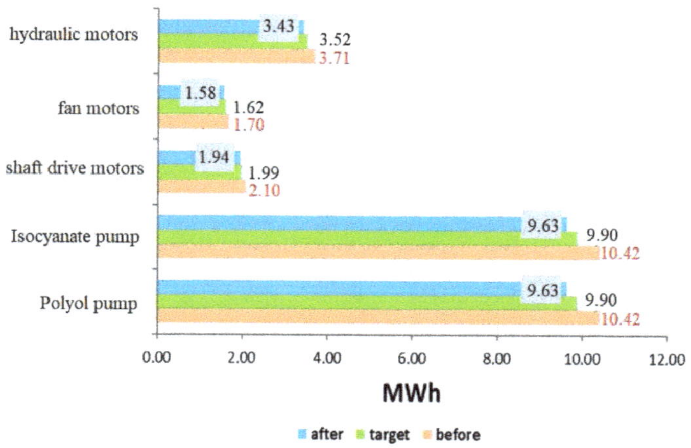

Figure 12. Energy consumption frame.

5. Application of Kaizen Methodology in Other Areas

The steps presented in the description of the Kaizen methodology are followed. This case study is dedicated to areas serving production lines and focuses on eliminating energy losses in the injection and extrusion areas.

5.1. Step 1: Determine the Next Project for Increasing Energy Consumption

In step 1 the energy consumption in the organization was determined for different areas. Figure 3 shows that the maintenance area is a high energy consumer. This area has the particularity of serving the whole organization. In order to apply the Kaizen methodology in the maintenance area, the management teams responsible for certain areas of the organization must cooperate. In this context, we have not yet been able to develop a plan whereby the managers responsible for the areas of the organization develop a project in which the Kaizen methodology can be applied. We can apply the Kaizen methodology to reduce energy consumption in the areas of injection and extrusion.

5.2. Step 2: Analysis of the Injection and Extrusion Area

In these areas, various parts are made that will be used in the assembly process for refrigerators. In these areas, we have 31 injection machines and 4 extruder machines. Table 7 shows the measurements for 3 injection machines and 3 extruders on 5 different days and the average consumption.

Table 7. Electricity consumption from different periods.

Machine	Day 1	Day 2	Day 3	Day 4	Day 5	Average
	kWh/Day					
Extruder 1	6602.17	6629.01	6656.46	6650.05	6643.26	6636.19
Extruder 2	2953.20	2969.47	2954.96	2860.56	2877.88	2923.21
Extruder 3	4557.50	4558.75	4548.42	4698.68	4674.21	4607.51
Injection 1	5163.63	5166.37	5073.52	4901.25	4840.20	5028.99
Injection 2	6103.74	5992.69	6163.59	5993.93	5665.92	5983.97
Injection 3	3016.12	3781.03	3981.64	3364.52	3331.28	3494.92

The highest energy consumption in the injection and extrusion processes has been identified in the regions where the machine heats the material (Figure 13).

Figure 13. Material heating zone on the injection machine.

5.3. Step 3: Identifying and Implementing Solutions

The material that will be injected/extruded must have a certain temperature. In the heating zone of the material, there are large temperature losses during the process which influence energy consumption. Hence, material heaters will be installed to reduce heat loss and will collect and heat the material to the required temperature. The material heaters are equipped with heaters and thermocouples that indicate the temperature of the heated material. The addition of insulation jackets increases the heating efficiency of the material by keeping it at the required temperature (Figure 14). To make this solution applicable for injection and extrusion machines, insulation jackets must be modular.

Figure 14. Insulation jackets for an extrusion machine.

5.4. Step 4: Check the Implemented Solutions

The aim is to save a significant percentage of electricity, with shorter cylinder heating times and maintaining a constant temperature of the material. With a budget of 24,500 €, the solution was initially applied to the 3 injection machines and 3 extruders. For this example, a new set of energy consumption measurements is performed (Table 8).

Comparing the average values obtained from the initial measurements with the ones obtained after the implementation of insulation jackets, a significant reduction in energy consumption is noted (Table 9).

In one year, by eliminating heat losses, a saving of 89,965.22 € will be made on electricity consumption. In view of the large reduction in energy consumption, this system will be implemented on all injection and extrusion machines.

Table 8. Electricity consumption from different periods after insulation jackets implementation.

Machine	Day 1	Day 2	Day 3	Average
	kWh/Day			
Extruder 1	6580.95	6530.10	6503.58	6538.21
Extruder 2	2861.73	2832.67	2855.62	2850.01
Extruder 3	4457.83	4444.87	4443.33	4448.68
Injection 1	4990.98	4992.11	4992.17	4991.75
Injection 2	5332.85	5357.66	5312.25	5334.25
Injection 3	2844.87	2803.20	2811.97	2820.01

Table 9. Average electricity consumption trends.

Machine	Before	After	Saving
	kWh/Day		
Extruder 1	6636.19	6503.58	132.61
Extruder 2	2923.21	2855.62	67.59
Extruder 3	4607.51	4443.33	164.18
Injection 1	5028.99	4992.17	36.82
Injection 2	5983.97	5312.25	671.72
Injection 3	3494.92	2811.97	682.95
	Total saved		1755.88

Energy consumption will be reduced by 400 MWh/year during the 228 working days. The greenhouse gas emission factor specific to the electricity grid in Romania is 0.26184 $kgCO_2e/kWh$ and was used to determine CO_2 emissions [78]. Based on this it has been estimated that the CO_2 emissions for the organization is 126,5630 tonnes.

In the next section, we make a brief presentation of the projects that can be carried out in the organization.

6. Future Research

The case studies show the use of the Kaizen approach to reduce energy consumption in 3 areas of an organization that produces refrigerating appliances. Kaizen's approach can be applied to other production lines and to the areas that serve the production lines. The research team is in discussion with the organization's management about developing new energy efficiency projects.

The organization is interested to develop Kaizen projects because there is already a culture in terms of application methodology. Based on the study already carried out, it was identified from the first step (Figure 3) that in the maintenance area, there is very high energy consumption. A proposal would be to carry out a Kaizen-type project on lighting in the organization. The aim is to create an interior lighting installation suitable for the production area, with LED lighting, in which the luminous flux is harmoniously distributed and ensures a special climate of comfort. With an investment of 270,000 € in such a project it is expected a reduction of energy consumption of 2,130,344 kWh per year. The lighting levels were calculated according to the regulations in force, by using lamps with higher or lower intensity, the circuits being dimensioned accordingly to allow this. Depending on the service area and the height of the room, projector type lighting will be used for very high heights (over 6 m) and linear lighting for heights below 6 m.

Increasing lighting efficiency in the organization will bring a profit of €167,535, with a payback time of 2 years and 6 months.

Investments in lighting can be made by replacing old lighting spans with the sun tracker mirror system. With an investment of 20,500 €, the annual energy consumption can be reduced by 344,461 kWh. This can lead to an annual financial efficiency of 7628 €/Year.

The objective of the organization is to continue reducing costs. We suggest to apply the Kaizen approach to analyze and discover the technical and technological changes needed to reduce other expenses such as the labor force, the product costs, or the costs of another production line.

In the future, we propose to develop research with the objective of reducing CO_2 emissions as a result of the technological changes proposed in all the production sectors of the analyzed organization. Thus, the total reduction of CO_2 at the organization level could be determined, as a result of the changes discovered and implemented with the help of the Kaizen methodology.

The changes identified and operationalized using Kaizen can be completed with the digitalization of organizational production and administrative processes [79,80]. These changes require extensive and in-depth research in advance to identify the appropriate solutions and the financial resources needed for their implementation as part of an extensive and systematic digital transformation process. Therefore, in future research, we propose to approach organizational digitalization and the increase of energy efficiency as real and necessary strategic options for a modern organization.

7. Conclusions

At the global level, in the current conditions related to energy shortage, but especially regarding the exponential increase of the energy price, we consider that steps are needed for organizations to find solutions to increase the energy efficiency of their production lines. Therefore, to remain in the market, increasing energy efficiency has become a strategic objective for any organization. Worldwide, this strategic organizational objective can lead to other beneficial effects for mankind, such as the reduction of pollution [81] and the decrease of the prices for the products obtained from the production processes carried out.

Kaizen is a process-centric approach towards people and the environment and has developed over time as a managerial strategic option, through which, with small steps and without high additional costs, a systematic analysis of a process can be made to achieve efficiency by detecting the losses (Muda) and by detecting the inefficient use of resources on the targeted production chain.

In this respect, the authors have systematically approached this problem, following the steps already established in the specialized literature of the Kaizen methodology.

By applying the Kaizen methodology in the organization, several areas could be identified for energy improvement. Our attention was focused on the areas with the highest energy consumption and the solutions identified for each studied area led to a reduction in energy consumption. Kaizen is a methodology that can be applied systematically and has led to the identification of losses and the most efficient solutions. The internal resources of the organization were used to implement the Kaizen methodology. Once a routine is established it is standardized and can be applied several times in the organization.

Kaizen methodology has not been applied in the maintenance area because it requires the involvement of managers responsible for all areas of the organization and at this moment there is no well-developed Kaizen culture.

We believe that the method used by us has achieved its intended goal of energy consumption in the organization where we made these case studies.

The use of the Kaizen methodology has a positive impact on the organization. It is a method that can be applied systematically and that can be adapted to the organization's needs.

The application of the Kaizen methodology in these case studies led to the identification of major energy losses and to quick and efficient solutions. The organization needs to ensure that investments made to reduce energy consumption will be recovered in a short period of time so that profits can be reinvested. In these case studies, applying the Kaizen methodology, we started from a general problem to find specific solutions. We can use Kaizen because it is a method for improving processes in an organization, available to everyone, that can be applied with internal resources and minimal costs.

The Kaizen approach can be successfully applied in other organizations to help them eliminate production line losses, product manufacturing process losses, or human resource time losses. Once these losses are eliminated, the total expenses are reduced and the profits of the organizations can increase.

The authors decided to publish this study in the context of the current energy crisis at the global level, to support organizations that want to increase their energy efficiency with minimal costs and pollution [82,83]. The price of the product is given by the production costs, and the customer should pay for costs that add value to the product, not for avoidable expenditures or losses.

Author Contributions: Conceptualization, A.A. and I.-C.E.; methodology, A.A. and E.-N.V.; validation, A.A., I.-C.E., E.-N.V. and F.-F.R.; formal analysis, A.A., I.-C.E., E.-N.V. and F.-F.R.; investigation, I.-C.E.; data curation, A.A., I.-C.E., E.-N.V. and F.-F.R.; writing—original draft preparation, A.A., I.-C.E., E.-N.V. and F.-F.R.; writing—review and editing, A.A., I.-C.E., E.-N.V. and F.-F.R.; supervision, A.A.; funding acquisition, A.A. All authors have read and agreed to the published version of the manuscript.

Funding: This research received no external funding.

Data Availability Statement: Not applicable.

Conflicts of Interest: The authors declare no conflict of interest.

References

1. Faiella, I.; Mistretta, A. The Net Zero Challenge for Firms' Competitiveness. *Environ. Resour. Econ.* **2022**, *83*, 85–113. [CrossRef]
2. Ren, X.; Liu, Z.; Jin, C.; Lin, R. Oil price uncertainty and enterprise total factor productivity: Evidence from China. *Int. Rev. Econ. Financ.* **2023**, *83*, 201–218. [CrossRef]
3. Zumian, X.; Juanhe, G.; Zongshu, W.; Zhichao, Y.; Lijin, X. Power shortage and firm productivity: Evidence from the World Bank Enterprise Survey. *Energy* **2022**, *247*, 123479. [CrossRef]
4. Hongming, N.; Jingchao, S.; Ziyang, Q.; Jianfei, H.; Yuxing, Y.; Tianyi, Y.; Tao, D. A novel evaluation method for energy efficiency of process industry—A case study of typical iron and steel manufacturing process. *Energy* **2021**, *233*, 121081. [CrossRef]
5. Zhengbiao, H.; Dongfeng, H.; Hongbo, Z. Multi-objective optimization of energy distribution in steel enterprises considering both exergy efficiency and energy cost. *Energy* **2023**, *263*, 125623. [CrossRef]
6. Nelfiyanti; Mohamed, N.; Rashid, M.; Ramadhan, A.I. Parameters of effects in decision making of automotive assembly line using the Analytical Hierarchy Process method. *CIRP J. Manuf. Sci. Technol.* **2022**, *37*, 370–377. [CrossRef]
7. EUR-Lex. DIRECTIVE 2012/27/Eu of the European Parliament and of the Council. Available online: http://data.europa.eu/eli/dir/2012/27/oj (accessed on 9 November 2022).
8. EUR-Lex. Directive (Eu) 2018/2002 of the European Parliament and of the Council. Available online: http://data.europa.eu/eli/dir/2018/2002/oj (accessed on 9 November 2022).
9. EUROSTAT. NRG_BAL_S. Available online: https://ec.europa.eu/eurostat/databrowser/view/ten00124/default/line?lang=en (accessed on 9 November 2022).
10. EUROSTAT. Energy Statistics—An Overview. Available online: https://ec.europa.eu/eurostat/statistics-explained/index.php?title=Energy_statistics_-_an_overview&oldid=557437#Final_energy_consumption (accessed on 9 November 2022).
11. García-Quevedo, J.; Jové-Llopis, E. Environmental policies and energy efficiency investments. An industry-level analysis. *Energy Policy* **2021**, *156*, 112461. [CrossRef]
12. Kalantzis, F.; Niczyporuk, H. Labour productivity improvements from energy efficiency investments: The experience of European firms. *Energy* **2022**, *252*, 123878. [CrossRef]
13. IEA. Global Energy Review: CO_2 Emissions in 2021—Analysis. Available online: https://www.iea.org/reports/global-energy-review-co2-emissions-in-2021-2 (accessed on 9 November 2022).
14. Geoffrey, M. *Kaizen Event Implementation Manual*, 5th ed.; Society of Manufacturing Engineers: Dearborn, MI, USA, 2006; p. 21.
15. Chan, D.Y.L.; Huang, C.F.; Lin, W.C.; Hong, G.B. Energy efficiency benchmarking of energy-intensive industries in Taiwan. *Energy Convers. Manag.* **2014**, *77*, 216–220. [CrossRef]
16. Shabbir, I.; Mirzaeian, M.; Sher, F. Energy efficiency improvement potentials through energy benchmarking in pulp and paper industry. *Clean. Chem. Eng.* **2022**, *3*, 100058. [CrossRef]
17. Gorus, M.S.; Karagol, E.T. Reactions of energy intensity, energy efficiency, and activity indexes to income and energy price changes: The panel data evidence from OECD countries. *Energy* **2022**, *254*, 124281. [CrossRef]

18. Boyd, G.A. Comparing the statistical distributions of energy efficiency in manufacturing: Meta-analysis of 24 Case studies to develop industry-specific energy performance indicators (EPI). *Energy Effic.* **2017**, *10*, 217–238. [CrossRef]
19. Soni, A.; Mittal, A.; Kapshe, M. Energy intensity analysis of Indian manufacturing industries. *Resour. Effic. Technol.* **2017**, *3*, 353–357. [CrossRef]
20. Vijayaraghavan, A.; Dornfeld, D. Automated energy monitoring of machine tools. *CIRP Ann.* **2010**, *59*, 21–24. [CrossRef]
21. Xiao, Q.; Li, C.; Tang, Y.; Li, L.; Li, L. A knowledge-driven method of adaptively optimizing process parameters for energy efficient turning. *Energy* **2019**, *166*, 142–156. [CrossRef]
22. Hrovatin, N.; Cagno, E.; Dolšak, J.; Zorić, J. How important are perceived barriers and drivers versus other contextual factors for the adoption of energy efficiency measures: An empirical investigation in manufacturing SMEs. *J. Clean. Prod.* **2021**, *323*, 129123. [CrossRef]
23. Cannata, A.; Karnouskos, S.; Taisch, M. Energy efficiency driven process analysis and optimization in discrete manufacturing. In Proceedings of the 35th Annual Conference of IEEE Industrial Electronics, Porto, Portugal, 3–5 November 2009; pp. 4449–4454. [CrossRef]
24. Dietmair, A.; Verl, A. Energy consumption modeling and optimization for production machines. In Proceedings of the IEEE International Conference on Sustainable Energy Technologies, Singapore, 24–27 November 2008; pp. 574–579. [CrossRef]
25. Cagno, E.; Accordini, D.; Trianni, A.; Katic, M.; Ferrari, N.; Gambaro, F. Understanding the impacts of energy efficiency measures on a Company's operational performance: A new framework. *Appl. Energy* **2022**, *328*, 120118. [CrossRef]
26. Petersen, J.P.; Winther, O.; Jacobsen, D.J. A Machine-Learning Approach to Predict Main Energy Consumption under Realistic Operational Conditions. *Ship Technol. Res.* **2012**, *59*, 64–72. [CrossRef]
27. Gupta, A.; Randhawa, P. *Implementing Industry 4.0 and Sustainable Manufacturing: Leading to Smart Factory. Industry 4.0 and Advanced Manufacturing*; Springer: Singapore, 2022; pp. 471–482. [CrossRef]
28. Saidur, R. A review on electrical motors energy use and energy savings. *Renew. Sustain. Energy Rev.* **2010**, *14*, 877–898. [CrossRef]
29. Waide, P.; Brunner, C.U. *Energy-Efficiency Policy Opportunities for Electric Motor-Driven Systems*; IEA Energy Papers; IEA: Paris, France, 2011; p. 7. [CrossRef]
30. Gonzalez-Abreu, A.-D.; Osornio-Rios, R.-A.; Jaen-Cuellar, A.-Y.; Delgado-Prieto, M.; Antonino-Daviu, J.-A.; Karlis, A. Advances in Power Quality Analysis Techniques for Electrical Machines and Drives: A Review. *Energies* **2022**, *15*, 1909. [CrossRef]
31. Tabora, J.M.; De Lima Tostes, M.E.; De Matos, E.O.; Bezerra, U.H.; Soares, T.M.; De Albuquerque, B.S. Assessing Voltage Unbalance Conditions in IE2, IE3 and IE4 Classes Induction Motors. *IEEE Access* **2020**, *8*, 186725–186739. [CrossRef]
32. European Union. Commission Regulation (EU) 2019/1781. Laying down ecodesign requirements for electric motors and variable speed drives pursuant to Directive 2009/125/EC of the European Parliament and of the Council, amending Regulation (EC) No 641/2009 with regard to ecodesign requirements for glandless standalone circulators and glandless circulators integrated in products and repealing Commission Regulation (EC) No 640/2009. *Off. J. Eur. Union* **2019**, *Ll272*, 74–94.
33. Atul, P.; Pankaj, D. Lean manufacturing a vital tool to enhance productivity in manufacturing. *Mater. Today Proc.* **2021**, *46*, 729–736. [CrossRef]
34. Baysan, S.; Kabadurmus, O.; Cevikcan, E.; Satoglu, S.; Durmusoglu, M.B. A simulation-based methodology for the analysis of the effect of lean tools on energy efficiency: An application in power distribution industry. *J. Clean. Prod.* **2019**, *211*, 895–908. [CrossRef]
35. Belekoukias, I.; Garza-Reyes, J.A.; Kumar, V. The impact of lean methods and tools on the operational performance of manufacturing organisations. *Int. J. Prod. Res.* **2014**, *52*, 5346–5366. [CrossRef]
36. Naeemah, A.J.; Wong, K.Y. Positive impacts of lean manufacturing tools on sustainability aspects: A systematic review. *J. Ind. Prod. Eng.* **2022**, *39*, 552–571. [CrossRef]
37. Matt, D.T.; Rauch, E. Implementation of Lean Production in Small Sized Enterprises. *Procedia CIRP* **2013**, *12*, 420–425. [CrossRef]
38. Tayal, A.; Kalsi, N.S. Review on effectiveness improvement by application of the lean tool in an industry. *Mater. Today Proc.* **2021**, *43*, 1983–1991. [CrossRef]
39. Omogbai, O.; Salonitis, K. The Implementation of 5S Lean Tool Using System Dynamics Approach. *Procedia CIRP* **2017**, *60*, 380–385. [CrossRef]
40. Ramos, A.R.; Espíndola-Ferreira, J.C.; Kumar, V.; Garza-Reyes, J.A.; Cherrafi, A. A lean and cleaner production benchmarking method for sustainability assessment: A study of manufacturing companies in Brazil. *J. Clean. Prod.* **2018**, *177*, 218–231. [CrossRef]
41. Sundararajan, N.; Terkar, R. Improving productivity in fastener manufacturing through the application of Lean-Kaizen principles. *Mater. Today Proc.* **2022**, *62*, 1169–1178. [CrossRef]
42. Abdulmalek, F.A.; Rajgopal, J. Analyzing the benefits of lean manufacturing and value stream mapping via simulation: A process sector case study. *Int. J. Prod. Econ.* **2007**, *107*, 223–236. [CrossRef]
43. Wang, P.-S.; Yang, T.; Yu, L.-C. Lean-pull strategy for order scheduling problem in a multi-site semiconductor crystal ingot-pulling manufacturing company. *Comput. Ind. Eng.* **2018**, *125*, 545–562. [CrossRef]
44. Singh, S.; Kumar, K. A study of lean construction and visual management tools through cluster analysis. *Ain Shams Eng. J.* **2021**, *12*, 1153–1162. [CrossRef]
45. Agyabeng-Mensah, Y.; Tang, L.; Afum, E.; Baah, C.; Dacosta, E. Organisational identity and circular economy: Are inter and intra organisational learning, lean management and zero waste practices worth pursuing? *Sustain. Prod. Consum.* **2021**, *28*, 648–662. [CrossRef]

46. Aksoy, A.; Öztürk, N. Supplier selection and performance evaluation in just-in-time production environments. *Expert Syst. Appl.* **2011**, *38*, 6351–6359. [CrossRef]
47. Brito, M.; Vale, M.; Leão, J.; Ferreira, L.P.; Silva, F.J.G.; Gonçalves, M.A. Lean and Ergonomics decision support tool assessment in a plastic packaging company. *Procedia Manuf.* **2020**, *51*, 613–619. [CrossRef]
48. Saurin, T.A.; Duarte Ribeiro, J.L.; Vidor, G. A framework for assessing poka-yoke devices. *J. Manuf. Syst.* **2012**, *31*, 358–366. [CrossRef]
49. Enache, I.C.; Simion, I.; Avramescu, A.-M.; Chiscop, F.; Chiscop, M.A. Using 6 Sigma in system capability diagnosis to increase productivity. *Univ. Politeh. Bucur. Sci. Bull. Ser. D* **2014**, *76*, 133–142.
50. Ortiz, C. Kaizen vs. Lean: Distinct but related. *Met. Finish.* **2010**, *108*, 50–51. [CrossRef]
51. Imai, M. *Gemba Kaizen: A Commonsense Approach to a Continuous Improvement Strategy*; McGraw-Hill: New York, NY, USA, 1997.
52. Harada, T. *Management Lessons from Taiichi Ohno*; McGraw-Hill: New York, NY, USA, 2015.
53. Kaplinsky, R.; Posthuma, A. *Easternization: The Spread of Japanese Management Techniques to Developing Countries*; Routledge: London, UK, 2004.
54. Otsuka, K.; Jin, K.; Sonobe, T. *Applying Kaizen in Africa: A New Avenue for Industrial Development*; Palgrave Macmillan: Cham, Switzerland, 2018. [CrossRef]
55. Maarof, M.G.; Mahmud, F. A Review of Contributing Factors and Challenges in Implementing Kaizen in Small and Medium Enterprises. *Procedia Econ. Financ.* **2016**, *35*, 522–531. [CrossRef]
56. Bellgran, M.; Kurdve, M.; Hanna, R. Cost driven Green Kaizen in pharmaceutical production—Creating positive engagement for environmental improvements. *Procedia CIRP* **2019**, *81*, 1219–1224. [CrossRef]
57. Radharamanan, R.; Godoy, L.P.; Watanabe, K.I. Quality and productivity improvement in a custom-made furniture industry using Kaizen. *Comput. Ind. Eng.* **1996**, *31*, 471–474. [CrossRef]
58. Pinto, F.L.; Silva, F.J.G.; Campilho, R.D.S.G.; Casais, R.B.; Fernandes, A.J.; Baptista, A. Continuous improvement in maintenance: A case study in the automotive industry involving Lean tools. *Procedia Manuf.* **2019**, *38*, 1582–1591. [CrossRef]
59. Orynycz, O.; Tucki, K.; Prystasz, M. Implementation of Lean Management as a Tool for Decrease of Energy Consumption and CO2 Emissions in the Fast Food Restaurant. *Energies* **2020**, *13*, 1184. [CrossRef]
60. Robinson, A.G.; Schroeder, D.M. *Ideas Are Free: How the Idea Revolution Is Liberating People and Transforming Organizations*; Berrett-Koehler Publishers: San Francisco, CA, USA, 2006.
61. García, J.L.; Rivera, D.G.; Iniesta, A.A. Critical success factors for Kaizen implementation in manufacturing industries in Mexico. *Int. J. Adv. Manuf. Technol.* **2013**, *68*, 537–545. [CrossRef]
62. Keating, E.; Oliva, R.; Repenning, N.; Rockart, S.; Sterman, J. Overcoming the improvement paradox. *Eur. Manag. J.* **1999**, *17*, 120–134. [CrossRef]
63. Meister, M.; Böing, T.; Batz, S.; Metternich, J. Problem-solving process design in production: Current progress and action required. *Procedia CIRP* **2018**, *78*, 376–381. [CrossRef]
64. Magnier-Watanabe, R. Getting ready for kaizen: Organizational and knowledge management enablers. *Vine* **2011**, *41*, 428–448. [CrossRef]
65. Miller, J.; Wroblewski, M.; Villafuerte, J. *Creating a Kaizen Culture: Align the Organization, Achieve Breakthrough Results, and Sustain the Gains*; McGraw-Hill Education: New York, NY, USA, 2013.
66. Janjić, V.; Bogicevic, J.R.; Krstić, B. Kaizen as a global business philosophy for continuous improvement of business performance. *Ekonomika* **2019**, *65*, 13–25. [CrossRef]
67. Chen, J.C.; Chen, K.M. Application of ORFPM system for lean implementation: An industrial case study. *Int. J. Adv. Manuf. Technol.* **2014**, *72*, 839–852. [CrossRef]
68. Marques, P.A.; Carvalho, A.M.; Santos, J.O. Improving Operational and Sustainability Performance in a Retail Fresh Food Market Using Lean: A Portuguese Case Study. *Sustainability* **2022**, *14*, 403. [CrossRef]
69. Rodrigues, J.; Sá, J.C.; Silva, F.J.G.; Ferreira, L.P.; Jimenez, G.; Santos, G. A Rapid Improvement Process through "Quick-Win" Lean Tools: A Case Study. *Systems* **2020**, *8*, 55. [CrossRef]
70. Souza, D.L.; Korzenowski, A.L.; Alvarado, M.M.; Sperafico, J.H.; Ackermann, A.E.F.; Mareth, T.; Scavarda, A.J. A Systematic Review on Lean Applications' in Emergency Departments. *Healthcare* **2021**, *9*, 763. [CrossRef]
71. Casey, J.T.; Brinton, T.S.; Gonzalez, C.M. Utilization of lean management principles in the ambulatory clinic setting. *Nat. Rev. Urol.* **2009**, *6*, 146–153. [CrossRef]
72. Palepu, S.; Nitsch, A.; Narayan, M.; Kim, S.; Osier, N. A Flat Organizational Structure for an Inclusive, Interdisciplinary, International, and Undergraduate-Led Team. *Front. Educ.* **2020**, *5*, 102. [CrossRef]
73. Salah, S.; Mustafa, A. Integration of energy saving with Lean production in a food processing company. *J. Mach. Eng.* **2021**, *21*, 118–133. [CrossRef]
74. Thiede, S.; Bogdanski, G.; Herrmann, C. A Systematic Method for Increasing the Energy and Resource Efficiency in Manufacturing Companies. *Procedia CIRP* **2012**, *2*, 28–33. [CrossRef]
75. Schneider Electric România. Available online: https://www.se.com/ro/ro/product-range/1155-altivar-71/ (accessed on 15 November 2022).
76. Balaji, M.; Dinesh, S.N.; Raja, S.; Subbiah, R.; Manoj Kumar, P. Lead time reduction and process enhancement for a low volume product. *Mater. Today Proc.* **2022**, *62*, 1722–1728. [CrossRef]

77. IEA. What Is Behind Soaring Energy Prices and What Happens Next?—Analysis. Available online: https://www.iea.org/commentaries/what-is-behind-soaring-energy-prices-and-what-happens-next (accessed on 15 November 2022).
78. Carbon Footprint. Available online: https://www.carbonfootprint.com/docs/2022_03_emissions_factors_sources_for_2021_electricity_v11.pdf (accessed on 7 January 2023).
79. Borowski, P.F. Digitization, Digital Twins, Blockchain, and Industry 4.0 as Elements of Management Process in Enterprises in the Energy Sector. *Energies* **2021**, *14*, 1885. [CrossRef]
80. Borowski, P.F. Digital Transformation and Prosumers Activities in the Energy Sector. In *Intelligent Systems in Digital Transformation. Lecture Notes in Networks and Systems*; Kahraman, C., Haktanır, E., Eds.; Springer: Cham, Switzerland, 2023; p. 549. [CrossRef]
81. Androniceanu, A.; Georgescu, I.; Sabie, O.M. Comparative research on government effectiveness and political stability in Europe. *Adm. Si Manag. Public* **2022**, *1*, 63–76. [CrossRef]
82. Androniceanu, A.; Sabie, O.M. Overview of green energy as a real strategic option for sustainable development. *Energies* **2022**, *15*, 857. [CrossRef]
83. Androniceanu, A.; Georgescu, I. The Impact of CO_2 Emissions and Energy Consumption on Economic Growth: A Panel Data Analysis. *Energies* **2023**, *16*, 1342. [CrossRef]

energies

MDPI

Article

Not All about the Money: The Role of Financial Information in Promoting Residential Rooftop Photovoltaics

David Boaz [1,*], Ruslana Rachel Palatnik [2,3] and Ofira Ayalon [1,3]

1 School of Environmental Sciences, University of Haifa, Haifa 3498838, Israel; oayalon@univ.haifa.ac.il
2 Department of Economics and Management, Yezreel Valley College, Emek Yezreel 1930600, Israel; rachelpa@yvc.ac.il
3 NRERC—Natural Resources and Environmental Research Center, University of Haifa, Haifa 3498838, Israel
* Correspondence: boaz1234@gmail.com; Tel.: +972-542448622

Abstract: This study investigates the role of financial information in shaping the intention to adopt (ITA) Residential Rooftop Photovoltaic (RPV) systems in Israel—a country with a surprisingly low RPV adoption rate given the high solar irradiance, and significant governmental support for RPV. We used an online survey of 899 respondents to examine how different presentations of RPV revenues impact both RPV deal comprehension and ITA. Our results align with existing literature on the roles of early technology adopters and pro-environmental characteristics while introducing two novel findings. First, households value their roofs—especially flat ones used for leisure. This negatively impacts ITA. Second, we found a positive correlation between trust in government entities and ITA, underscoring the importance of trust in institutions. Although financial information does enhance RPV deal comprehension, it does not significantly influence ITA. This study concludes that RPV penetration should not be promoted solely by financial incentives, but also by other motivations like environmental concerns, novelty seeking, and improved institutional trust.

Keywords: Residential Photovoltaics; financial information; photovoltaic adoption; willingness to accept

1. Introduction

In the milieu of rising global apprehension over climate change, nations are increasingly pivoting towards renewable energy sources. Residential Photovoltaic (RPV) systems, characterized by their favorable energy life cycle, have become central to this environmental mitigation strategy. Despite governmental incentives and the demonstrated financial viability of these systems, their adoption remains disappointingly low. We hypothesized that one of the barriers for RPV adoption is the lack of understanding of the financial benefits of the RPV deal by roof owners' households. This study, conducted in the Israeli context, examines the question whether clarifying the financial aspects of the RPV deal would increase RPV adoption among households.

PV systems can be incorporated in a centralized or distributed fashion, including dual-land-usage sites, such as rooftops, parking lots, cemeteries, and water reservoirs. The interplay between ground and dual-use facilities presents a complex quandary: larger, centralized ground facilities offer cheaper electricity production through economies of scale, albeit with significant land usage [1]. Conversely, dual-use PV facilities are costlier but less land intensive. Decentralized small PV sources also have several distinct advantages, including reliable power supply, minimized energy transmission losses due to proximity to consumption areas, the potential for deferring costly transmission network upgrades, distributed profit-sharing opportunities, job creation, and enhanced energy-saving awareness [2,3].

Considering these benefits and notwithstanding higher initial costs, several countries actively encourage residents to become prosumers (producers–consumers of energy) by installing small, distributed RPV systems on their rooftops. Feed-in tariffs, which are pre-agreed, long-term rates that state entities commit to buying electricity, are the most common incentive for prosumers [4,5]. These tariffs are set based on system size, technology, quality, and grid connection type, and are designed to ensure a satisfactory return on investment.

From the household's point of view, installing RPV systems requires substantial initial capital expenditure, making it a high-stakes decision. Potential adopters are confronted with a complex decision matrix, balancing the promised financial returns and environmental contributions against the high initial costs of the system, long payback periods, perceived risks associated with innovative technology, trust in governmental entities, and the lack of knowledge and uniformity, often leading individuals to default to the status quo in times of uncertainty [6]. Tables 1 and 2 outline the literature on household motivations and barriers to RPV adoption.

Table 1. Motivations to adopt RPV.

Motivation/Factor	Reason	References
Perceived financial benefits, due to subsidies or electricity bill savings	There is a consensus that this is the most important factor	[7–20]
Self-sufficiency	People aspire to supply themselves with their own energy. It fulfills households' desire to be more independent from the state, their utility provider, conventional energy sources, and rising energy prices	[18,21,22]
Novelty seeking, being a technology early adaptor	People with higher novelty seeking have greater interest in pursuing RPV	[12,19]
Pro-environmental behavior	Among early market stages	[12,17–20,22,23]
Peers effects and subjective norms	Among later adopters	[12,15,17,19,20,22,24–27]
Age	Younger tend more to adopt RPV	[12,15,23]
Gender	Men declared more ITA. But women found to actually adopt	[12,20,23]
Education	Education is needed to understand the financial cost and benefits of RPV	[12,20,28]
Rural location	Explained by higher distribution tariffs in rural areas of Finland	[20]
Socio-economic level, income	High income correlates with actual adoption Low- and moderate-income levels have similar ITA	[8,17,24,28–31]
Information campaigns	Information campaign had a significantly positive effect on PV adoption rates	[32]
System and installer reliabilities	Homeowners care about system reliability and ease of maintenance. In- staller's reliability is potentially key to the long-term savings	[14]

Table 2. Barriers to adopting RPV.

Barrier	Reason	References
High upfront cost	An RPV system costs tens of thousands of USD—a significant amount	[7,11,13,17]
Perceived as an innovation	Lack of trust in the new technology, fear of unexpected failures and high maintenance costs	[11,12,17]
Long-term deal and time discounting	The payback period is 5 to 10 years. People value today's money more than future money	[5,20,33–35]
Uncertainty in government or market policies	Fear that the government will change the feed-in tariffs in the future, offer better deals, or that the price of energy will decrease	[7,9,15,33,36]
Lack of knowledge and poor information	People are not aware to the option of installing RPV	[12,20,37]
Lack of uniformity among system installers	A variety of technologies, layouts, and system sizes, making it hard to compare among offers	[38,39]

Our research offers a unique dimension by investigating the role of visual financial cues in shaping ITA RPV, presented through artificial fictive electricity bills. Grounded in Larrick et al.'s [40] recommendations of using expanded scales when presenting metrics of expensive energy-related products and drawing inspiration from Allcott's [41] behavioral inquiries using electricity bills, this study hypothesizes that incorporating long-term financial benefits into monthly bills may enhance prospective adopters' perception of the RPV investment.

Additionally, we scrutinize an overlooked factor in RPV adoption: the perceived value of existing rooftop usages [19]. We hypothesize that the type and utilization of a rooftop—whether for leisure, gardening, or storage—could influence its perceived value, thereby negatively affecting the ITA RPV.

Yet another unique dimension of our study is the exploration of the link between households' trust in institutional entities, specifically the government and the electricity company, and ITA RPV systems. This extends the findings of Petrovich et al. [36], who noted that the ITA RPV declines when there's a perception of increased policy risk. Our research investigates whether trust in these critical institutions could significantly influence long-term investment decisions in this area.

This study employs a stated preference approach, utilizing an online survey to evaluate whether the presentation of a revenue graph, detailed financial information, or a combination thereof impacts ITA RPV. These innovative methods aim to provide answers to pivotal queries from potential RPV adopters, such as the expected total revenue, return on investment, and payback period.

Situated in Israel—a country with ambitious renewable energy targets but low RPV adoption rates—this research holds implications for energy policy in other countries. Through the focus on financial nudges and the under-investigated variable of perceived rooftop value, this study offers a nuanced understanding of the complex decision-making processes behind RPV adoption.

As presented in Figure 1, the state of Israel has incentivized small PV systems since 2008 (the tariff declines in correspondence with the decline in PV panel prices—aiming to be attractive for the households, and yet to minimize its effect on the cost of living). Yet, till 2022, approximately 3% of Israel's attached houses are covered by PV panels (see calculation in Appendix A). The Israeli backdrop serves as a compelling case study given its combination of high solar irradiance, and significant governmental backing for renewable energy. Government Decision 465 (2020) underscores Israel's targets: 20% renewable energy by 2025 and 30% by 2030. Therefore, understanding the behavioral nuances that influence RPV adoption in such an environment is not just nationally relevant, but holds wider implications for countries with similar characteristics or aspirations.

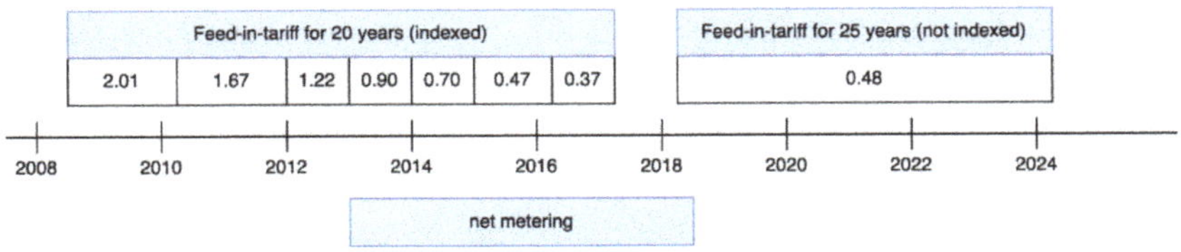

Figure 1. History of feed-in tariffs for small PV systems (in ILS, per generated kWh. One ILS is lower than approximately a-quarter of a USD).

Feed-in tariffs in Israel are designed to incentivize the adoption of smaller RPV systems, encouraging residents to become energy producers. However, despite these tariffs, there is a significant gap between policy goals and the actual adoption rates, prompting a need for further investigation. We conclude that the decision on RPV adoption is "not all about the money", and other factors are playing important role. The study conclusions might also apply to other countries willing to promote RPV penetration without harming the cost of living.

This study also embarks few methodological innovations. We employed fictive electricity bills to simulate real-world conditions, thus bypassing the limitations associated with an inability to alter actual utility bills. By doing so, it aims to gauge the influence of financial nudges more accurately on ITA, offering a robust methodological framework for similar studies in the future. This research also contributes to the field by focusing on the often-neglected aspect of perceived rooftop value. With its multifaceted approach, this study enriches the scholarly conversation on renewable energy adoption and provide practical insights for policy-making and public engagement strategies.

The rest of this paper is structured as follows: First, we describe the variables and the examined hypotheses on the relations among them. Then, we present the methods we used to collect and test our hypotheses and a description of the sample. This is followed by the results, a discussion, and a conclusion section summarizing the main research findings, recommendations for action, research limitations, and future research.

2. Materials and Methods

To evaluate whether explicitly presenting financial information could accelerate the diffusion of the innovation to potential adopters, this study utilized a stated preference approach. The section outlines the variables and hypotheses, methodology, and data.

2.1. Variables and Hypotheses

In this study, we expanded upon the model proposed by Wolske et al. [12] for intent to install RPV systems. We incorporated additional independent variables related to the respondents' characteristics, inspired by the model's original variables: novelty seeking, pro-environmental behavior, trust in government and electricity companies, gender, income, education, age, urban or rural residence, proximity to other PV systems, and acquaintances with PV. We further introduced two additional independent variables related to roof features: roof type (e.g., flat or tiles) and roof usage (e.g., storage or leisure). We hypothesized that these variables could significantly influence the decision-making process for RPV adoption.

Our study examined two dependent variables:

1. The level of financial comprehension of the deal was measured through three quantitative open-ended questions regarding average monthly revenue, total accumulated revenues over the system's lifetime, and the payback period in years.

2. The ITA an RPV system, measured similarly to [12,15], as the average of three 1 to 5 Likert scale questions. Similarly to Mundaca and Samahita [15] and Ahmar et al. [42], we constructed an ordinal logistic regression model to predict the ITA of a roof owner given the explanatory variables. The model is defined as:

$$ITA_i = \alpha + \sum_k \beta_k \times x_{k,i} + e_{k,i} \quad (1)$$

where ITA_i is the ITA RPV of household i; α is the constant regression cut; $x_{k,i}$ is the value of explanatory variable k for household i, with the coefficient β_k, and the stochastic standard error $e_{k,i}$.

Our research, presented in Figure 2 investigates the following hypotheses:

- **H1.** *The presentation of accumulated financial information about the RPV deal directly increases the ITA RPV.*
- **H2.** *The presentation of accumulated financial information affects the comprehension of the RPV deal.*
- **H3.** *For a financially viable RPV deal, better comprehension of the deal increases the ITA of an RPV project.*
- **H4.** *RPV-specific beliefs and attitudes have an impact on the ITA an RPV project.*
- **H5.** *Respondents' characteristics influence the ITA an RPV project.*
- **H6.** *Roof characteristics influence the ITA an RPV project.*
- **H7.** *Respondents' characteristics have an impact on the perceived roof value.*
- **H8.** *Roof characteristics have an impact on the perceived roof value.*
- **H9.** *Households that perceive high value to their roof have lower ITA an RPV project.*

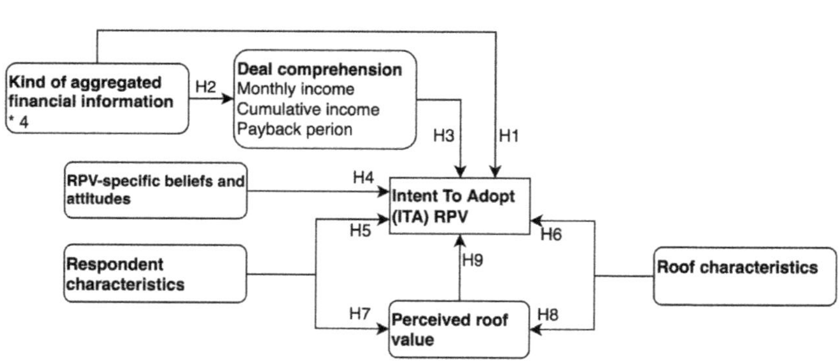

Figure 2. Research hypotheses.

2.2. Data Acquisition

To test these hypotheses, we conducted an online survey. The target population consisted of households in Israel that own detached houses and have not installed a PV system. The survey flow included stratified sampling according to known demographic distribution, consent to participate, screening questions, socio-demographic characteristics, roof characteristics, and a series of questions related to RPV adoption, financial comprehension, and RPV-specific beliefs and attitudes.

We designed a fictive April 2022 electricity bill simulating a resident with a 15 kW PV system installed 25 months ago. The hypothetic purchase is based on characteristics of an actual system: Purchasing and installing such a system cost approximately ILS 80,000 (approximately 23,000 USD, in 2022 prices). Assuming 1700 effective sunlight hours per year, a conservative degradation rate of 1% per year in panels efficiency [43], maintenance costs of 1% per year, and a constant feed-in tariff (FIT) of ILS 0.48 per generated kWh, with 3% annual real interest rate and project duration of 25 years. The conservative estimation reveals RPV as a highly profitable endeavor. The expected return on investment (ROI) is estimated to reach 140%, a net present value (NPV) of 112,000 ILS (more than 30,000 USD), an internal rate of return (IRR) of 12.7%, and a payback period of approximately 9 years. Yet this information is not evident to the average household. A typical electricity bill summarizes the periodic (monthly) electricity consumption and electricity generation, but it does not contain the accumulated revenues from the PV system.

To examine H1, we designed four electricity bills, with various types of information regarding revenues from RPV. Each respondent was randomly assigned to one of four groups and was presented with a corresponding electricity bill (see Figure S1 in the Supplementary Materials):

1. *Control*: as-is electricity bill, with no accumulated revenue information.
2. *Graph*—as-is bill + past and future revenues flow graph, as presented in Figure 3.
3. *Text of financial information*—as-is bill + the following concise text description summarizing the revenues in accepted financial indices:

   "The contract with the electricity company is for 25 years in which it undertakes to purchase the electricity you will generate for 0.48 ILS per kWh. Your system capacity is 15 kW. The system was installed 25 months ago. Till today, the system revenues are 23,507 ILS. The system is expected to create a total revenue of 178,000 ILS. Assuming the system cost was 80,000 ILS, the payback period is 9 years, the expected return on investment is 140%, and the internal return rate is 12.7% per year".

4. *Both*—combined display of graph and financial information.

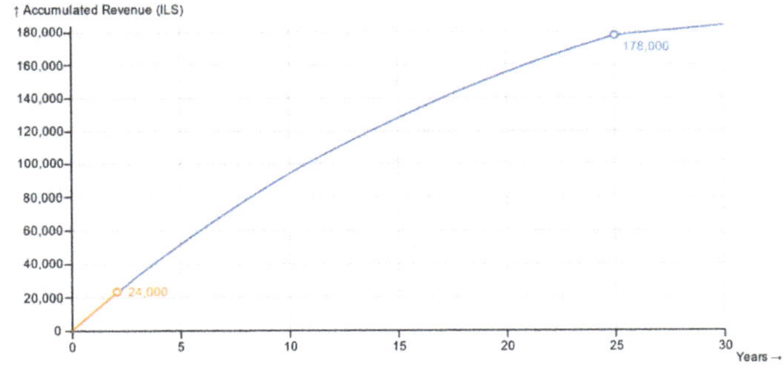

Figure 3. Revenues graph we presented in the experiment. Red: past revenues, Blue: future revenues.

To test the hypotheses, we designed a questionnaire inspired by similar previous research [12,15,44]. The target population is residential roof owners who don't have a PV system. Hence, we screened out participants that did not satisfy these two conditions. We made sure to keep the questionnaire at a reasonable length and not tedious. When translating or adapting to Hebrew, we paid attention to wording that is short, specific, neutral, and gender-agnostic. Once we had an initial draft, we validated the questionnaire by ten scholar experts in the fields of economic and environmental behavior. After we compiled the experts' feedback, we conducted a pilot among ten respondents in the target population. The full questionnaire and the conditional information displays are shown in the Supplementary Materials S1 and S2.

Due to budget constraints, two online surveys were conducted in March 2022 and January 2023 (see Table 3). We worked with a market research company that manages a panel of members willing to participate in questionnaires for a nominal fee (shopping vouchers worth approximately ILS 12). The online panel contains approximately 50,000 active members aged 18 and over. The panel is a representative sample of the Jewish population in Israel. The panel administrators invite members to participate in the sample and validate that there will be no significant deviation from four controlled variables whose distribution has been defined in advance according to their distribution in the population:

gender, age group (6 age groups), geographic area code (5 areas) and proximity to religion (4 categories—to avoid overrepresentation of Jews ultra-Orthodox Jews). The average survey response time was 13 min. Participants who completed the questionnaire in less than 5 min were filtered out to avoid guesswork, yielding a unified dataset of 899 valid cases. The two surveys were mostly identical, except few differences, as described below.

Table 3. Summary of samples.

Sample	Date	No. of Participants	No. Passed Screening Questions	No. Duration > 5 min
A	March 2022	980	301	270
B	January 2023	2779	822	629

Tables 4 and 5 show descriptive statistics of the sample and compare them to the general Israeli population in the census. The sample distribution does not deviate from the census in terms of gender, age, and the number of persons in a household. The distribution of urban respondents in the sample is smaller compared to the population, as the target population is roof owners who are typically located in rural areas. The household income average was significantly lower than the population mean, possibly due to respondent's confusion between individual vs. household incomes. Supplementary Material S3 presents the distribution of the respondents among the four condition groups.

Table 4. Distribution of categorical variables in the sample compared to the Israeli Census.

		Sample N = 899, n (%)	Census [45] (%)
Gender (% Men)		477 (53.1)	(49.6)
Up to 35	Age, years (%)	244 (27.1)	(32.1)
35 to 61		477 (53.1)	(49.1)
61 and above		178 (19.8)	(18.7)
Urban (% urban)		655 (72.9)	(91)
Education (% academic)		568 (63.2)	(50.1)
Household Income (% High = over the median 15,000 ILS), N = 814		147 (18.1)	(50) [46]
Tiles	Roof type (%) *	407 (45.3)	
Flat concrete		453 (50.4)	
Flat floored		91 (10.1)	
No use	Roof usage (%) *	272 (30.3)	
Facilities		545 (60.6)	
Storage		114 (12.7)	
Leisure		103 (11.5)	

* Roof type and roof usage were multiple-choice questions. Hence, the sums are greater than 100%.

Table 5. Distribution of numeric variables in the sample.

	Range	Mean	SD	Median
Household size (census = 3.23)	>1	3.9	1.9	4.0
PV in neighborhood	≥0	1.7	0.6	2.0
Acquaintances with PV	≥0	1.4	3.8	0.0
Trust gov.	1 to 5	2.7	1.0	3.0
Trust IEC	1 to 5	2.6	1.0	3.0
Novelty seeking	1 to 5	3.7	0.8	4.0
Pro environmental	1 to 5	3.0	0.8	3.0

3. Results

The report of findings of our research starts with discussing the valuation of perceived roof value and how it correlates with the characteristics of the participants. Next, we delve into the primary reasons cited by participants for not yet RPV systems. Subsequently, we explore the impact of how information was presented on participants' comprehension of RPV deals. Lastly, we scrutinize the interrelationships between diverse factors and the ITA RPV.

3.1. *Perceived Roof Value*

We applied a willingness-to-accept (WTA) methodology [47,48] to calculate the value households attribute to their "free" roof. This approach allows us to conduct a comprehensive cost-benefit analysis, factoring in the alternative value of the roof's use under current conditions. We employed a single-bounded dichotomous choice model for this study. Each participant was presented with a random compensation bid to consider in exchange for relinquishing his current use of the roof. This WTA compensation format was chosen as the most fitting way to guide participants in quantifying the value they place on their current rooftop usage.

The WTA-specific bids to compensate for relinquishing the current use of the roof and the correspondent ITA RPV are presented in Table 6. In all bids, the mean ITA RPV was higher among those who accepted the bid. This difference was statistically significant in four of five bids. Table 7 presents a logistic regression analysis predicting the WTA ("yes" answers) ratio. The model accounted for 69.2% of the variation (C statistic = 0.692, R^2 = 11.48%).

Table 6. Willingness to accept a bid for giving up the roof and its relation to ITA RPV.

Bid (ILS)	Answer	N	Ratio "Yes"	ITA RPV Mean (SD)
10,000	"Yes"	64	54.7%	3.1 (0.9) **
	"No"	53		2.8 (0.7)
50,000	"Yes"	91	66.0%	2.9 (0.8)
	"No"	48		2.7 (0.9)
100,000	"Yes"	103	75.7%	3.1 (0.8) ***
	"No"	33		2.1 (1.1)
150,000	"Yes"	106	75.7%	3.1 (0.7) **
	"No"	34		2.7 (0.9)
200,000	"Yes"	71	73.2%	2.9 (0.9) ***
	"No"	26		2.2 (0.9)

** $p < 0.01$, and *** $p \leq 0.001$.

Table 7. Logistic regression predicting the willingness to accept bid compensation for current roof services.

	b (S.E.)
Bid	0.29 (0.07) ***
Gender (Men)	
Urban	
Education	
Income	
Household size	
Age	
Roof type—Tiles	0.40 (0.20) *
Roof type—Flat concrete	
Roof type—Flat floored	
Roof Usage—No use	1.01 (0.38) **
Roof usage: Storage	
Roof Usage—Facilities	
Roof Usage—Leisure	−1.32 (0.32) ***

* $p < 0.05$, ** $p < 0.01$, and *** $p \leq 0.001$.

Following Hanemann [47], we utilized the regression equation to estimate the value of having a "free" roof (Figure 4). The estimated roof value is computed by the area enclosed between the distribution function and the Y-axis, which yielded 62,134 ILS. The significantly high roof value exceeds four times the median household monthly net income [46].

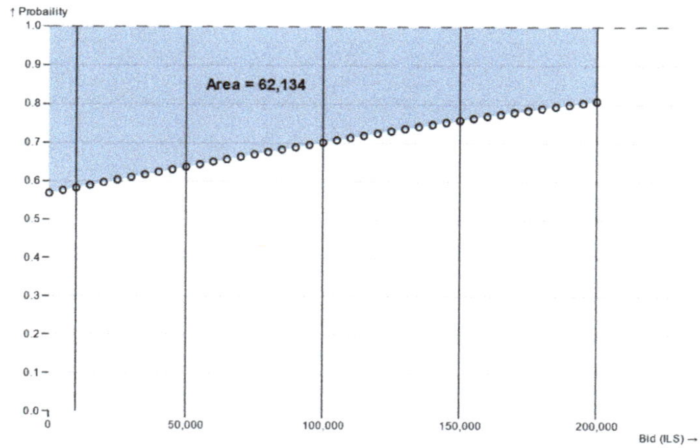

Figure 4. The value of current roof services. The circles indicate the estimated WTA for a given bid. The blue area represents the estimated value (in NIS) households assign to their "free" roofs.

3.2. Stated Reasons for Non-Adoption

At the beginning of the survey, participants were prompted to "choose the reasons you have not yet installed a PV system" from a pre-established list of options. The results are presented in Figure 5. The most selected reasons were the need for initial capital investment, chosen by 54% of respondents, and a lack of information, selected by 42%. Notably, 17% of participants cited a fear of electromagnetic radiation as a deterrent, a factor that, to the best of our knowledge, has not been previously highlighted in existing literature.

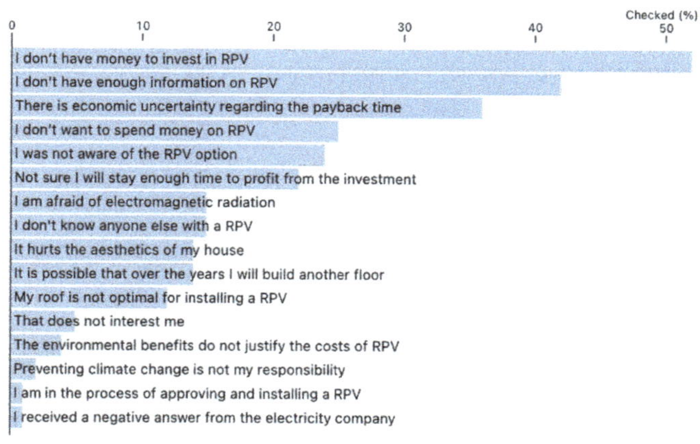

Figure 5. Selected reasons for not installing RPV.

Grouping the selected options according to their main reasons revealed that economic considerations (82% of respondents) and lack of information (52% of respondents) were the leading factors. The high weight of economic reasons supports our research hypothesis that improving financial comprehension of the deal may enhance the ITA a PV system.

3.3. Comprehension of the RPV Deal

We administered three questions that necessitated respondents to scrutinize the displayed electricity bill to provide accurate answers. These questions aimed to gauge the participants' understanding of three essential aspects of the RPV deal. For each of these questions, we established a range of values that would be considered as correct answers. Table 8 outlines these acceptable ranges and analyzes the variations in comprehension across the four different condition groups using a Chi-square test for statistical evaluation.

Table 8. Effect of presented information on deal comprehension.

	Correct Answer Range	**Total N = 899 (100%)**	**Graph N = 217 (24.1%)**	**Fin N = 221 (24.6%)**	**Both N = 228 (25.4%)**	**Control N = 233 (25.9%)**	***p* Value**
Monthly revenue (ILS)	500 to 1500	654 (72.8)	157 (72.4)	162 (73.3)	165 (72.4)	170 (73.0)	0.995
Cumulative revenue (ILS)	150,000 to 300,000	337 (37.6)	72 (33.3)	96 (43.4)	94 (41.2)	75 (32.5)	0.032
Payback period (years)	6 to 9	453 (50.4)	103 (47.5)	115 (52.0)	117 (51.3)	118 (50.6)	0.785

A statistically significant difference was observed among the treatment vs. control groups. Respondents in the "Fin" (financial description) and "Both" (financial description and graph) groups provided more accurate answers to the question concerning the cumulative revenue. Consequently, we can confirm Hypothesis 2, which posits that the presentation of cumulative financial information has a meaningful impact on respondents' comprehension of the RPV deal.

3.4. Intent-to-Adopt RPV

The ITA variable was measured using three items (1 to 5 Likert scale, Cronbach's alpha = 0.88), as shown in Table S4 (in Supplementary Materials). We found no significant difference between the four groups on any item (Kruskal–Wallis Test).

As the average of the ITA items did not pass normal distribution tests (visual, Shapiro–Wilk, Kolmogorov–Smirnov, Anderson–Darling tests), we transformed the continuous variable (average) into an ordinal variable with four bins: Low (1–2), Medium-low (2–3), Medium-high (3–4), and High (4–5). We then used a-parameterized statistical tests. Tables S4 and S5 (in the Supplementary Materials) present the relationships between the categorical and numeric variables to the ITA.

We applied a stepwise selection method and *p*-value threshold of 5%. We repeated the analysis in three hierarchy levels of the independent variables, as shown in Table 9. In the first level, we entered only the presented information group (3 condition groups and the control group as a reference) and the three comprehension items. In the second level, we added socio-economic characteristics and roof characteristics. In the third level, we introduced the RPV-specific beliefs and attitudes.

The most robust model for predicting the ITA the RPV systems accounted for 41% of the variance. The analysis underscores the significance of various factors, presented here in descending order of importance. The perception of RPV systems offering personal and financial advantages over the current situation emerged as particularly significant. This was followed by the subjective belief that using RPV is a social norm. The characteristics of the roof, specifically having a pitched, tiled surface that is currently idle, also played a role. For households with lower incomes, the potential revenues from the system carry more weight in the decision-making process. Since RPV is seen as a novel innovation, many prospective users are inclined to experiment on a small scale before fully committing. Understanding RPV's personal financial and broader environmental benefits necessitates a certain educational background. Additionally, some potential adopters feel hindered by perceived behavioral controls, thinking, "I want to, but I can't". The inclination to be an early adopter of new technology, trust in the electric company, existing environmental behavior, and apprehensions about unpredictable maintenance costs due to the technology's novelty also factored into the intention to adopt RPV systems.

Table 9. Ordinal logistic regression predicting ITA from different input variable sets.

	I	II	III
	b (S.E.)	b (S.E.)	b (S.E.)
Information: Graph			
Information: Finance			
Information: Both			
Information: Control			
Comprehension: Monthly revenue			
Comprehension: Cumulative revenue			
Comprehension: Payback period	0.45 (0.13) ***	0.42 (0.14) **	
Gender (Men)		0.28 (0.14) *	
Age			
Urban			
Education			0.40 (0.15) **
Income			−0.46 (0.19) *
Novelty seeking		0.46 (0.10) ***	0.27 (0.10) **
Trust Gov.		0.17 (0.08) *	
Trust electricity company		0.17 (0.08) *	0.24 (0.08) ***
Pro-environmental		0.46 (0.09) ***	0.23 (0.10) *
Household size			
PV in Neighborhood			
Acquaintances with PV		0.05 (0.02) **	
Roof type: Tiles		0.30 (0.14) *	0.48 (0.15) ***
Roof type: Flat concrete		0.51 (0.23) *	
Roof type: Flat floored			
Roof usage: Facilities			
Roof usage: Leisure			
Roof usage: Storage			
Roof usage: None			
Relative advantage			1.31 (0.13) ***
Expense concerns			
Maintenance costs			
Trialability			0.45 (0.11) ***
Subjective norms			0.93 (0.11) ***
Perceived behavioral control			−0.28 (0.09) **
C statistics †	0.55	0.69	0.81
adjR^2	0.12	0.15	0.41

* $p < 0.05$, ** $p < 0.01$, and *** $p \leq 0.001$. † C statistics represent the area under the Receiver Operating Characteristic (ROC) curve with values that range between 0.5 to 1.

4. Discussion

Figure 6 provides a summary of the statistically significant findings related to our initial hypotheses. Contrary to our central hypothesis (H1), this study found no direct relationship between the way accumulated financial information about RPV systems is presented and the intention to adopt for this sample. Partial support was found for Hypothesis 2 (H2), indicating that presenting accumulated financial data does enhance the precision of understanding cumulative revenue, though it does not necessarily affect the ITA. This lends partial credence to Hypothesis 3 (H3), as a better comprehension of the payback period and monthly revenue correlates positively with ITA. The findings highlight a disconnect between how the RPV deal's information is presented and understood, and the ensuing ITA, explaining why H1 could not be confirmed in the sample despite partial confirmations for H2 and H3.

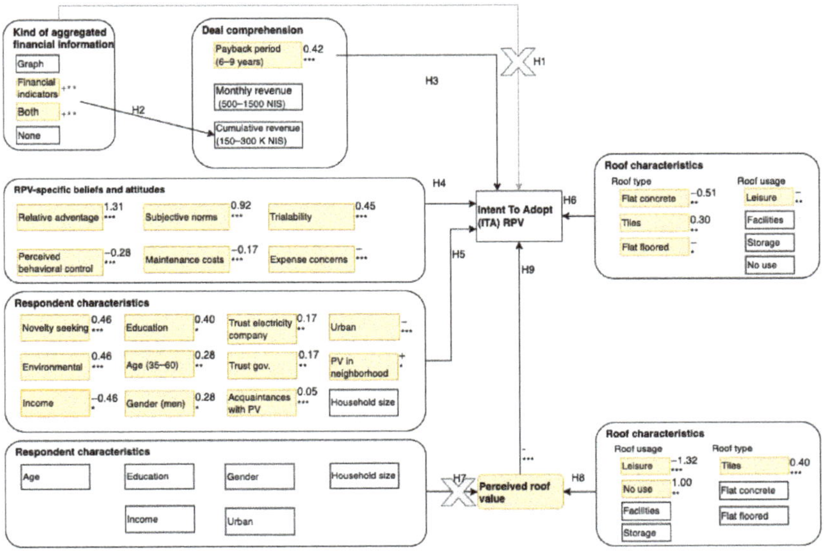

Figure 6. Summary of statistical tests for hypotheses. Statistically significant affecting variables are marked in color and stars * $p < 0.05$, ** $p < 0.01$, and *** $p \leq 0.001$. Variables are sorted by their strength of effect. Numbers indicate regression coefficients, and +/− signs indicate a positive or negative difference or correlation.

For Hypothesis 4 (H4), all RPV-specific beliefs and attitudes stemming from Wolske et al. [12] were confirmed. In terms of Hypothesis 5 (H5), several variables were identified that positively correlate with the ITA, such as early adopters, pro-environmental behavior, men, age between 35 and 60 years, higher education, trust in authorities (governmental and electrical entities) and acquaintances with PV. We confirmed Hypothesis 6 (H6) and found that flat roofs and roofs used for leisure have a negative correlation with ITA, while tiled roofs have a positive correlation. We could not confirm Hypothesis 7 (H7) as we did not find a significant relationship between the respondent characteristics and the "perceived value of the free roof". We confirmed Hypothesis 8 (H8), showing that both roof usage and roof type affect the perceived roof value which presents the opportunity cost for the RPV system. Notably, leisure usage was the most valuable, discouraging people from adopting RPVs. Hypothesis 9 (H9) revealed that in four out of five bids, significantly higher ITA was observed among those who accepted the offered WTA bid for roof services compared to those who refused. This suggests that the "perceived roof value" as an opportunity cost is an important factor affecting ITA that was previously overlooked.

To summarize, this study confirms the complexity of factors that contribute to RPV adoption, including financial comprehension, social norms, and institutional trust. This study confirmed previous research findings on the relationship between innovative behavior, pro-environmental stance, and ITA for RPV systems [12,15,19]. It also added contextual insights, indicating that the observed lack of trust in governmental and electrical institutions in the Israeli setting [49] could hinder ITA. This aligns with the national political instability and broken promises from politicians.

RPV systems on other roofs in the neighborhood contribute to the ITA RPV. The subjective perception explains that RPV is a norm [12,19,22,50]. This is also aligned with [26,27], which found that the best predictor for RPV is the presence of other systems nearby.

5. Conclusions

Our research contributes unique and significant insights into the realm of RPV system adoption in Israel, a context where government incentives exist but uptake remains surprisingly low. Despite our conservative estimates revealing a strong financial case for RPV adoption—with a net present value exceeding 100,000 ILS and an internal rate of return of approximately 13% per year—household adoption is not commensurate with these promising financial metrics.

One intriguing finding is the high value that households assign to their roof amenities, calculated as much as 62,134 ILS in our WTA analysis. This consideration substantially erodes the net present value to approximately 40,000 ILS—still financially viable but less attractive if the opportunity costs of roof usage are factored in.

We explored the influence of presenting cumulative financial information to potential adopters, presuming that a clearer depiction of long-term gains would nudge decisions in favor of RPV adoption. Contrary to our hypothesis, the presentation of financial data had no significant impact on the intention to adopt (ITA) an RPV system. While financial considerations are indeed important in making the RPV decision, and the way the financial information was presented did affect the financial comprehension of the RPV deal, we found no relationship between the presented financial information and the ITA an RPV system.

This gap between our hypothesis and the actual results warrants further discussion. Wolske et al. [50] found that customers were not sensitive to financial information in the initial marketing stages of solar systems. This apparent paradox—that financial motivations are crucial yet insufficient—raises critical questions. It suggests that financial metrics may only come into play in later stages of decision making. Indeed, in the Israeli context, with its unique set of circumstances like high living costs and security concerns, households lack an immediate "need or problem recognition" [51] to install solar systems. Climate and renewable energy discussions are notably absent in both social and official channels, leaving renewable energy, quite literally, off the rooftop and out of mind.

Aligned with Petrovich et al. [36] conclusion that mitigating RPV policy risks could be more effective than increasing the level of financial incentives, we recommend enhancing trust in RPV deals through solid contracts or trusted third-party involvement (e.g., banks or insurance companies which are trusted for the long-term). These could be an important step in overcoming skepticism. Policy makers in Israel must also demonstrate unwavering commitment to climate action and sustainable energy, providing consistent policies that cultivate an individual sense of urgency to act.

We acknowledge that this study has some limitations. The interest rate environment at the time of our study in early 2022 was quite different from current conditions. Changes in economic variables like interest rates could affect future research findings. Moreover, our research highlighted a disconnection between financial comprehension and RPV adoption in the sample. Subsequent studies should explore how tweaking the presentation of financial metrics can affect RPV uptake.

To robustly address these issues, we propose future research utilizing experimental economic methods to mimic "real-world" conditions. Specifically, natural experiments where participants are provided with specific, personalized information about their own rooftops, and where their actual decisions are monitored over time. These could offer a more nuanced understanding of the myriad factors influencing RPV adoption and could inform more effective strategies to bolster its uptake.

To conclude, RPV adoption is a high involvement consumer decision. Households' considerations are complex. Financial aspects are important, but the decision is "not all about the money". These insights provide valuable guidance for policy makers, researchers, and industry stakeholders, all striving to expedite the shift towards sustainable energy solutions.

Supplementary Materials: The following supporting information can be downloaded at: https://www.mdpi.com/article/10.3390/en17092043/s1, S1. Survey Questionnaire; S2. Controlled Information Presentation; S3. Descriptive Statistics of Four Condition Groups in the Sample; S4. Univariate Analysis of ITA RPV.

Author Contributions: Conceptualization, D.B.; Writing—original draft, D.B.; Writing—review & editing, R.R.P. and O.A.; Supervision, R.R.P. and O.A. All authors have read and agreed to the published version of the manuscript.

Funding: This research was funded by a grant from the rector of the University of Haifa for contributing to the UN's Sustainable Development Goals (SDGs), particularly to SDG 7: affordable and clean energy (date: February 2022).

Data Availability Statement: The original contributions presented in the study are included in the article and Supplementary Materials, further inquiries can be directed to the corresponding author.

Conflicts of Interest: The authors declare no conflicts of interest.

Appendix A

Estimation of the RPV Implementation Ratio in Israel

We estimated the ratio of residential roofs covered with PV systems. According to the electricity company (in reply to an information request), in July 2023, there were 16,290 PV systems with a capacity of less than 20 kW. To estimate the number of residential roofs in Israel, we found that, in 2018, 24% of households lived in detached houses (https://www.cbs.gov.il/he/mediarelease/doclib/2021/445-1/15_21_445t3.pdf, accessed on 1 September 2023), and the number of households was 2,587,100 (https://www.cbs.gov.il/he/publications/LochutTlushim/2020/mb_years.xlsx, accessed on 1 September 2023). Hence, the number of detached houses in 2018 was 620,904. Assuming that the number of detached houses remained constant till 2023, only 2.6% of detached residences are covered by PV systems.

References

1. Lakhani, R.; Doluweera, G.; Bergerson, J. Internalizing land use impacts for life cycle cost analysis of energy systems: A case of California's photovoltaic implementation. *Appl. Energy* **2014**, *116*, 253–259. [CrossRef]
2. Kleiner, A. *The Added Benefit of Solar Facilities on Roofs*; Israel Minister of Energy: Jerusalem, Israel, 2017. (In Hebrew)
3. Israeli Electricity Authority. *Increasing the Electricity Production Targets in Renewable Energies for 2030*; Israeli Electricity Authority: Haifa, Israel, 2020. Available online: https://www.gov.il/he/departments/publications/reports/2030_final (accessed on 1 September 2023).
4. Cots, F. *Engaging Citizens and Local Communities in the Solar Revolution*; Rooftop solar PV Country Comparison Report; Climate Action Network: Ixelles, Belgium, 2022.
5. Dijkgraaf, E.; Van Dorp, T.P.; Maasland, E. On the Effectiveness of Feed-in Tariffs in the Development of Solar Photovoltaics. *Energy J.* **2018**, *39*, 81–100. [CrossRef]
6. Samuelson, W.; Zeckhauser, R. Status quo bias in decision making. *J. Risk Uncertain.* **1988**, *1*, 7–59. [CrossRef]
7. Li, L.; Dai, C. Internal and External Factors Influencing Rural Households' Investment Intentions in Building Photovoltaic Integration Projects. *Energies* **2024**, *17*, 1071. [CrossRef]
8. Stanisławska, M. An Examination of Households' Attitudes towards Renewable Energy Source Investments in Lower Silesian Voivodeship. *Energies* **2023**, *16*, 7100. [CrossRef]
9. Karneyeva, Y.; Wüstenhagen, R. Solar feed-in tariffs in a post-grid parity world: The role of risk, investor diversity and business models. *Energy Policy* **2017**, *106*, 445–456. [CrossRef]
10. Klein, M.; Deissenroth, M. When do households invest in solar photovoltaics? An application of prospect theory. *Energy Policy* **2017**, *109*, 270–278. [CrossRef]
11. Moezzi, M.; Ingle, A.; Lutzenhiser, L.; Sigrin, B. *A Non-Modeling Exploration of Residential Solar Photovoltaic (PV) Adoption and Non-Adoption*; National Renewable Energy Laboratory: Golden, CO, USA, 2017; NREL/SR-6A20-67727, 1379469. [CrossRef]
12. Wolske, K.S.; Stern, P.C.; Dietz, T. Explaining interest in adopting residential solar photovoltaic systems in the United States. *Energy Res. Soc. Sci.* **2017**, *25*, 134–151. [CrossRef]
13. Jacksohn, A.; Grösche, P.; Rehdanz, K.; Schröder, C. Drivers of renewable technology adoption in the household sector. *Energy Econ.* **2019**, *81*, 216–226. [CrossRef]
14. Bao, Q.; Sinitskaya, E.; Gomez, K.J.; MacDonald, E.F.; Yang, M.C. A human-centered design approach to evaluating factors in residential solar PV adoption: A survey of homeowners in California and Massachusetts. *Renew. Energy* **2020**, *151*, 503–513. [CrossRef]

15. Mundaca, L.; Samahita, M. What drives home solar PV uptake? Subsidies, peer effects and visibility in Sweden. *Energy Res. Soc. Sci.* **2020**, *60*, 101319. [CrossRef]
16. Stauch, A.; Gamma, K. Cash vs. solar power: An experimental investigation of the remuneration-related design of community solar offerings. *Energy Policy* **2020**, *138*, 111216. [CrossRef]
17. Zander, K.K. Unrealised opportunities for residential solar panels in Australia. *Energy Policy* **2020**, *142*, 111508. [CrossRef]
18. Colasante, A.; D'Adamo, I.; Morone, P. Nudging for the increased adoption of solar energy? Evidence from a survey in Italy. *Energy Res. Soc. Sci.* **2021**, *74*, 101978. [CrossRef]
19. Schulte, E.; Scheller, F.; Sloot, D.; Bruckner, T. A meta-analysis of residential PV adoption: The important role of perceived benefits, intentions and antecedents in solar energy acceptance. *Energy Res. Soc. Sci.* **2022**, *84*, 102339. [CrossRef]
20. Ruokamo, E.; Laukkanen, M.; Karhinen, S. Innovators, followers and laggards in home solar PV: Factors driving diffusion in Finland. *Energy Res. Soc. Sci.* **2023**, *102*, 103183. [CrossRef]
21. Karakaya, E.; Sriwannawit, P. Barriers to the adoption of photovoltaic systems: The state of the art. *Renew. Sustain. Energy Rev.* **2015**, *49*, 60–66. [CrossRef]
22. Engelken, M.; Römer, B.; Drescher, M.; Welpe, I. Why homeowners strive for energy self-supply and how policy makers can influence them. *Energy Policy* **2018**, *117*, 423–433. [CrossRef]
23. Karytsas, S.; Vardopoulos, I.; Theodoropoulou, E. Factors Affecting Sustainable Market Acceptance of Residential Microgeneration Technologies. *Energies* **2019**, *12*, 3298. [CrossRef]
24. Dharshing, S. Household dynamics of technology adoption: A spatial econometric analysis of residential solar photovoltaic (PV) systems in Germany. *Energy Res. Soc. Sci.* **2017**, *23*, 113–124. [CrossRef]
25. Palm, A. Peer effects in residential solar photovoltaics adoption—A mixed methods study of Swedish users. *Energy Res. Soc. Sci.* **2017**, *26*, 1–10. [CrossRef]
26. Curtius, H.C.; Hille, S.L.; Berger, C.; Hahnel, U.J.J.; Wüstenhagen, R. Shotgun or snowball approach? Accelerating the diffusion of rooftop solar photovoltaics through peer effects and social norms. *Energy Policy* **2018**, *118*, 596–602. [CrossRef]
27. Barton-Henry, K.; Wenz, L.; Levermann, A. Decay radius of climate decision for solar panels in the city of Fresno, USA. *Sci. Rep.* **2021**, *11*, 8571. [CrossRef] [PubMed]
28. Lukanov, B.R.; Krieger, E.M. Distributed solar and environmental justice: Exploring the demographic and socio-economic trends of residential PV adoption in California. *Energy Policy* **2019**, *134*, 110935. [CrossRef]
29. Reames, T.G. Distributional disparities in residential rooftop solar potential and penetration in four cities in the United States. *Energy Res. Soc. Sci.* **2020**, *69*, 101612. [CrossRef]
30. Wolske, K.S. More alike than different: Profiles of high-income and low-income rooftop solar adopters in the United States. *Energy Res. Soc. Sci.* **2020**, *63*, 101399. [CrossRef]
31. O'Shaughnessy, E.; Barbose, G.; Wiser, R.; Forrester, S.; Darghouth, N. The impact of policies and business models on income equity in rooftop solar adoption. *Nat. Energy* **2021**, *6*, 84–91. [CrossRef]
32. Palm, A.; Lantz, B. Information dissemination and residential solar PV adoption rates: The effect of an information campaign in Sweden. *Energy Policy* **2020**, *142*, 111540. [CrossRef]
33. Bauner, C.; Crago, C.L. Adoption of residential solar power under uncertainty: Implications for renewable energy incentives. *Energy Policy* **2015**, *86*, 27–35. [CrossRef]
34. Sussman, R.; Chikumbo, M.; Gifford, R. Message framing for home energy efficiency upgrades. *Energy Build.* **2018**, *174*, 428–438. [CrossRef]
35. De Groote, O.; Verboven, F. Subsidies and time discounting in new technology adoption: Evidence from solar photovoltaic systems. *Am. Econ. Rev.* **2019**, *109*, 2137–2172. [CrossRef]
36. Petrovich, B.; Carattini, S.; Wüstenhagen, R. The price of risk in residential solar investments. *Ecol. Econ.* **2021**, *180*, 106856. [CrossRef]
37. Sommerfeldt, N.; Lemoine, I.; Madani, H. Hide and seek: The supply and demand of information for household solar photovoltaic investment. *Energy Policy* **2021**, *161*, 112726. [CrossRef]
38. Gillingham, K.; Deng, H.; Wiser, R.; Darghouth, N.; Nemet, G.; Barbose, G.; Rai, V.; Dong, C. Deconstructing Solar Photovoltaic Pricing. *Energy J.* **2016**, *37*, 231–251. [CrossRef]
39. O'Shaughnessy, E.; Margolis, R. The value of price transparency in residential solar photovoltaic markets. *Energy Policy* **2018**, *117*, 406–412. [CrossRef]
40. Larrick, R.P.; Soll, J.B.; Bruner Keeney, R.L. Designing better energy metrics for consumers. *Behav. Sci. Policy* **2015**, *1*, 63–75. [CrossRef]
41. Allcott, H. Social norms and energy conservation. *J. Public Econ.* **2011**, *95*, 1082–1095. [CrossRef]
42. Ahmar, M.; Ali, F.; Jiang, Y.; Alwetaishi, M.; Ghoneim, S.S.M. Households' Energy Choices in Rural Pakistan. *Energies* **2022**, *15*, 3149. [CrossRef]
43. Deline, C.; Jordan, D.; Sekulic, B.; Parker, J.; McDanold, B.; Anderberg, A. *PV Lifetime Project—2021 NREL Annual Report*; National Renewable Energy Laboratory: Golden, CO, USA, 2021; NREL/TP-5K00-81172. Available online: https://www.nrel.gov/docs/fy22osti/81172.pdf (accessed on 1 September 2023).
44. Abreu, J.; Wingartz, N.; Hardy, N. New trends in solar: A comparative study assessing the attitudes towards the adoption of rooftop PV. *Energy Policy* **2019**, *128*, 347–363. [CrossRef]

45. Israel Census Bureau. Population—Statistical Abstract of Israel 2022—No. 73. 2022. Available online: https://www.cbs.gov.il/en/publications/Pages/2022/Population-Statistical-Abstract-of-Israel-2022-No.73.aspx (accessed on 1 September 2023).
46. Israel Census Bureau. Household Income and Expenditure, Data from the 2020 Household Expenditure Survey—General Summary. 2023. Available online: https://www.cbs.gov.il/en/publications/Pages/2023/Household-Income-and-Expenditure-Data-From-the-2020-Household-Expenditure-Survey%E2%80%93General-Summary.aspx (accessed on 1 September 2023).
47. Hanemann, M. Welfare Evaluations in Contingent Valuation Experiments with Discrete Responses. *Am. J. Agric. Econ.* **1984**, *66*, 332–341. [CrossRef]
48. Whittington, D.; Adamowicz, W.; Lloyd-Smith, P. Asking Willingness-to-Accept Questions in Stated Preference Surveys: A Review and Research Agenda. *Annu. Rev. Resour. Econ.* **2017**, *9*, 317–336. [CrossRef]
49. Michaels, L.; Parag, Y. Motivations and barriers to integrating 'prosuming' services into the future decentralized electricity grid: Findings from Israel. *Energy Res. Soc. Sci.* **2016**, *21*, 70–83. [CrossRef]
50. Wolske, K.S.; Todd, A.; Rossol, M.; McCall, J.; Sigrin, B. Accelerating demand for residential solar photovoltaics: Can simple framing strategies increase consumer interest? *Glob. Environ. Chang.* **2018**, *53*, 68–77. [CrossRef]
51. Bruner, G.C.; Pomazal, R.J. Problem recognition: The crucial first stage of the consumer decision process. *J. Serv. Mark.* **1988**, *2*, 43–53. [CrossRef]

energies

MDPI

Article

European Climate Policy in the Context of the Problem of Methane Emissions from Coal Mines in Poland

Bożena Gajdzik [1,*], Katarzyna Tobór-Osadnik [2], Radosław Wolniak [3,*] and Wiesław Wes Grebski [4]

[1] Department of Industrial Informatics, Silesian University of Technology, 44-100 Gliwice, Poland
[2] Department of Safety Engineering, Silesian University of Technology, 44-100 Gliwice, Poland; katarzyna.tobor-osadnik@polsl.pl
[3] Faculty of Organization and Management, Silesian University of Technology, 44-100 Gliwice, Poland
[4] Penn State Hazleton, Pennsylvania State University, 76 University Drive, Hazleton, PA 18202-8025, USA; wxg3@psu.edu
* Correspondence: bozena.gajdzik@polsl.pl (B.G.); radoslaw.wolaniak@polsl.pl or radoslaw.wolniak@polsl.pl (R.W.)

Abstract: This paper presents a thorough examination of methane capture from Polish coal mines, contextualized within the framework of the European Union's (EU) climate policy objectives. Through a strategic analysis encompassing the interior of coal mines, the surrounding environment, and the macro environment, this study elucidates the complex dynamics involved in methane emissions and capture initiatives. The key findings include a declining trend in absolute methane emissions since 2008, despite fluctuations in coal extraction volumes, and a relatively stable level of methane capture exceeding 300 million m^3/year since 2014. The analysis underscores the critical role of government support, both in terms of financial incentives and streamlined regulatory processes, to facilitate the integration of methane capture technologies into coal mining operations. Collaboration through partnerships and stakeholder engagement emerges as essential for overcoming resource competition and ensuring the long-term success of methane capture projects. This paper also highlights the economic and environmental opportunities presented by methane reserves, emphasizing the importance of investment in efficient extraction technologies. Despite these advancements, challenges persist, particularly regarding the low efficiency of current de-methanation technologies. Recommendations for modernization and technological innovation are proposed to enhance methane capture efficiency and utilization.

Keywords: methane capture; climate policy; energy utilization; environmental sustainability; coal mining

1. Introduction

Increased interest in the effects of human activity on the environment can be traced back to the late 1960s and early 1970s. They became the main theme of the First UN Conference Environment and Development, held in Stockholm in 1972. This event gave rise to international environmental cooperation. At this conference, "the outline of a new idea of development was put forward, taking into account the negative dimension of anthropocentric pressures on nature" [1]. When the Paris Agreement came into force in 2016, many countries around the world, including Poland, pledged to reduce greenhouse gas emissions by 2050. The European Union's climate policy, reflecting the UN Paris Agreement, implies many legislative and technological changes in many sectors of the economy, including mining. Among the objectives of the European Green Deal is the one to significantly reduce greenhouse gas emissions by 2030 and 2050. To achieve the EGD targets, the European Commission has prepared a set of legislative proposals whose implementation should contribute to a zero-carbon economy [2]. In the communication 'Fit for 55', it shows the way to achieve these reduction targets. The EU's 'Fit for 55' package is a unique set of policies and instruments aimed at reducing greenhouse gas emissions by

55% in 2030 compared to 1990 and achieving net zero emissions by 2050 [3]. The package prepared includes a set of policies addressing every aspect of society and the operation of the economy [4].

The capturing of methane from coal mines is closely connected to the climate policy objectives of the European Union (EU). The EU has been at the forefront of global efforts to combat climate change and transition towards a low-carbon economy [5]. Methane is a potent greenhouse gas, and its capture aligns with the EU's commitment to reducing emissions and mitigating the impacts of climate change [6]. The EU's climate policy framework, including the European Green Deal and the 2030 Climate Target Plan, emphasizes the need to decarbonize various sectors, including energy production. Coal mines are significant sources of methane emissions, especially during extraction and processing activities [7]. By capturing methane from coal mines, the EU aims to minimize the release of this potent greenhouse gas into the atmosphere, contributing to the overall reduction in emissions, in line with its climate goals [8]. It can be stated that the EU has set ambitious targets to achieve climate neutrality by 2050. This involves not only reducing carbon dioxide emissions but also addressing other potent greenhouse gases like methane [9]. Methane has a higher warming potential than carbon dioxide in the short term, making its capture an important aspect of the EU's broader strategy to combat climate change. The EU has established regulations and directives to address methane emissions, and these policies encourage the adoption of technologies and practices that capture methane from various sources, including coal mines [10–12]. Companies operating within the EU are subject to stringent environmental standards and regulations, creating a regulatory environment that promotes the implementation of methane capture technologies [13,14].

The novelty of this paper lies in its comprehensive examination of methane capture from Polish coal mines within the context of the European Union's climate policy objectives. While the existing literature on methane emissions from coal mining operations often focuses on individual aspects such as technological solutions or regulatory challenges, this paper offers a holistic analysis that considers various interconnected factors.

An important aspect of novelty is the strategic analysis framework adopted in the study. By examining methane emissions at multiple levels—the interior of coal mines, the surrounding environment, and the macro environment—this paper provides a nuanced understanding of the complexities involved. This strategic approach allows for a more comprehensive assessment of the challenges and opportunities associated with methane capture initiatives. Also, this paper's empirical analysis of methane emissions trends over a significant timeframe (2008–2022) provides valuable insights into the effectiveness of current mitigation efforts. By analyzing data from Polish coal mines, this study offers empirical evidence to support its findings and recommendations. This empirical approach adds credibility to the paper's conclusions and contributes to the scientific understanding of methane capture in coal mining operations.

It can be stated that the paper's emphasis on the need for collaboration and stakeholder engagement underscores its practical relevance. By advocating for partnerships between mining companies, government agencies, research institutions, and the private sector, this paper recognizes the importance of collective action in addressing methane emissions from coal mines.

2. Methane as One of the Main Greenhouse Gases

It has been 200 years since the discovery of the greenhouse effect. The greenhouse effect was described by French mathematician Joseph Fourier in 1824 [15]. The greenhouse effect is caused by a group of gases called greenhouse gases. Three main categories of gases are responsible, to a very large extent, for this effect, with these gases being carbon dioxide (CO_2), methane (CH_4), and nitrous oxide (N_2O). Among these gases, carbon dioxide ranks first in terms of negative environmental impact, followed by methane. Carbon dioxide (CO_2) is scientifically estimated to be responsible for approximately 66% of global warming and methane for 30% of global warming (IEA data) [16,17]. In EU countries, methane is estimated to contribute more than 12% to global warming [18], and in Poland, it is 10.7% [19]. The relative contribution to the greenhouse effect of this gas over the last 100 years has been 15%. Methane (CH_4) is a potent GHG but has a short lifetime compared with CO_2 (9.1 ± 0.9 years, total atmospheric lifetime) [20,21]. At the molecular level, this gas is a much more potent driver of climate change (with a global warming potential over a 100-year period 28 times higher than that of carbon dioxide and 86 times higher over a 20-year period) [22]. Methane is a much more poisonous gas than carbon dioxide. It makes warming happen much faster.

Since the beginning of the industrial revolution, an increase in atmospheric methane concentrations has been noted in global reports. Increases in atmospheric methane concentrations have increased as economies have grown. Although atmospheric methane levels increased more slowly in the 1990s than in the 1980s (methane concentrations were below 1780 ppb for several years at the turn of the 20th century), atmospheric CH_4 concentrations rose again around 2007. According to data published by the US National Oceanic and Atmospheric Administration (NOAA) [23], there is more and more methane in the atmosphere. Only a few hundred years ago, the background concentration of methane (CH_4) in the Earth's atmosphere was around 700 ppb (parts per billion), and in 1920, it was still below 1000 ppb [22]. On a global scale, the increasing trends in methane emissions over the last few years are a cause for concern [15]. In recent years, methane production has increased most sharply in Africa and the Middle East; China; and South Asia and Oceania. Each of these three regions emitted 10–15 million tonnes more methane per year between 2000 and 2017. They were immediately followed by the United States, which emitted about 4.5 million tonnes more methane each year [16]. In 2021 (after the pandemic), methane concentrations reached a record high of 1889 ppb (parts per billion)—11 ppb more than the year before [24,25]. According to the NOAA report, methane concentrations in September 2022 were 1915.44 ppb and in September 2023, 1927.35 ppb [14]. The NOAA website provides a table (Table 1 in our article) with the annual increase in globally averaged atmospheric methane.

Table 1. Annual increase in globally averaged atmospheric methane [16].

Year	1984	1985	1986	1987	1988	1989	1990	1991	1992	1993
Growth (ppb)	12.87	12.20	12.89	11.43	10.68	11.20	8.66	14.07	2.39	3.87
Uncertainty (ppb)	0.84	0.84	0.72	0.73	0.47	0.57	0.49	0.56	0.47	0.63
Year	1994	1995	1996	1997	1998	1999	2000	2001	2002	2003
Growth (ppb)	7.24	3.83	2.48	6.36	12.12	2.38	−1.45	−0.72	3.24	4.84
Uncertainty (ppb)	0.62	0.52	0.42	0.69	0.64	0.61	0.53	0.45	0.56	0.60
Year	2004	2005	2006	2007	2008	2009	2010	2011	2012	2013
Growth (ppb)	−4.84	0.32	1.91	7.85	6.51	4.71	5.21	4.92	4.98	5.64
Uncertainty (ppb)	0.42	0.43	0.51	0.64	0.44	0.56	0.70	0.59	0.46	0.57
Year	2014	2015	2016	2017	2018	2019	2020	2021	2022	
Growth (ppb)	12.78	10.00	7.03	6.88	8.76	9.67	15.13	17.98	13.22	
Uncertainty (ppb)	0.43	0.72	0.64	0.75	0.62	0.65	0.45	0.45	0.77	

Source: [16].

Additional information on Table 1: The Global Monitoring Division of NOAA's Earth System Research Laboratory has measured methane since 1983 at a globally distributed network of air sampling sites [26]. A global average is constructed by first smoothing the data for each site as a function of time, and then, smoothed values for each site are plotted as a function of latitude for 48 equal time steps per year. Global means are calculated from the latitude plot at each time step [27].

High atmospheric methane concentrations pose a very serious threat to global efforts to keep warming below 1.5–2 °C. The global warming impact of 1 tonne of atmospheric methane over a 20-year horizon is comparable to the footprint of more than 80 tonnes of CO_2. Methane is also responsible for at least 25% of the increase in the Earth's temperature since the beginning of the industrial revolution [28–30].

Approximately 60 per cent of global methane emissions are anthropogenic, the largest sources of which, based on estimates, are fossil fuel production and use (one-quarter to one-third); waste (about one-quarter); and agriculture (about half of total methane emissions), particularly agriculture associated with intensive production [31]. According to IEA data [16], the source structure of methane emissions in 2021 was as follows: (1) natural from wetlands (more than 190 Mt methane/contents), (2) agriculture (about 140 Mt) (3) energy (about 135 Mt including oil, gas, coal), (4) waste (73 Mt), (5) other (about 40 MT), and (6) biomass burning (less than 10 Mt).

The rapid increase in atmospheric methane concentrations in recent years has worried the world, and the problem of methane was put on the agenda of the COP26 summit. The Global Methane Pledge is a multi-country pact to reduce methane emissions by 30% by 2030 from 2020 levels. Over 100 countries (representing 50% of anthropogenic methane emissions) recently signed up for the Global Methane Pledge (globalmethanepledge.org), and there is extensive potential for mitigation [32–34]. These countries agree that moves towards decarbonization must (should) run in parallel with moves towards de-methanization. The term 'de-methanization' may not yet be as popular as decarbonization, but it too will accelerate in the coming years. The Intergovernmental Panel on Climate Change (IPCC) notes that in order for the world to meet the global goal of limiting temperature increase by 1.5 °C (or even 2 °C) by 2050, there must be sharp reductions in methane emissions by 2030 [35]. In September 2021, the European Union and the United States announced the Global Methane Commitment, which is a political commitment to reduce global methane emissions in all methane-emitting sectors by at least 30% (relative to 2020 levels) by 2030, at the United Nations Climate Change Conference (COP 26) in Glasgow in November 2021. The Global Methane Commitment includes a move towards the use of the best available inventory methods to quantify methane emissions with a focus on high emission sources. The International Panel (IPCC) recommends reducing its emissions as one of the best ways to tackle global warming in the coming years. Methane (CH_4) emissions not only are a factor in global warming but also negatively affect human health—particularly respiratory diseases. Methane also contributes to the formation of ozone, which is a potent air pollutant. Methane is an important precursor of tropospheric ozone production and is responsible for 50% of the increase in tropospheric ozone levels [36]. Ozone is not only a greenhouse gas but also a major component of photochemical smog (UNEP, 2011) [37]. It contributes to one million premature deaths from respiratory diseases worldwide each year [37,38]. The role of methane as a lever to achieve climate goals is an active topic [25–27,33,39].

Reducing atmospheric emissions of methane is a priority strategic direction today, all the more so because predictions for atmospheric emissions of this gas are poor. Allen et al. [40] showed that by 2050, global average atmospheric methane concentrations will decrease by 26%, resulting in a 9.7% decrease in surface ozone concentrations and a 0.39 ± 0.05 K decrease in temperature compared to SSP3-7.0. Shindell et al. [32] examined a 30% reduction in methane concentrations from a specific day to present day (2015) using an ensemble of models. They attributed a global, population-weighted decrease in ozone exposure of 2–2.5 ppb and a temperature decrease of 0.18 ± 0.02 K to the methane change.

However, a global policy to reduce methane emissions must take into account the specificities of methane emissions in different countries. Poland, which is the subject of the research in this article, is a country where more than 90% of harmful methane emissions come from coal mines [41].

When analyzing the structure of methane emission sources in Poland and the EU, it can be seen that in Poland, the main emitters of methane are mines. So much of Poland's methane emissions come from coal mining, much higher in relation to the European Union as a whole, where the share of emissions from fossil fuel extraction is 38% [29]. Ember (2020) points out that in Poland, about 70% of methane emissions comes from the mining industry (hard coal, oil, and gas mining) [28,29]. In the total makeup of greenhouse gases in Poland, the share of methane is 10.7% [19].

After the pandemic (2021), Polish coal mines emitted more than 420,000 tonnes of methane, 28% more than a decade ago. Between 2013 and 2021, methane emissions from active coal mines in Poland remained stable at 450,000–525,000 tonnes per year [29]. Despite the fact that between 2010 and 2022, the amount of annual hard coal production in Poland fell by more than 20 million tonnes, or 28%, methane emissions increased by 26% over the same period—to 426,000 tonnes in 2021 [29]. This situation is due to the fact that, despite the steady decline in hard coal mining in Poland, coking coal mining has remained stable—around 12 million tonnes. In 2021, coking coal extraction in Poland amounted to 12.6 million tonnes and accounted for 23% of total hard coal extraction in that year [19]. As coking coal is a critical raw material of the European Union from 2020 (European Commission), restrictions on methane emissions from these deposits have not been planned. However, work is currently underway in Poland to detail how methane emissions are reported. The problem lies in a good estimation of how much methane there is per 1 tonne of thermal coal mined and how much per 1 tonne of coking coal. In Poland, it is assumed that more than 60% of methane emissions come from mines, with both thermal and coking coal in their deposits. At the same time, the attribution of methane emissions to the extraction of a specific type of coal is important because of the new requirements set out in the EU policy on methane emission reduction. However, as these provide the methane emission charges for 2027 only from mined thermal coal, this lack of differentiation makes it impossible today to assess the impact of the regulation as a whole. An important aspect of the new regulation is that emissions will be billed per operator (coal company) and not per mine. However, it will still be difficult to account for methane emissions accurately. There is a gap in Poland at the level of conscious state policy to reduce methane emissions. The social and environmental costs of methane emissions are not reflected either in the Polish environmental fee system or in the EU ETS emissions pricing mechanism. The environmental fee in Poland in 2023 was only 34 gr per tonne of methane emissions and was only revalued in line with inflation [29]. The government program on ending the exploitation of the most emitting coal deposits is therefore of key importance for reducing methane emissions from Polish mines. This is due to the fact that Poland has committed to reducing its emissions by 30% by 2030, as part of the Global Methane Pledge. It is assumed that as part of the planned transformation process in the Silesian Voivodship in Poland, further mines will be liquidated by 2030. By 2030, hard coal production will fall by 7.1 million tonnes per year—from 30 million tonnes per year in 2021 to 23 million tonnes per year by 2030. The reduction in coal production will represent 62% of the planned national reduction in coal production by 2030 according to Polish policy (the government program KPEiK) and will contribute to a reduction in the share of coal in electricity generation [29,42,43]. The decommissioning of mines is also expected to contribute to the reduction in methane emissions into the atmosphere. However, the amount of methane per extraction level is important. The mines selected for decommissioning currently emit 62.4 million m^3/year of methane. The closure of more mines is planned for after 2030, and it will be gradual (up to 2049), which will translate into a reduction in hard coal production and, as a result, will additionally contribute to a decrease in the production of raw material and methane emissions (Table 2).

Table 2. Schedule of liquidation of thermal power plants and mines in Poland.

Type of Plants	Thermal Power Plant			Hard Coal Mines					Lignite Coal Mines
year	2030	2035	2036	to 2030	to 2035	to 2040	to 2045	to 2049	2036
number	3	1	1	2	2	4	2	4	21

Source: [44].

The release of gas from coal occurs not only in the process of mining but also during processing, during transportation or storage, as well as after mine closure. Methane escapes into the atmosphere from active underground coal mines in two different ways:

- Through de-methanation systems;
- Through ventilation shafts.

In Polish mining, the key sources of methane emissions from mines are unclear installations (35%), ventilation shafts (22%), and the methane from methane installations (43%) [19].

It is assumed that Polish mines will increasingly switch to a reserve-peak operation system during decommissioning periods (until 2049). However, this is not always possible due to the safety of plant operation and the requirements of mining and geological law. On the other hand, due to the destabilization of the energy market (high energy prices), and the lack of nuclear power in the Polish electricity system, coal will be needed in Poland until 2030. In 2018, the Program for the hard coal mining sector in Poland was adopted, which covers the period until 2030. The goal of the Program is to improve the situation of the hard coal mining sector and make efficient use of its resources. Implementation of the Program is monitored in annual cycles. In 2022, the Program was updated, taking into account the assumptions of PEP2040 [45] and the provisions of the Social Agreement on the Transformation of the Coal Mining Sector and Selected Transformation Processes of the Silesian Voivodeship, which set out the planned dates for coal mine shutdowns. As amended, the Program pursues an additional objective of equitable transformation, taking into account activities aimed at winding down the mining sector and mechanisms to support this process. According to the indications of the Program, the termination of thermal coal mining from hard coal mines should take place by 2049 [46,47].

Methane emissions into the atmosphere will continue to be a problem during the transition. To reduce these emissions, the emitter (mine) should use captured methane. It should also do this for its own energy economy. A solution is also to sell the methane on the open market [48]. Ways of managing methane-rich mine gases from mine de-methanation in Polish conditions are described in [49–51]. Currently, methane from many mines in Poland is lost due to release into the atmosphere. A second source of methane may also be mines that are no longer in operation. Preliminary analyses conducted by PIG-PIB have shown that it is inefficient to extract methane from the Lower Silesian Mining Basin, but it is possible to extract it from the recently abandoned deposits of the Upper Silesian Mining Basin, according to ongoing analyses. As part of the ongoing work, a method for its extraction has also been developed. Therefore, the capture of methane from closed mines can—and should—complement the national energy mix. The capture of methane from mines in Poland is, next to the decarbonization of steelworks, a key industrial problem in the radical climate policy [52,53]. In Poland, the extraction of methane from decommissioned coal mines has so far been marginal. It is carried out in only two deposits, and the annual volume of extraction is about 5 million cubic meters (data for 2020), which is 1 per mille of national gas extraction and about 1.7% of methane captured during coal mining [54]. It has been assumed that the most important criterion indicating the possibility of extracting methane from a discontinued deposit is the total amount of methane emissions (methane yield) recorded during coal mining [55,56]. In the case of closed mines, the amount of methane decreases from year to year. According to estimates by the Central Mining Institute, during the first year, the amount of methane emissions from a single longwall drops to less than 20% of the average methane yield of that longwall from the period of its operation [54]. After a few years, the amount of methane emitted is only a few percent of the initial volume of emissions, and after a dozen years (estimated to be about 15 years), emissions virtually disappear [54–56].

The process of capturing methane involves employing advanced technologies to trap and extract the gas before it is released into the atmosphere [57–59]. One method involves the installation of methane capture systems directly within coal mines. These systems utilize a combination of ventilation and gas extraction techniques to capture methane emissions at their source [60–63]. By implementing these technologies, coal mining operations can minimize their environmental impact and simultaneously harness methane as a valuable energy resource [64,65].

The captured methane can be utilized for various purposes, such as power generation or as a feedstock for industrial processes [66]. This not only reduces the reliance on traditional fossil fuels but also provides an economic incentive for coal mining companies to adopt methane capture technologies [67].

Also, capturing methane from coal mines aligns with global efforts to transition towards cleaner and more sustainable energy sources [68]. Governments, environmental organizations, and industries are increasingly recognizing the importance of reducing methane emissions to combat climate change. Implementing methane capture technologies in coal mines represents a tangible and effective step towards achieving these environmental goals [69–72].

In addition to environmental benefits, the reduction in methane emissions enhances safety within coal mines [73]. Methane is not only a potent greenhouse gas but also poses a significant explosion risk in underground coal mines. By capturing and utilizing methane, the likelihood of accidental releases leading to explosions is mitigated, promoting a safer working environment for miners [74].

In Table 3, there is a juxtaposition of the main advantages of using methane from coal mines.

Table 3. Advantages of using methane from coal mines.

Advantage	Description
Mitigation of Greenhouse Gas Emissions	Prevents its release into the atmosphere, reducing the greenhouse effect and mitigating climate change.
Utilization as a Clean Energy Source	Allows for its use in power generation or as a feedstock for industrial processes, contributing to cleaner energy production.
Economic Incentives for Coal Mining Companies	Provides an economic incentive for coal mining companies to adopt methane capture technologies by offering a valuable energy resource.
Alignment with Global Sustainability Goals	Supports international efforts to reduce methane emissions and the transition towards more sustainable and environmentally friendly practices.
Enhanced Safety in Coal Mines	Reduces the risk of methane-related explosions in underground coal mines, promoting a safer working environment for miners.
Responsible Resource Extraction and Management	Demonstrates a commitment to responsible mining practices by addressing environmental concerns associated with coal extraction.
Environmental Stewardship	Contributes to environmental conservation by preventing the release of methane, a potent greenhouse gas, into the atmosphere.
Reduction in Carbon Footprint	Significantly lowers the carbon footprint of coal mining operations, aligning with global efforts to combat climate change.
Renewable Energy Resource	Allows for it to serve as a renewable energy resource, reducing dependence on non-renewable fossil fuels for energy production.
Economic Diversification for Coal Industry	Offers economic diversification for the coal industry by turning methane emissions into a valuable energy commodity.
Green Energy Portfolio Enhancement	Enhances the green energy portfolio of coal mining companies, promoting a positive image and attracting environmentally conscious stakeholders.
Compliance with Environmental Regulations	Meets and exceeds regulatory requirements for reducing methane emissions, ensuring adherence to environmental standards.
Technological Innovation and Development	Drives the development and implementation of advanced technologies for methane capture, fostering innovation within the mining industry.
Global Leadership in Sustainable Practices	Positions coal mining companies as leaders in sustainable practices, contributing to a positive reputation within the global market.
Community Relations and Social Responsibility	Demonstrates social responsibility by addressing environmental concerns, improving community relations, and fostering a positive relationship with local stakeholders.
Long-term Viability of Coal Industry	Enhances the long-term viability of the coal industry by integrating environmentally friendly practices, adapting to changing energy trends.

Source: Authors own work based on [54–57,68,70–85].

Capturing methane from coal mines, while offering significant environmental benefits, is not without its challenges. One prominent issue is the substantial initial investment costs associated with the installation of methane capture systems [86]. Coal mining operations face financial constraints when implementing these technologies, hindering widespread adoption. To address this, governments can play a pivotal role by providing financial incentives and subsidies, thereby alleviating the financial burden and making methane capture more economically feasible for mining companies [87].

Operational and maintenance expenses pose an ongoing challenge for coal mining operations that have implemented methane capture systems. The sustained operation of these systems requires consistent financial resources, necessitating long-term financial planning to manage these costs effectively. Without proper planning, the economic viability of methane capture projects may be compromised over time [88]. Technological challenges further impede the successful implementation of methane capture initiatives. Developing and implementing effective capture technologies requires substantial research and development investment to overcome technical hurdles. Increased funding in this area can lead to the creation of more efficient and cost-effective methane capture technologies, fostering technological advancements within the mining industry [57,89].

The variability in methane emissions from coal mines presents another challenge. Designing capture systems that are consistently effective across different mining conditions proves to be a complex task. Adaptive system designs are essential in addressing this variability, ensuring the systems can adapt to the changing emission levels encountered in different mining scenarios [58,90,91]. Land use and infrastructure constraints can hinder the implementation of methane capture systems, often leading to resistance and logistical challenges. Government infrastructure support becomes crucial in overcoming these obstacles, facilitating the necessary modifications and ensuring a smoother integration of methane capture technologies into existing coal mining operations [92]. Collaboration through partnerships and funding initiatives with industry stakeholders, research institutions, and environmental organizations becomes essential to share the financial load and overcome resource competition [58,93]. Regulatory compliance and reporting burden can also impede progress. Simplifying and streamlining regulatory processes are necessary to reduce the administrative burden on coal mining companies, ensuring they can meet regulatory requirements for methane capture without undue strain [57].

Also, the perception of continued fossil fuel use and potential economic viability concerns in the short term may deter investment in methane capture projects. Integrating these projects with broader renewable energy initiatives can enhance their economic viability and contribute to a more sustainable energy portfolio, making them more appealing in the short term [57,58]. Public resistance and perception issues within local communities pose a final challenge. Proactive stakeholder engagement, clear communication about the benefits of methane capture, and addressing concerns are crucial in overcoming resistance and gaining community support for these projects [64,91,92].

In Table 4, there is an analysis of the main problems connected to using methane from coal mines.

Incentives and funding programs provided by the EU also play a role in encouraging the adoption of methane capture technologies. Financial support for research and development, as well as subsidies for companies investing in cleaner technologies, further facilitate the integration of methane capture systems in coal mining operations [57–59,88,91]. For good preparation of a strategy for methane management from Polish mines, a comprehensive analysis of the current status of emissions of this gas from mines is necessary. This article will present data analyses to illustrate the directions of activities in this field to date.

Table 4. Problems of using methane from coal mines.

Problem	Description	Methods of Overcoming
Initial Investment Costs	The installation of methane capture systems requires a substantial upfront investment, which can be a financial challenge for coal mining operations.	Government incentives and subsidies: governments can provide financial incentives and subsidies to alleviate the initial investment burden, making methane capture more economically viable for coal mining companies.
Operational and Maintenance Expenses	Ongoing operational and maintenance costs associated with maintaining methane capture systems may strain financial resources over time.	Long-term financial planning: implementing long-term financial planning can help coal mining companies manage operational and maintenance costs effectively, ensuring the sustained operation of methane capture systems.
Technological Challenges	Developing and implementing effective methane capture technologies can pose technical challenges, leading to delays and potential setbacks.	Research and development investment: increased investment in research and development can address technological challenges, fostering the development of more efficient and cost-effective methane capture technologies.
Variability in Methane Emissions	Methane emissions from coal mines can vary, making it challenging to design capture systems that are consistently effective across different conditions.	Adaptive system designs: designing methane capture systems with adaptability to varying emissions can enhance effectiveness across different mining conditions, ensuring a more consistent performance.
Land Use and Infrastructure Constraints	Implementing methane capture systems may require significant land use and infrastructure modifications, facing resistance and logistical challenges.	Government infrastructure support: government support for infrastructure modifications can facilitate the implementation of methane capture systems, addressing land use and infrastructure constraints.
Competition for Limited Resources	Coal mining companies may face competition for limited resources, diverting attention and funding away from methane capture initiatives.	Collaborative funding and partnerships: establishing partnerships and collaborative funding initiatives with industry stakeholders, research institutions, and environmental organizations can help share the financial load and overcome resource competition.
Regulatory Compliance and Reporting Burden	Meeting regulatory requirements for methane capture may impose additional administrative burdens, leading to compliance challenges.	Streamlined regulatory processes: simplifying and streamlining regulatory processes can reduce the administrative burden on coal mining companies, encouraging compliance with methane capture requirements.
Economic Viability in the Short Term	The economic viability of methane capture projects may be questioned in the short term, potentially hindering widespread adoption.	Integration with renewable energy initiatives: integrating methane capture projects with broader renewable energy initiatives can enhance the economic viability and contribute to a more sustainable energy portfolio, making them more attractive in the short term.
Perception of Continued Fossil Fuel Use	Critics argue that investing in methane capture may perpetuate the use of fossil fuels, diverting focus from renewable energy alternatives.	Public awareness campaigns and education: conducting public awareness campaigns and education initiatives can address misconceptions, emphasizing the transitional nature of methane capture and its contribution to cleaner fossil fuel use.
Public Resistance and Perception Issues	Local communities and the public may resist methane capture projects due to concerns about safety, aesthetics, or perceptions of environmental impact.	Stakeholder engagement and communication: proactive engagement with local communities, clear communication about the benefits, and addressing concerns can help overcome resistance and gain community support for methane capture projects.

Source: Authors own work based on [57–64,69,86–92].

3. Study Method

It should be noted that the nature of this analysis is primarily a review. However, based on an analysis of the historical data over a period of more than 20 years, hypotheses can be formulated to show that limiting methane emissions from Polish thermal coal mines is not a straightforward subject and the solutions proposed so far are not unambiguous. Its aim is to analyze the data on methane production levels and the use of methane released from Polish hard coal mines for the years 2008–2022 from the perspective of building a decarbonization strategy for the Polish mining industry. Therefore, the authors would like to draw attention to the strategic position of this gas within the policy of decarbonization and diversification of market offerings. The management of methane from existing and closed hard coal mines in Poland should be one of the goals of strategic plans for mining companies, especially in relation to EU policy and accounting for emissions of this greenhouse gas. Therefore, the authors chose a three-area strategic analysis as the method [92]. The strategic analysis [94,95] was conducted for three areas, i.e., the interior of Polish hard coal mines, the close environment, and the macro environment (Figure 1).

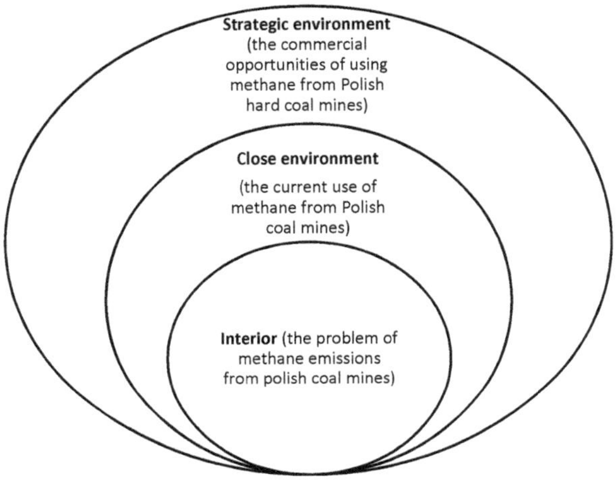

Figure 1. Methodology—three-area strategic analysis of the use of methane from coal mines in Poland. Source: individual study.

The research hypotheses (RH) were formulated on the basis of the source data and literature research conducted:

RH$_1$. *Reducing the volume of extraction does not significantly reduce methane emissions per tonne of thermal coal.*

RH$_2$. *The level of methane capture from hard coal mines is still dictated in particular by safety considerations and is not adapted to EU climate legislation.*

RH$_3$. *Capturing methane from Polish coal mines should be a long-term goal in the development and innovation strategies of these companies.*

The three-area strategic analysis applied allows the problem of methane emissions from Polish coal mines to be looked at from different points of view and to prove or deny the hypotheses assumed.

The interior of hard coal mines in Poland was analyzed to understand the problem of methane emissions from coal mines. The situation in the close environment was analyzed to understand the methods used to utilize methane from actively decommissioned coal mines. Then, from a strategy-building perspective (third level), the possible methods to improve methane management in response to the Regulation of the European Parliament and of the Council on the reduction in methane emissions in the energy sector and amending Regulation (EU) 2019/94 were discussed.

A set of official data published annually by the State Mining Authority in Katowice (WUG) was selected for the analysis. The division into methane emissions from thermal and coking coals is difficult, as it is still captured on an estimated basis. As mentioned in Section 2, it can be assumed that it accounts for 60% of all emitted gas. To simplify the analysis, we decided to assume that the trends and proportionality of the captures remain the same for methane from thermal coal deposits as for mining as a whole. For the purpose of inference in the applied strategic analysis, such an approach will make it possible to show the trends of change and to determine the current state of methane management from Polish mines. The results of the analyses are presented in graphical form. According to the authors, the use of complex statistical analyses not only is unnecessary but also would not contribute anything to the inference process. After analyzing the preliminary database from WUG, it was decided that the time period adopted would not be the whole of 2008–2022, but primarily 2014–2022. The justification and presentation of the results are included in next section.

4. Results of the Analysis

4.1. The Problem of Methane Emissions from Coal Mines in Poland

As is well known, the fuel–energy situation in the world is difficult, especially due to the war in Ukraine, the post COVID-19 situation, and the turbulence on fuel markets. At the same time, the European Union's climate policy is very demanding, strongly one-sided, and costly for many economies. This is due to legal regulations and financial burdens (e.g., EU ETS). The Polish hard coal mining industry not only faces quality and price requirements in relation to the needs of the Polish energy sector but also, as an emitter of the greenhouse gas methane, is subject to the rules of the EU climate policy in its assumptions of zero emissions by 2025. The amount of methane emitted from the Polish coal mines is mainly due to the characteristics of Polish deposits and the mining method used in Polish mines. Thus, the question of the analysis was as follows:

QR: *Does limiting the amount of hard coal mining have a direct impact on the amount of methane emissions into the air?*

From the point of view of emission charges, it is not so much the volume of thermal coal extraction that is important, but the amount of methane per 1 tonne of extraction. Coking coal, which is on the EU's list of critical raw materials, is not affected by the restrictions at this point.

Figure 2, based on data obtained from the WUG, presents the quantity of absolute methane emissions from 2008 to 2022, and is followed by Figure 3, which presents the variation in coal extraction to absolute methane emissions from 2008 to 2022.

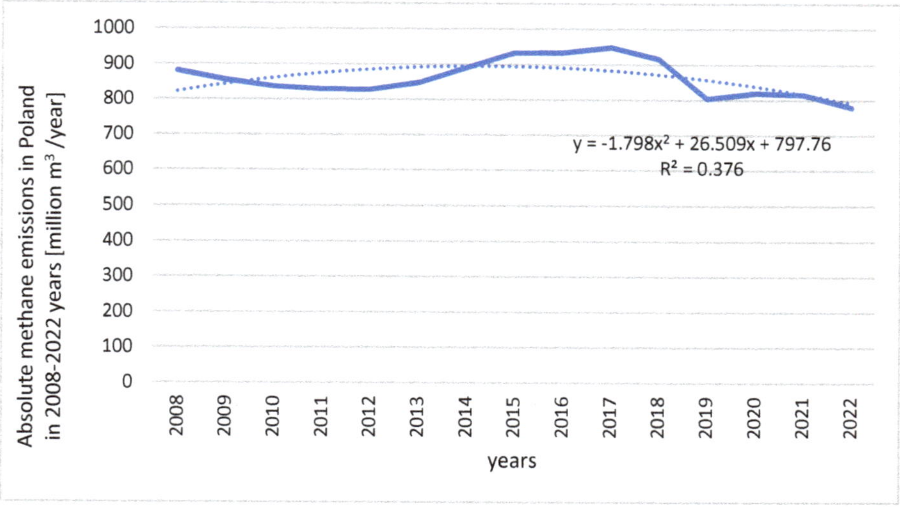

Figure 2. Absolute methane emissions in Poland in the years 2008–2022 [million m^3/year]. Source: individual study based on the data of WUG.

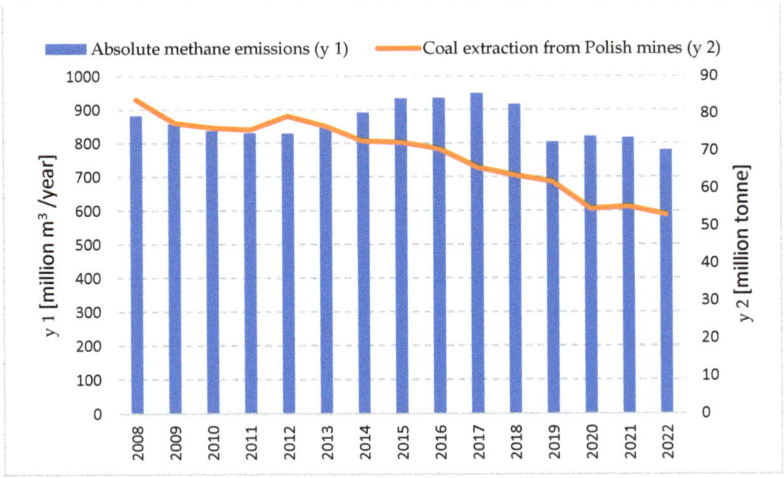

Figure 3. Variability in coal extraction volumes to absolute methane emissions in Poland in the years 2008–2022. Source: individual study based on the data of WUG.

As can be observed, the volume of absolute methane emissions in the years 2008–2022 is slightly decreasing. Coefficient m, the slope as a function of trend, is negative (Formula (1)):

$$y = -1.798x^2 + 26.509x + 797.76, \tag{1}$$

However, when analyzing this development against the background of extraction volumes for the period in the years 2008–2022 (Figure 3), it can be seen that hard coal extraction volumes in Poland are gradually declining, with absolute methane emissions continuing to oscillate between 778.9 [million m^3/year] in 2022 and 948.5 [million m^3/year] in 2017.

The analysis of the quantity of methane captured between 2008 and 2022 is presented in Figure 4.

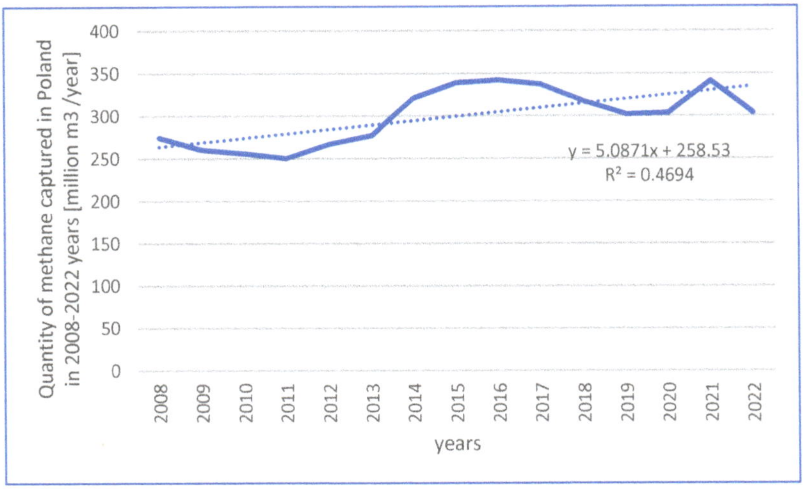

Figure 4. Quantity of methane captured in Poland in the years 2008–2022 [million m^3/year]. Source: individual study based on the data of WUG.

The trend for the quantity of methane captured between 2008 and 2022 was determined using Excel spreadsheet tools and has the form of (2):

$$y = 5.0871x + 258.53 \quad (2)$$

Analysing the available data from the WUG, it was noted that the cut-off year for further analysis should be 2014 (Figure 5). When we narrowed the scope of the assessment of the quantity of methane captured to the years 2014–2022, the trend function changed its characteristics and had the form of (3):

$$y = -2.9717x + 337.63 \quad (3)$$

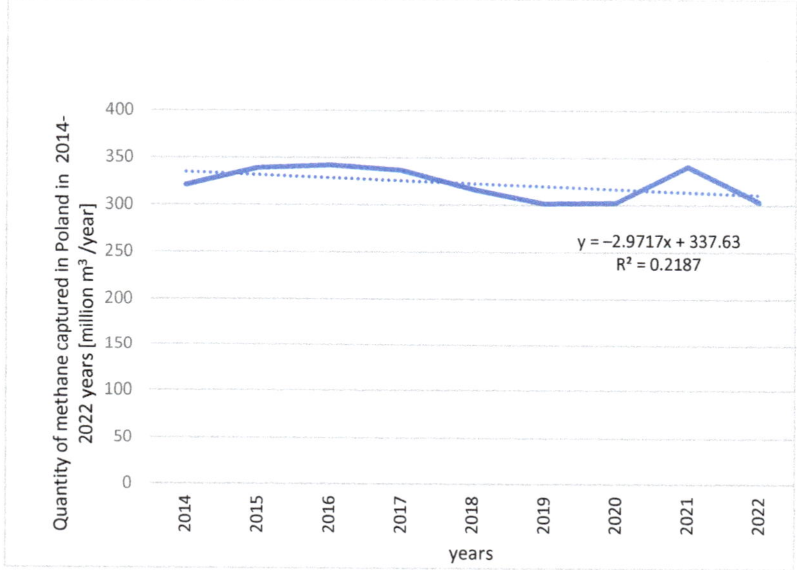

Figure 5. Quantity of methane captured in Poland in the years 2014–2022 [million m^3/year]. Source: individual study based on the data of WUG.

The slope m indicator thus changed to a negative sign. This means that, in the long term, it can be said that the amount of methane captured will increase. Unfortunately, this is most likely due to several factors: the number of mines working; the amount of coal extraction; and, above all, the level of methane in the exploited deposits. Since 2014, we have seen a relatively stabilized situation. Although m has a negative sign and unfortunately worsens our methane capture results, looking at the characteristics of the curve in Figure 5, it can be cautiously concluded that the level of methane capture from Polish coal mines is at a fairly stable level and is over 300 [million m^3/year]. Until 2013, the level was below this value.

As can be concluded from Figure 6, there is no clear correlation between coal extraction volumes and the quantity of methane captured. This finding is important for building a future methane management strategy in line with the EU climate policy.

When analyzing the level of de-methanation resulting from the absolute value of methane and the quantity of methane captured, first, the range of years 2008–2022 was analyzed (Figure 7). The analysis period was then shortened to the years from 2014 to 2022 (Figure 8). In both cases, slight upward trends of $m_1 = 0.7046$ and $m_2 = 0.5150$ were observed. However, taking into account the previous analyses, it has to be said that such a result does not allow for the formulation of the conclusion that the level of methane capture from Polish mines is improving. Thus, improvements to these results are ongoing and are the subject of much research work.

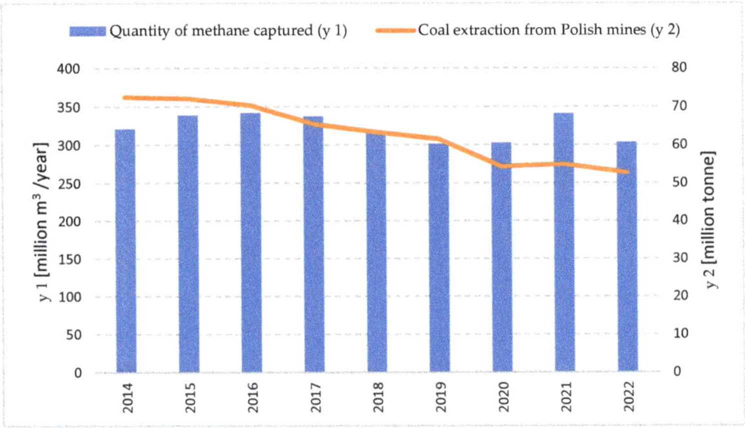

Figure 6. Variability in coal extraction volumes to quantity of methane captured in Poland in the years 2014–2022. Source: individual study based on the data of WUG.

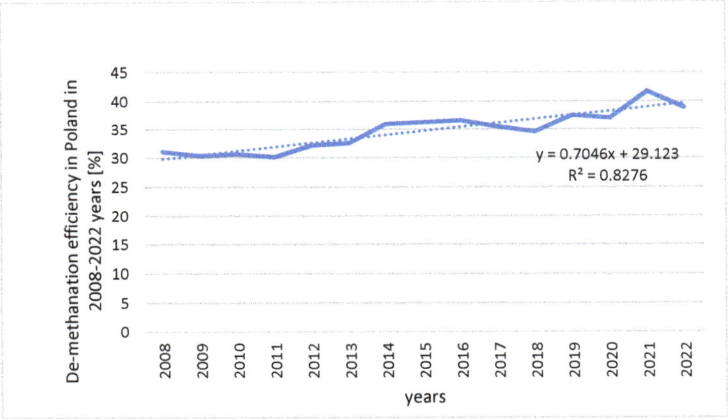

Figure 7. De-methanation efficiency in Poland in the years 2008–2022 [%]. Source: individual study based on the data of WUG.

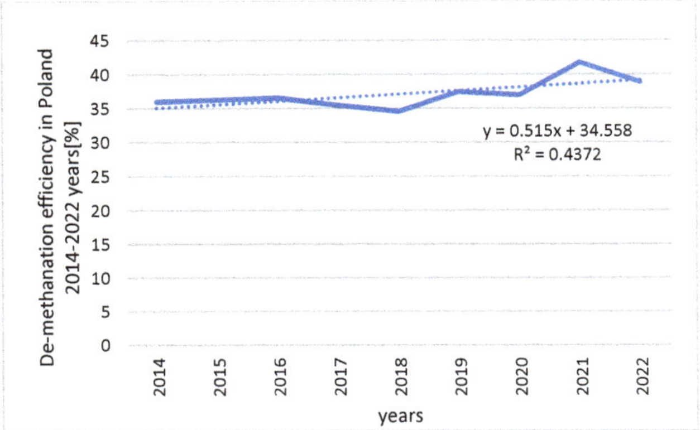

Figure 8. De-methanation efficiency in Poland the years 2014–2022 [%]. Source: individual study based on the data of WUG.

4.2. The Current Use of Methane from Coal Mines in Poland

Currently, the extraction of methane from mines is driven by safety regulations. However, as a result of changes in regulations resulting from the implementation of the Green Deal, methane must be looked at as an additional product from mines. Therefore, research is being conducted on extracting methane from mine air or mine de-methanation systems. Figure 9 presents the variation in the amount of use of captured methane in the Polish coal mines for absolute methane. As before, only data from 2014–2022 were analyzed. Currently, only about 25% of the methane emitted from Polish mines is used and this trend has no significant increase, m = 0.0053 (Figure 10). This situation requires decisive action on the part of mining companies if they are to prepare for the implementation of the EU climate policy.

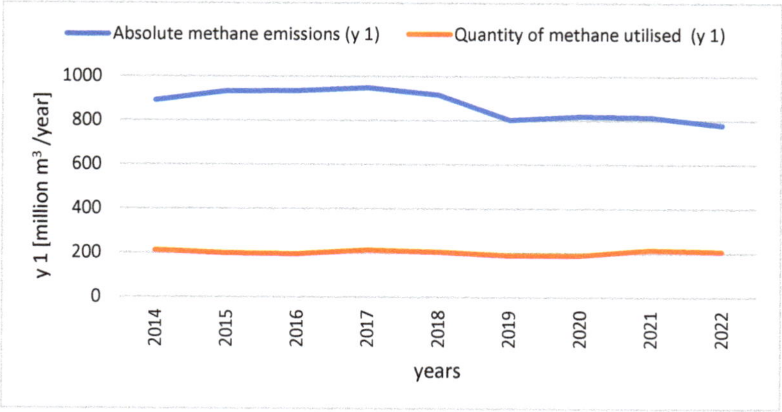

Figure 9. Absolute methane emissions and quantity of methane utilized in Poland in the years 2014–2022 [million m^3/year]. Source: individual study based on the data of WUG.

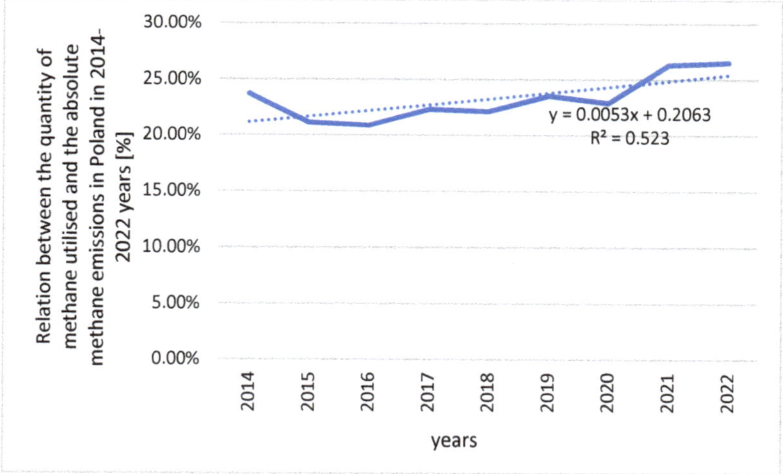

Figure 10. Relation between utilized and absolute methane emissions in Poland in the years 2014–2022 [%]. Source: individual study based on the data of WUG.

Officially, the efficiency of captured methane utilization is determined for reporting reasons (Figure 11), but this value is determined only from the value of captured methane and the level of its utilization. According to the authors, it is more appropriate to consider the efficiency of captured methane utilization to the value of absolute methane in terms of the tasks of the decarbonization policy of the Polish energy industry.

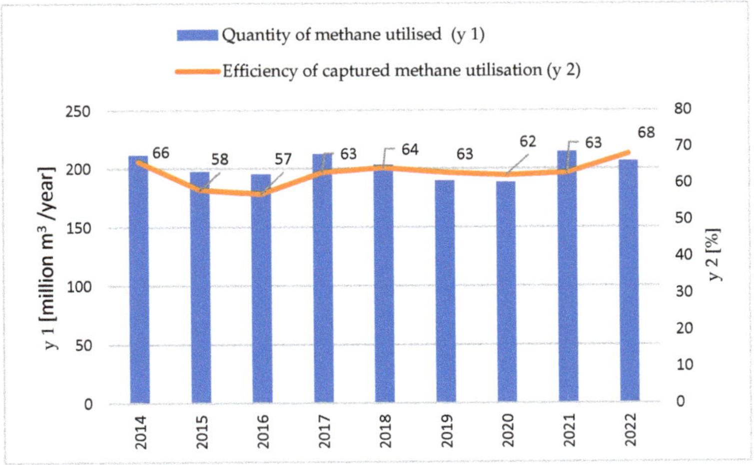

Figure 11. Variability in quantity of methane utilized and efficiency of captured methane utilization based the Polish database. Source: individual study based on the data of WUG.

The authors of this paper (analysis) compared the absolute methane emissions to the number of active longwalls (Figure 12). Although the Pearson correlation coefficient for these data was r = 0.761, it cannot be assumed that a reduction in the number of active longwalls will lead to a reduction in methane emissions in Polish mines.

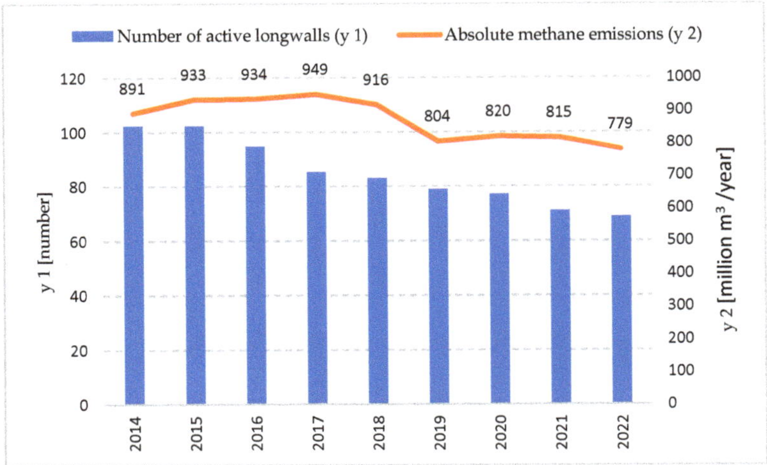

Figure 12. Variability in absolute methane emissions to number of active longwalls based on the Polish database. Source: individual study based on the data of WUG.

Analyzing the relative methane emissions first from 2008 to 2022 and then from 2014 to 2022, it was noted that the trend of this value is unchanged (m_1 = 0.3564 and m_2 = 0.2933)—Figures 13 and 14. Unfortunately, it is upward. This is not a good result in the context of per-tonne charges for methane emissions from Polish mines. The increase in relative methane emissions is due to mining at increasing depths and entering new seams with higher methane contents.

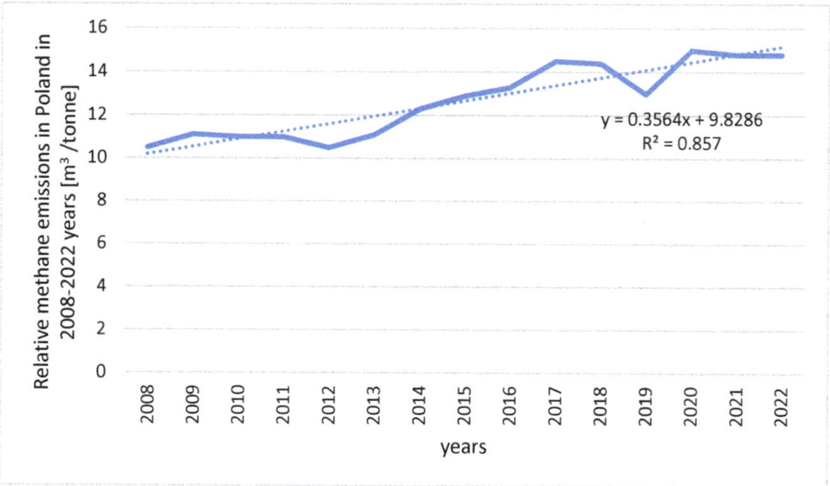

Figure 13. Relative methane emissions in Poland in the years 2008–2022 [m^3/tonne]. Source: individual study based on the data of WUG.

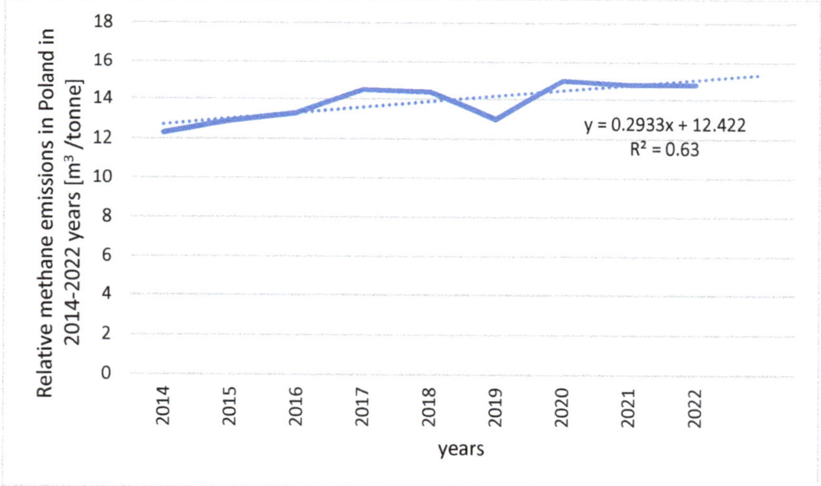

Figure 14. Relative methane emissions in Poland in the years 2014–2022 [m^3/tonne]. Source: individual study based on the data of WUG.

Therefore, the large reserves of coal bed methane in active and decommissioned Polish mines represent an economic and environmental opportunity for mining companies. This opportunity should be exploited through appropriate investment and state policies supporting the development of energy installations based on extracted methane. Unfortunately, a prerequisite for this is the development of investments in more efficient extraction of methane from ventilation air and de-methanation installations.

4.3. Commercial Opportunities of Using Methane from Polish Hard Coal Mines

JSW SA is already taking a number of measures to build a climate-friendly enterprise. Since 2018, the Budryk Mine has been implementing a 300 million project entitled Economic Use of Methane. It aims to reduce its carbon footprint and obtain energy for its own needs. As part of these measures, the installation of further de-methanisation gas engines at the

Budryk and Knurów-Szczygłowice mines continued in 2020, which, when completed, will reduce methane emissions by a total of approximately 1.3 million Mg CO_2 e.

Unfortunately, as demonstrated earlier, the efficiency of de-methanation is low. Therefore, mine de-methanation technologies need to be modernized. Modern drilling rigs that enable long and directional drilling must be introduced. It should be noted here that the nature of methane fixation in Polish deposits makes it difficult to apply solutions from other countries. The methane in Polish deposits is basically released only when the head is excavated. Therefore, de-methanation technologies dedicated to Polish deposits should be applied. The purpose of methane extraction also needs to be looked at differently. Currently, de-methanation is carried out to ensure the safety of coal mining, and not for energy, environmental, or economic purposes.

Mine de-methanation gas is mainly a mixture of air and methane with a concentration of 30% to 75% and a calorific value of 15 MJ/kg to 37.5 MJ/kg. The efficiency of methane utilization could be increased by building surface and underground reservoirs. This would improve the situation resulting from uneven methane capture during de-methanation.

Investing in the construction of gas engines will not only allow the extraction of own energy, but the heat from the flue gas and cooling can also be used for municipal purposes. If investments are made in the construction of switchgear and electrical grids to enable power to be taken out of the mine, electricity would become a new product of the mining company.

The second issue is the use of methane from mine air released into the atmosphere. Methane concentrations discharged through mine exhaust shafts are low and variable over time, ranging from 0.00% to 0.50–0.3% on average.

Depending on the density of methane in the mine air (VAM—Ventilation Air Methane), it can

- Be added (auxiliary) to power plants where the high combustion temperature is provided by another primary fuel;
- Be used in installations where VAM is the primary, or even the only, fuel [96,97].

Currently, in the literature on VAM utilization, most attention is paid to reversible reactors. Noteworthy is the innovative method developed by a team of scientists from the Institute of Chemical Engineering of the Polish Academy of Sciences in Gliwice for the utilization of methane from the ventilation air of coal mines. The TFRR (Thermal Flow Reversal Reactor) technology developed by the team does not generate toxic gases such as nitrogen oxides or carbon monoxide [53]. Under EU regulations, thermal coal mines will have to use such solutions.

The third issue is the extraction of methane from decommissioned mines. The GZW area is by far the most promising in terms of extracting methane. Several licenses have already been granted for the exploration, prospecting, and exploitation of methane from coal seams. An example of how this is already being applied is the use of methane from the "Morcinek" deposit.

5. Discussion

The analysis presented in this paper provides valuable insights into the challenges and opportunities associated with capturing methane from coal mines, particularly in the context of the European Union's (EU) climate policy objectives. One of the key findings is the substantial initial investment required for methane capture systems in coal mining operations. This paper suggests that government incentives and subsidies can play a crucial role in alleviating this financial burden, making methane capture economically viable for mining companies.

Operational and maintenance expenses are identified as ongoing challenges, requiring long-term financial planning for effective management. Additionally, this paper highlights the significance of addressing technological challenges through increased research and development investment to create more efficient and cost-effective methane capture technologies.

The variability in methane emissions from coal mines is acknowledged, emphasizing the need for adaptive system designs to ensure consistent effectiveness across different mining conditions [57,58]. Furthermore, land use and infrastructure constraints are identified as impediments, necessitating government infrastructure support for smoother integration of methane capture technologies [59].

Resource competition among coal mining companies and regulatory compliance burdens are discussed as potential obstacles, with collaboration and streamlined regulatory processes proposed as solutions [65]. This paper also emphasizes the importance of integrating methane capture projects with broader renewable energy initiatives for enhanced economic viability.

The study method involved a three-area strategic analysis, covering the interior of Polish hard coal mines, the close environment, and the macro environment. The results of the analysis indicate a gradual decline in absolute methane emissions from 2008 to 2022, despite varying coal extraction volumes. The quantity of methane captured has shown a positive trend, reaching a relatively stable level of over 300 million m^3/year since 2014.

The efficiency of methane capture, represented by de-methanation efficiency, exhibits a slight upward trend, indicating potential improvements. However, this paper notes that ongoing research is necessary to enhance these results further.

The current use of methane from the Polish coal mines is primarily driven by safety regulations, with about 25% of emitted methane utilized. The efficiency of methane utilization remains relatively constant, highlighting the need for more decisive actions to align with EU climate policy objectives.

The commercial opportunities of using methane from Polish hard coal mines are discussed, with JSW SA cited as an example of taking measures to build a climate-friendly enterprise. This paper suggests modernizing mine de-methanation technologies, introducing efficient drilling rigs, and exploring surface and underground reservoirs to increase methane utilization.

Improving methane utilization requires a multifaceted approach encompassing technological innovation, regulatory support, and strategic investment [98]. Enhancing the efficiency of de-methanation technologies is paramount. This could involve modernizing existing methods and introducing innovative techniques tailored to the unique characteristics of Polish coal deposits. Additionally, investing in advanced drilling rigs capable of long and directional drilling can facilitate more effective methane extraction [99]. Surface and underground reservoirs should be explored to optimize methane storage and distribution, ensuring consistent utilization [100].

There is a need to reframe the purpose of methane extraction from a safety-driven measure to an energy and environmental opportunity. This shift in perspective can incentivize mining companies to invest in methane capture systems and prioritize methane utilization [101]. Government incentives and subsidies can play a crucial role in making methane capture economically viable for mining operations. Also, expanding the infrastructure for methane utilization is essential [102]. This includes the construction of gas engines and electrical grids within mines to harness methane for energy production. Heat generated from flue gas and cooling processes can be repurposed for municipal use, contributing to energy efficiency and reducing environmental impact [103].

Innovative approaches to utilizing methane from mine air, such as reversible reactors or thermal flow reversal reactor technology, should be explored [104]. These solutions offer environmentally friendly alternatives while complying with EU regulations. It should be mentioned that fostering collaboration between stakeholders, including mining companies, government agencies, research institutions, and the private sector, is crucial [105,106]. By working together, these entities can pool resources, share expertise, and drive innovation in methane utilization. This collaborative effort is essential for realizing the economic and environmental potential of methane from Polish coal mines [101–104,107].

The findings of this paper shed light on the multifaceted challenges and potential avenues for mitigating methane emissions from coal mines, with a particular emphasis on the European Union's climate policy objectives [8–11,91,92]. The identified issues align with existing theoretical frameworks related to environmental sustainability, technology adoption, and policy implementation.

This paper underscores the substantial initial investment costs associated with methane capture systems in coal mining operations. This financial hurdle resonates with economic theories highlighting barriers to the adoption of environmentally friendly technologies [61]. The proposed solution involving government incentives and subsidies draws from theories of policy intervention to address market failures, promoting the adoption of sustainable practices.

Operational and maintenance expenses, recognized as ongoing challenges, reflect the importance of long-term financial planning. This aligns with theories emphasizing the need for businesses to incorporate environmental considerations into their strategic planning for sustainable operations [14,108,109]. The suggestion of increased research and development investment to tackle technological challenges resonates with innovation theories, acknowledging the role of technology in overcoming barriers to sustainable practices.

The acknowledgement of variability in methane emissions from coal mines underscores the complexities of designing effective capture systems. The proposed adaptive system designs align with ecological theories emphasizing the importance of resilience and adaptability in addressing environmental challenges [93,110]. Additionally, this paper's emphasis on government infrastructure support reflects the theories related to the role of institutions in facilitating sustainable practices.

Resource competition among coal mining companies, regulatory compliance burdens, and potential economic viability concerns are discussed in the context of broader environmental policies [81–95]. Collaborative funding initiatives and streamlined regulatory processes, proposed as solutions, draw from theories related to collaborative governance and policy design for sustainability. From the macro-economic level, this cooperation moves to the sectoral level, as mines are strongly linked to metallurgy, and methane can be used in metallurgy, heat, and energy management [111,112]. CO_2 emissions are a key negative aspect of steelworks in Poland, as more than 50% of steel is produced using high CO_2 emission technology [113,114].

This paper's discussion of integrating methane capture projects with broader renewable energy initiatives reflects a strategic approach grounded in theories of sustainability transitions [115,116]. This aligns with the recognition that a comprehensive shift toward cleaner energy requires the integration of various technologies and practices [117,118].

The analysis of absolute methane emissions, quantities of methane captured, and de-methanation efficiency trends provides empirical evidence supporting the theoretical frameworks discussed. The decline in absolute methane emissions, despite varying extraction volumes, aligns with theories of efficiency gains in resource use [119,120]. The positive trend in the quantity of methane captured, coupled with a relatively stable de-methanation efficiency since 2014, suggests ongoing efforts to improve methane capture technologies.

6. Conclusions

This paper provides a comprehensive analysis of the challenges, opportunities, and strategies associated with capturing methane from the Polish coal mines, with a focus on aligning with the European Union's climate policy. The identified challenges, ranging from initial investment costs to technological complexities and public perceptions, underscore the multifaceted nature of implementing methane capture initiatives in the coal mining industry.

This study emphasizes the critical role of government support, both in terms of financial incentives and streamlined regulatory processes, to alleviate the economic burden on coal mining companies and facilitate the integration of methane capture technologies. Collaboration through partnerships, funding initiatives, and stakeholder engagement emerges as a crucial approach to overcoming resource competition and ensuring the long-term success of methane capture projects.

The main scientific value of this paper lies in its comprehensive examination of the challenges associated with capturing methane from coal mines and its alignment with the climate policy objectives of the European Union. By delving into issues such as initial investment costs, operational challenges, technological hurdles, and regulatory complexities, this paper provides a nuanced understanding of the multifaceted nature of methane capture initiatives within the coal mining industry.

The identification of these challenges is complemented by a thoughtful analysis of potential solutions, grounded in both practical considerations and theoretical frameworks. The integration of economic, ecological, and governance theories adds depth to the proposed strategies for addressing obstacles to widespread adoption of methane capture systems.

A comprehensive analysis of both the source data over the period 1988–2023 and the literature data carried out made it possible not only to prove complex hypotheses but also show directions for building future strategies to reduce methane emissions per tonne of thermal coal from Polish mines.

While this paper provides a comprehensive analysis of the challenges and opportunities associated with capturing methane from coal mines, it is important to acknowledge certain limitations in its scope and methodology. This study predominantly focuses on the European context, specifically emphasizing the European Union's climate policy objectives. This geographical specificity may limit the generalizability of the findings to other regions with different regulatory frameworks, economic conditions, and technological landscapes. Consequently, this paper may not fully capture the diverse challenges faced by coal mining operations in other parts of the world. The empirical analysis, particularly the examination of trends in absolute methane emissions, quantities of methane captured, and de-methanation efficiency, relies on data from coal mines in Poland. The applicability of these findings to mines in other countries might be constrained by variations in mining practices, geological conditions, and the extent of technological adoption.

Author Contributions: Conceptualization, B.G., R.W. and K.T.-O.; methodology, K.T.-O.; software, R.W. and B.G.; validation, B.G. and R.W.; formal analysis, K.T.-O. and B.G.; investigation, B.G., R.W. and W.W.G.; resources, R.W. and B.G.; data curation, B.G. and K.T.-O.; writing—original draft preparation, B.G., R.W. and K.T.-O.; writing—review and editing, R.W., B.G. and W.W.G.; visualization, B.G.; supervision, B.G., R.W. and K.T.-O.; project administration, B.G., R.W. and W.W.G.; funding acquisition, R.W. All authors have read and agreed to the published version of the manuscript.

Funding: This research received no external funding.

Data Availability Statement: The data are contained within the article.

Conflicts of Interest: The authors declare no conflicts of interest.

References

1. Bohumil, H.; Joanna, P.; Łukasz, W.; Jakub, S.; Petra, M.; Patrycja, P.; Iveta, S.; Katarzyna, T.-O.; Jiri, K.; Barbora, H.; et al. *Low Carbon Economy in the Context of Developing the Automotive, Construction, and Tourism Sector*; Akademia WSB Publisher: Dąbrowa Górnicza, Poland, 2022.
2. Bluszcz, A.; Manowska, A. Differentiation of the Level of Sustainable Development of Energy Markets in the European Union Countries. *Energies* **2020**, *13*, 4882. [CrossRef]
3. Ovaere, M.; Proost, S. Cost-effective reduction of fossil energy use in the European transport sector: An assessment of the Fit for 55 Package. *Energy Policy* **2022**, *168*, 113085. [CrossRef]
4. Markowska, K.; Sękala, A.; Stecuła, K.; Kawka, T.; Sirovitskiy, K.; Pankova, O.; Vnukova, N.; Shulyak, M.; Kharchenko, S.; Shchur, T.; et al. Comparison of the Sustainability and Economic Efficiency of an Electric Car and an Aircraft—A Case Study. *Sustainability* **2023**, *15*, 1238. [CrossRef]
5. Proost, S. Looking for winning policies to address the climate issue in EU-aviation. *J. Air Transp. Manag.* **2024**, *115*, 102534. [CrossRef]
6. Dupont, C.; Moore, B.; Boasson, E.L.; Gravey, V.; Jordan, A.; Kivimaa, P.; Kulovesi, K.; Kuzemko, C.; Oberthür, S.; Panchuk, D.; et al. Three decades of EU climate policy: Racing toward climate neutrality? *Wiley Interdiscip. Rev. Clim. Chang.* **2024**, *15*, e863. [CrossRef]

7. Paleari, S. The EU policy on climate change, biodiversity and circular economy: Moving towards a Nexus approach. *Environ. Sci. Policy* **2024**, *151*, 103603. [CrossRef]
8. Jensen, A.; Nielsen, H.Ø.; Russel, D. Diffusion of climate policy integration in adaptation strategies: Translating the EU mandate into UK and Danish national contexts. *Reg. Environ. Chang.* **2023**, *23*, 130. [CrossRef]
9. Brad, A.; Schneider, E. Carbon dioxide removal and mitigation deterrence in EU climate policy: Towards a research approach. *Environ. Sci. Policy* **2023**, *150*, 103591. [CrossRef]
10. Korosuo, A.; Pilli, R.; Abad Viñas, R.; Blujdea, V.N.; Colditz, R.R.; Fiorese, G.; Rossi, S.; Vizzarri, M.; Grassi, G. The role of forests in the EU climate policy: Are we on the right track? *Carbon Balance Manag.* **2023**, *18*, 15. [CrossRef]
11. Weishaar, S.E.; Kim, E.; Tiche, F.G. *Climate and Energy Law and Policy in the EU and East Asia: Transition and Policy Cooperation*; Edward Elgar Publishing: Cheltenham, UK, 2023; pp. 1–220.
12. Vogler, J. Global dimensions of EU climate, energy and transport policies. In *Handbook on European Union Climate Change Policy and Politics*; Edward Elgar Publishing: Cheltenham, UK, 2023; pp. 144–157.
13. Zhang, X.-B.; Xu, J. Optimal policies for climate change: A joint consideration of CO_2 and methane. *Appl. Energy* **2018**, *211*, 1021–1029. [CrossRef]
14. Quelopana, A.; Órdenes, J.; Araya, R.; Navarra, A. Geometallurgical Detailing of Plant Operation within Open-Pit Strategic Mine Planning. *Processes* **2023**, *11*, 381. [CrossRef]
15. Greenhouse Effect. Wikipedia. Available online: http://en.wikipedia.org/wiki/Greenhouse_effect (accessed on 4 February 2024).
16. IEA Report: Global Methane Tracker. 2022. Available online: https://www.iea.org/reports/global-methane-tracker-2022/methane-and-climate-change (accessed on 4 February 2024).
17. Shindell, D.; Ravishankara, A.R.; Kuylenstierna, J.C.I.; Michalopoulou, E.; Höglund-Isaksson, L.; Zhang, Y.; Seltzer, K.; Ru, M.; Castelino, R.; Faluvegi, G.; et al. *Global Methane Assessment: Benefits and Costs of Mitigating Methane Emissions*; United Nations Environment Programme: Nairobi, Kenya, 2021.
18. CIREPL Paper: Świat OZE (Global RES). Available online: https://swiatoze.pl/metan-odpowiada-za-30-globalnego-ocieplenia-skad-wzial-sie-w-atmosferze-i-czy-jest-niebezpieczny/ (accessed on 4 February 2024).
19. KOBiZE. *KOBiZE 2023, Krajowy Raport Inwentaryzacyjny 2023. Warsow, 2023, Marc., Report Prepared by: Krajowy Ośrodek Bilansowania i Zarządzania Emisjami*; Ministry of Climate: Warsaw, Poland, 2023.
20. IPCC. *Climate Change 2021: The Physical Science Basis. Contribution of Working Group I to the Sixth Assessment Report of the Intergovernmental Panel on Climate Change*; Cambridge University Press: Cambridge, UK, 2021.
21. Nisbet, E.G.; Manning, M.R.; Dlugokencky, E.J.; Fisher, R.E.; Lowry, D.; Michel, S.E.; Myhre, C.L.; Platt, S.M.; Allen, G.; Bousquet, P.; et al. Very strong atmospheric methane growth in the 4 years 2014–2017: Implications for the Paris agreement. *Glob. Biogeochem.* **2019**, *33*, 318–342. [CrossRef]
22. Report: Our World in Data. Available online: https://ourworldindata.org/atmospheric-concentrations (accessed on 4 February 2024).
23. Report NOAA. Trends in Atmospheric Global Monitoring Laboratory Earth System Research Laboratories. Methane. Available online: https://gml.noaa.gov/ccgg/trends_ch4/#:~:text=Global%20CH4%20Monthly%20Means&text=The%20Global%20Monitoring%20Division%20of,et%20al.,%201994 (accessed on 4 February 2024).
24. CIRE.PL. Centrum Informacji o Rynku Energii, Metan Jest Większym Problemem Niż CO_2, Jest Plan Redukcji Emisji Metanu o 30 Proc. 3.11.2021, 13.00. Available online: https://www.cire.pl/artykuly/o-tym-sie-mowi/metan-jest-wiekszym-problemem-niz-co2-jest-plan-redukcji-emisji-metanu-o-30-proc (accessed on 4 February 2024).
25. Kuśmierczyk-Michulec, J.; Baré., J. Global atmospheric monitoring of radionuclides and noble gases: Insights into the interhemispheric transport of anthropogenic emissions. *Atmos. Environ.* **2024**, *319*, 120304. [CrossRef]
26. Dlugokencky, E.J.; Steele, L.P.; Lang, P.M.; Masarie, K.A. The growth rate and distribution of atmospheric methane. *J. Geophys. Res.* **1994**, *99*, 17021–17043. [CrossRef]
27. Masarie, K.A.; Tans, P.P. Extension integration of atmospheric carbon dioxide data into a globally consistent measurement record. *J. Geophys. Res.* **1995**, *100*, 11593–11610. [CrossRef]
28. COM202. IPCC. Summary for policymakers. In *Climate Change 2021: The Physical Science Basis. Contribution of Working Group I to the Sixth Assessment Report of the Intergovernmental Panel on Climate Change*; Cambridge University Press: Cambridge, UK, 2021; pp. 3–32.
29. Balcerowski, J.; Charkowska, Z. W poszukiwaniu źródeł Emisje metanu w polskim górnictwie węgla kamiennego i systemy raportowania. Instrat Policy Paper 01/2023, Warsow, Feb. 2023. Available online: https://instrat.pl/wp-content/uploads/2023/02/Instrat-Policy-Paper-01-2023-W-poszukiwaniu-zrodel.-Emisje-metanu-....pdf (accessed on 4 February 2024).
30. Ocko, I.B.; Sun, T.; Shindell, D.; Oppenheimer, M.; Hristov, A.N.; Pacala, S.W.; Mauzerall, D.L.; Xu, Y.; Hamburg, S.P. Acting rapidly to deploy readily available methane mitigation measures by sector can immediately slow global warming. *Environ. Res. Lett.* **2021**, *16*, 054042. [CrossRef]
31. Saunois, M.; Stavert, A.R.; Poulter, B.; Bousquet, P.; Canadell, J.G.; Jackson, R.B.; Raymond, P.A.; Dlugokencky, E.J.; Houweling, S.; Patra, P.K.; et al. The global methane budget 2000–2017. *Earth Syst. Sci. Data* **2020**, *12*, 1561–1623. [CrossRef]
32. Liu, Y.; Gao, H.; Yu, Z.; Li, C.; Ren, H.; Zhao, Y.; Huang, F.; Guo, Y.; Wan, M.; Tian, J. Managing Methane Emissions in Abandoned Coal Mines: Comparison of Different Recovery Technologies by Integrating Techno-Economic Analysis and Life-Cycle Assessment. *Environ. Sci. Technol.* **2023**, *56*, 13900–13908. [CrossRef]

33. Harmsen, M.; van Vuuren, D.P.; Bodirsky, B.L.; Chateau, J.; Durand-Lasserve, O.; Drouet, L.; Fricko, O.; Fujimori, S.; Gernaat, D.E.H.J.; Hanaoka, T.; et al. The role of methane in future climate strategies: Mitigation potentials and climate impacts. *Clim. Chang.* **2020**, *163*, 1409–1425. [CrossRef]
34. Shindell, D.; Fuglestvedt, J.; Collins, W. The social cost of methane: Theory and applications. *Faraday Discuss.* **2017**, *200*, 429–451. [CrossRef]
35. IPCC. *"Sumary for Policymakers" ["Podsumowanie dla Decydentów"] w: "Climate Change 2021: The Physical Science Basis"*; IPCC: Geneva, Switzerland, 2021.
36. Royal Society, Ground-Level Ozone in the 21st Century: Future Trends, Impacts and Policy Implications (RS1276 ed., 2008, Vol. 15/08). The Royal Society. Available online: https://royalsociety.org/topics-policy/publications/2008/ground-level-ozone/ (accessed on 4 February 2024).
37. UNEP. *Integrated Assessment of Black Carbon and Tropospheric Ozone*; United Nations Environmental Programme: Nairobi, Kenya, 2011.
38. Pujade-Lauraine, E.; Ledermann, J.A.; Selle, F.; Gebski, V.; Penson, R.T.; Oza, A.M.; Korach, J.; Huzarski, T.; Poveda, A.; Pignata, S.; et al. Olaparib tablets as maintenance therapy in patients with platinumsensitive, relapsed ovarian cancer and a BRCA1/2 mutation (SOLO2/ENGOT-Ov21): A double-blind, randomised, placebo-controlled, phase 3 trial. *Lancet Oncol.* **2017**, *18*, 1274–1284. [CrossRef]
39. Shoemaker, J.K.; Schrag, D.P.; Molina, M.J.; Ramanathan, V. What role for shortlived climate pollutants in mitigation policy? *Science* **2013**, *342*, 1323–1324. [CrossRef] [PubMed]
40. Allen, R.J.; Horowitz, L.W.; Naik, V.; Oshima, N.; O'Connor, F.M.; Turnock, S.; Shim, S.; Le Sager, P.; van Noije, T.; Tsigaridis, K.; et al. Significant climate benefits from near-term climate forcer mitigation in spite of aerosol reductions. *Environ. Res. Lett.* **2021**, *16*, 034010. [CrossRef]
41. Odpowiedzialny Biznes. Raport. Available online: https://odpowiedzialnybiznes.pl/aktualno%C5%9Bci/raport-90-emisji-szkodliwego-metanu-w-polsce-pochodzi-z-kopalni-wegla-kamiennego/ (accessed on 4 February 2024).
42. Ember. Drugi Bełchatów. Metan Kopalniany w Polsce. Ember Climate. 2020. Available online: https://ember-climate.org/app/uploads/2022/01/Polish-Polands-Second-Belchatow.pdf (accessed on 4 February 2024).
43. Krajowy Program Energetyki i Kopalń KPEiK/*Energy Policy of Poland (EPP2040) until 2040*, (Title in Polish) *Polityka Energetyczna Polski do 2040 r.* (PEP2040). Available online: https://www.gov.pl/web/klimat/polityka-energetyczna-polski (accessed on 4 February 2024).
44. Terytorialny Plan Sprawiedliwej Transformacji Województwa Śląskiego 2030 i Umowy Społecznej dotyczącej transformacji sektora górnictwa węgla kamiennego oraz wybranych procesów transformacji woj. śląskiego. In *Terytorialny Plan Sprawiedliwej Transformacji Województwa Śląskiego*; Urząd Marszałkowski Województwa Śląskiego: Katowice, Poland, 2022. Available online: https://transformacja.slaskie.pl/images/Dokumenty/1646921120_terytorialny_plan_spr.04.pdf (accessed on 25 May 2023).
45. Forum Energii, PEP 2040–Rekomendacje Eksperckiej Rady ds. Bezpieczeństwa Energetycznego. Available online: https://forum-energii.eu/pl/blog/pep-2040-rekomendacje-rady (accessed on 3 February 2024).
46. Polski Komitet Energii Elektrycznej Polska ścieżka Transformacji Energetycznej Warszawa, Październik 2022. Available online: https://pkee.pl/publikacje/raport-ey-i-pkee-polska-sciezka-transformacji-energetycznej/ (accessed on 3 February 2024).
47. Stasińska, B. Ograniczenie emisji metanu z kopalń węglowych poprzez katalityczne oczyszczanie powietrza wentylacyjnego. *Polityka Energetyczna* **2009**, *12*, 123–132.
48. Nawrat, S.; Stopa, J. Pozyskanie i Wykorzystanie Energetyczne Metanu z Kopalń Węgla Kamiennego Szansą Ekonomiczną i Ekologiczną Dla Śląska i Polski (Capturing and Using Coal-Mine Methane for Energy Production—An Economic and Ecological Opportunity for Silesia and Poland), Jastrzębie Zdrój, 17 October 2022. Available online: https://orka.sejm.gov.pl/opinie9.nsf/nazwa/650_20221017_4/$file/650_20221017_4.pdf (accessed on 4 February 2024).
49. Badyda, K. Możliwości zagospodarowania gazu kopalnianego w Polsce dla celów energetycznych. *Energetyka* **2008**, *6*, 416–423.
50. Nawrat, S.; Gantnar, K. Ocena stanu i możliwości utylizacji metanu z powietrza wentylacyjnego podziemnych kopalń węgla kamiennego. *Polityka Energetyczna* **2008**, *11*, 69–84.
51. Jureczka, J. Metan z Zaniechanych Złóż Węgla Kamiennego–Wsparcie Dla Transformacji Energetycznej. Państwowy Instytut Geologiczny (02 Listopad 2021). Available online: https://www.pgi.gov.pl/aktualnosci/display/13314-metan-z-zaniechanych-zloz-wegla-kamiennego-wsparcie-dla-transformacji-energetycznej.html (accessed on 27 May 2023).
52. Gajdzik, B.; Wolniak, R.; Grebski, W. Process of Transformation to Net Zero Steelmaking: Decarbonisation Scenarios Based on the Analysis of the Polish Steel Industry. *Energies* **2023**, *16*, 3384. [CrossRef]
53. Available online: https://informacje.pan.pl/71-media/materialy-prasowe/3136-innowacyjna-metoda-utylizacji-metanu-doceniona-na-swiecie (accessed on 2 February 2024).
54. Tobór-Osadnik, K.; Gajdzik, B.; Strzelec, G. Configurational Path of Decarbonisation Based on Coal Mine Methane (CMM): An econometric model for the Polish mining industry. *Sustainability* **2023**, *15*, 9980. [CrossRef]
55. Brodny, J.; Tutak, M. Analysis of methane emission into the atmosphere as a result of mining activity. *Int. Multidiscip. Sci. GeoConf. SGEM* **2016**, *3*, 83–90.
56. Kaczmarek, J. The Balance of Outlays and Effects of Restructuring Hard Coal Mining Companies in Terms of Energy Policy of Poland PEP 2040. *Energies* **2022**, *15*, 1853. [CrossRef]

57. Lv, D.; Wu, Y.; Chen, J.; Li, Z.; Xia, Q. Improving CH4/N2 selectivity within isomeric Al-based MOFs for the highly selective capture of coal-mine methane. *AIChE J.* **2020**, *66*, e16287. [CrossRef]
58. Niu, Z.; Cui, X.; Pham, T.; Lan, P.C.; Xing, H.; Forrest, K.A.; Wojtas, L.; Space, B.; Ma, S. A Metal–Organic Framework Based Methane Nano-trap for the Capture of Coal-Mine Methane. *Angew. Chem.-Int. Ed.* **2019**, *58*, 10138–10141. [CrossRef]
59. Li, Q.; Ruan, M.; Lin, B.; Zhao, M.; Zheng, Y.; Wang, K. Molecular simulation study of metal organic frameworks for methane capture from low-concentration coal mine methane gas. *J. Porous Mater.* **2016**, *23*, 107–122. [CrossRef]
60. Koroviaka, Y.; Pinka, J.; Tymchenko, S.; Rastsvietaiev, V.; Astakhov, V.; Dmytruk, O. Elaborating a scheme for mine methane capturing while developing coal gas seams. *Min. Miner. Depos.* **2020**, *14*, 21–27. [CrossRef]
61. Siemek, J.; Macuda, J.; Łukańko, Ł.; Hendel, J. The technology of drilling wells for capturing methane from abandoned coal mines. *Arch. Min. Sci.* **2020**, *65*, 89–101.
62. Szlązak, N.; Swolkień, J. Possibilities of capturing methane from hard coal deposits lying at great depths. *Energies* **2021**, *14*, 3542. [CrossRef]
63. Duda, A. The Impact of Atmospheric Pressure Changes on Methane Emission from Goafs to Coal Mine Workings. *Energies* **2024**, *17*, 173. [CrossRef]
64. Liu, Q.; Teng, F.; Zhang, L.-Y. The improvement and recalculation of China's coal methane emission inventory based on a dynamic mine-level database. *Clim. Chang. Res.* **2023**, *19*, 704–713.
65. Zhang, X.; Zhu, T.; Yi, N.; Yuan, B.; Li, C.; Ye, Z.; Zhu, Z.; Zhang, X. Study on Characteristics and Model Prediction of Methane Emissions in Coal Mines: A Case Study of Shanxi Province, China. *Atmosphere* **2023**, *14*, 1422. [CrossRef]
66. Konkol, I.; Cebula, J.; Świerczek, L.; Sopa, J.; Sopa, J.; Cenian, A. Characterization of Deposits Formed in Gas Engines Fuelled by Coal Mine Methane. *Materials* **2023**, *16*, 2517. [CrossRef] [PubMed]
67. Li, S.; Gong, Y.; Fu, S.; Wang, L.; Cheng, Y. Research progress of methane purification from coal mine gas by variable pressure adsorption. *Sci. Sin. Chim.* **2023**, *53*, 992–1007. [CrossRef]
68. Zhou, Y.; Liu, Y.; Shi, Y.; Zhang, Y.; Shi, J.; Sun, P. Strategies for heat recovery in a reverse flow reactor designed to lessen coal mine methane emissions test, Energy Sources. *Part A Recovery Util. Environ. Eff.* **2023**, *45*, 8665–8685.
69. Su, X.B.; Zhao, W.Z.; Wang, Q.; Yu, S.Y.; Song, J.X.; Wang, X.M.; Xia, D.P.; Fu, H.J.; Sun, C.Y.; Guo, H.G.; et al. Conception of key technologies for low-negative carbon emission reduction in the process of coalbed methane development from the CBM well, coal mine and goaf. *Meitan Xuebao/J. China Coal Soc.* **2023**, *48*, 335–356.
70. Swolkień, J.; Fix, A.; Gałkowski, M. Factors influencing the temporal variability of atmospheric methane emissions from Upper Silesia coal mines: A case study from the CoMet mission. *Atmos. Chem. Phys.* **2022**, *22*, 16031–16052. [CrossRef]
71. Smirnova, A.; Varnavskiy, K.; Nepsha, F.; Kostomarov, R.; Chen, S. The Development of Coal Mine Methane Utilization Infrastructure within the Framework of the Concept "Coal-Energy-Information". *Energies* **2022**, *15*, 8948. [CrossRef]
72. Meyer, A.G.; Lindenmaier, R.; Heerah, S.; Peischl, J.; Dubey, M.K. Using Multiscale Ethane/Methane Observations to Attribute Coal Mine Vent Emissions in the San Juan Basin From 2013 to 2021. *J. Geophys. Res. Atmos.* **2022**, *127*, e2022JD037092. [CrossRef]
73. Wu, X.; Cui, J.; Tong, R.; Li, Q. Research on Methane Measurement and Interference Factors in Coal Mines. *Sensors* **2022**, *22*, 5608. [CrossRef]
74. Sadavarte, P.; Pandey, S.; Maasakkers, J.D.; van der Gon, H.D.; Houweling, S.; Aben, I. A high-resolution gridded inventory of coal mine methane emissions for India and Australia. *Elementa* **2022**, *10*, 10. [CrossRef]
75. Yin, F.; Nie, B.; Wei, Y.; Lin, S. Co-Production System Based on Lean Methane and Biogas for Power Generation in Coal Mines. *Atmosphere* **2022**, *13*, 803. [CrossRef]
76. Bolt, A.; Dincer, I.; Agelin-Chaab, M. Investigation of a renewable energy based integrated system with carbon capturing for hydrogen, methane and other useful outputs. *J. Clean. Prod.* **2023**, *422*, 138572. [CrossRef]
77. Anyanwu, U.O.; Okafor, O.C.; Nkwor, C.A. A Scheduling Model in Capturing Methane Gas from Methane Clathrates Deposits. *J. Eng. Sci.* **2023**, *10*, G1–G13. [CrossRef]
78. Fukuhara, C.; Matsui, Y.; Tanebayashi, M.; Watanabe, R. A novel catalytic reaction system capturing solid carbon from greenhouse gas, combined with dry reforming of methane. *Chem. Eng. J. Adv.* **2021**, *5*, 100057. [CrossRef]
79. Liu, B.; Yu, X.; Shi, W.; Shen, Y.; Zhang, D.; Tang, Z. Two-stage VSA/PSA for capturing carbon dioxide (CO_2) and producing hydrogen (H2) from steam-methane reforming gas. *Int. J. Hydrogen Energy* **2020**, *45*, 24870–24882. [CrossRef]
80. Ghannadi, M.; Dincer, I. A new system for hydrogen and methane production with post-combustion amine-based carbon capturing option for better sustainability. *Energy Convers. Manag.* **2024**, *301*, 118026. [CrossRef]
81. Kumar, V.; Tiwari, A.K. Performance assessment of green hydrogen generation via distinct electrolytes dedicated to renewable energy. *Desalination* **2024**, *582*, 117651. [CrossRef]
82. Sofiiskyi, K.; Petukh, O. The results of experimental research of the parameters of methane capturing by local degassing wells in the undermining area. *E3S Web Conf.* **2019**, *109*, 00097. [CrossRef]
83. Nekhubvi, V.; Tinarwo, D. Methane capturing using anaerobic digestion technology: A way towards mitigating greenhouse gases in Limpopo Province, South Africa. *Nat. Environ. Pollut. Technol.* **2017**, *16*, 1243–1247.
84. Cluff, D.L.; Kennedy, G.A.; Bennett, J.G.; Foster, P.J. Capturing energy from ventilation air methane a preliminary design for a new approach. *Appl. Therm. Eng.* **2015**, *90*, 1151–1163. [CrossRef]
85. Wang, L.; Sun, Y.; Zheng, S.; Shu, L.; Zhang, X. How efficient coal mine methane control can benefit carbon-neutral target: Evidence from China. *J. Clean. Prod.* **2023**, *424*, 138895. [CrossRef]

86. Su, Z.; Xing, L.; Ali, H.E.; Alkhalifah, T.; Alturise, F.; Khadimallah, M.A.; Assilzadeh, H. Latest insights on separation and storage of carbon compounds in buildings towards sustainable environment: Recent innovations, challenges, future perspectives and application of machine learning. *Chemosphere* **2023**, *329*, 138573. [CrossRef] [PubMed]
87. Teng, Y.; Zhao, W.; Tang, K.; Yang, K.; Zhao, G.; Zhu, S.; Ye, J.; Gu, S. High efficiency of boron doping and fast growth realized with a novel gas inlet structure in diamond microwave plasma chemical vapor deposition system. *Carbon Lett.* **2023**, *34*, 1115–1128. [CrossRef]
88. Subbian, S.; Natarajan, P.; Murugan, C. Circular economy-based multi-objective decentralized controller for activated sludge wastewater treatment plant. *Front. Chem. Eng.* **2023**, *5*, 1235125. [CrossRef]
89. Smirnova, A.D.; Shevtsov, A.G.; Chen, S. Analysis of factors affecting the coal seams distribution permeability in the southern part of the Tutuyasskaya Area of the Kuznetsk Basin, Russia. *Sustain. Dev. Mt. Territ.* **2022**, *14*, 657–665. [CrossRef]
90. Wang, C.; Wang, R.; Zhou, W.; Li, F. The use of a novel spacer and ultralow-density cement system to control lost circulation in coalbed-methane wells. *SPE Drill. Complet.* **2015**, *30*, 76–85. [CrossRef]
91. Qi, Q.-X.; Ji, W.-B.; Yuan, J.-H.; Cheng, Z.-H.; Liu, X.-G. In-situ observation on the floor-connected crack field and its effects on methane extraction. *Meitan Xuebao/J. China Coal Soc.* **2014**, *39*, 1552–1558.
92. Puchkov, L.A.; Kaledina, N.O.; Kobylkin, S.S. System solutions of provision of methane safety of coal mines. *Gorn. Zhurnal* **2014**, *5*, 12–16.
93. Tang, X.; Hu, M. On the resilience of bio-cemented silica sands in chemically reactive environment. *Geomech. Energy Environ.* **2024**, *37*, 100527. [CrossRef]
94. Gierszewska, G.; Romanowska, M. *Analiza Strategiczna Przedsiębiorstwa*; Polskie Wydawnictwo Ekonomiczne: Warszawa, Poland, 2017.
95. Tyrańska, M.; Walas-Trębacz, J. *Wykorzystanie Metod Analizy Strategicznej w Przedsiębiorstwie*; Wydawnictwo Uniwersytetu Ekonomicznego w Krakowie: Kraków, Poland, 2010.
96. Korski, J.; Tobór–Osadnik, K.; Wyganowska, M. Reasons of problems of the Polish hard coal mining in connection with restructuring changes in the period 1988–2014. *Resour. Policy* **2016**, *48*, 25–31. [CrossRef]
97. Gosiewski, K.; Pawlaczyk-Kurek, A.; Manfred, J.; Wojdyła, A.; Kleszcz, T.; Machej, T.; Michalski, L. Realne możliwości utylizacji metanu z powietrza wentylacyjnego kopalń węgla kamiennego z odzyskiem energii. *Pr. Nauk. Inst. Inżynierii Chem. Pol. Akad. Nauk* **2021**, 75–119. Available online: https://integro.biblioteka.sosnowiec.pl/ici/recorddetail?id=oai:bibliotekanauki.pl:2175685 (accessed on 4 February 2024).
98. Sun, D.; Liang, Y.; Huang, X.; Ran, Q. Progress and prospects of coalbed methane development and utilization in coal mining areas with large dip angle and multiple coal groups in Xinjiang. *Meitan Kexue Jishu/Coal Sci. Technol.* **2023**, *51*, 162–172.
99. Kong, X.; Tang, Y.; Li, P.; Chen, B.; Ye, M. Thingking and utilization technology of coalbed methane in soft and low permeability coal seams in Huainan Mining Area. *Meitan Kexue Jishu/Coal Sci. Technol.* **2022**, *50*, 26–35.
100. Qin, K.; Hu, W.; He, Q.; Lu, F.; Cohen, J.B. Individual coal mine methane emissions constrained by eddy covariance measurements: Low bias and missing sources. *Atmos. Chem. Phys.* **2024**, *24*, 3009–3028. [CrossRef]
101. Varnavskiy, K.A.; Nepsha, F.S.; Kostomarov, R.V.; Chen, Q. Business Diversification of Coal Mining Enterprises Based on the Development of Coal Mine Methane Utilization Infrastructure. In Proceedings of the 2022 IEEE International Multi-Conference on Engineering, Computer and Information Sciences, SIBIRCON 2022, Yekaterinburg, Russia, 11–13 November 2022; pp. 2060–2063.
102. Liu, H.; Zhao, M.; Zhao, K.; Zhang, Q.; Zhang, X. Fine calculation of methane emissions and utilization amount from coal mines in Shanxi Province. *Nat. Gas Ind.* **2022**, *42*, 179–185.
103. Ponnudurai, V.; Rajarathinam, R.; Muthuvelu, K.S.; Nalajala, R.K.; Arumugam, L. Investigation on the utilization of coal washery rejects by different microbial sources for biogenic methane production. *Chemosphere* **2022**, *287*, 132165. [CrossRef] [PubMed]
104. Wang, X.; Zhou, F.; Ling, Y.; Wei, K.; Kang, J. Overview and Outlook on Utilization Technologies of Low-Concentration Coal Mine Methane. *Energy Fuels* **2021**, *35*, 15398–15423. [CrossRef]
105. Blinova, E.S.; Nevskaya, M.A. Methane utilization as a resource-saving method inthe coal industry. *E3S Web Conf.* **2021**, *266*, 06002. [CrossRef]
106. Mahdevari, S. Coal mine methane: Control, utilization, and abatement. In *Advances in Productive, Safe, and Responsible Coal Mining*; Woodhead Publishing: Sawston, UK, 2018; pp. 179–198.
107. Liu, J.; Sun, H.; Lei, Y.; Cao, J. Current situation and development trend of coalbed methane development and utilization technology in coal mine area. *Meitan Xuebao/J. China Coal Soc.* **2020**, *45*, 258–267.
108. Noriega, R.; Pourrahimian, Y. A systematic review of artificial intelligence and data-driven approaches in strategic open-pit mine planning. *Resour. Policy* **2023**, *77*, 102727. [CrossRef]
109. Pouresmaieli, M.; Ataei, M.; Nouri Qarahasanlou, A.; Barabadi, A. Integration of renewable energy and sustainable development with strategic planning in the mining industry. *Results Eng.* **2023**, *20*, 101412. [CrossRef]
110. Pape, U.; Brandsen, T.; Pahl, J.B.; Pieliński, B.; Baturina, D.; Brookes, N.; Chaves-Ávila, R.; Kendall, J.; Matančević, J.; Petrella, F.; et al. Changing Policy Environments in Europe and the Resilience of the Third Sector. *Voluntas* **2020**, *31*, 238–249. [CrossRef]
111. Gajdzik, B.; Sroka, W.; Vveinhardt, J. Energy intensity of steel manufactured utilising EAF technology as a function of investments made: The case of the steel industry in Poland. *Energies* **2021**, *14*, 5152. [CrossRef]

112. Gajdzik, B.; Wolniak, R.; Grebski, W.W. An Econometric Model of the Operation of the Steel Industry in Poland in the Context of Process Heat and Energy Consumption. *Energies* **2022**, *15*, 7909. [CrossRef]
113. Gajdzik, B. Environmental aspects, strategies and waste logistic system based on the example of metallurgical company. *Metalurgija* **2009**, *48*, 63–67.
114. Gajdzik, B. Comprehensive classification of environmental aspects in a manufacturing enterprise. *Metalurgija* **2012**, *51*, 541–544.
115. Stuikyte, R.; Varentsov, I.; Cook, C.; Dvoriak, S. Measuring sustainability of opioid agonist therapy programs in the context of transition from global fund support. *Harm Reduct. J.* **2024**, *21*, 7. [CrossRef] [PubMed]
116. Mahmood, H.; Irshad, A.R.; Tanveer, M. Do innovation and renewable energy transition play their role in environmental sustainability in Western Europe? *Humanit. Soc. Sci. Commun.* **2024**, *11*, 22. [CrossRef]
117. Bogner, K.; Kump, B.; Beekman, M.; Wittmayer, J. Coping with transition pain: An emotions perspective on phase-outs in sustainability transitions. *Environ. Innov. Soc. Transit.* **2024**, *50*, 100806. [CrossRef]
118. Chen, Y.; Yu, Z. Digitalization, trust, and sustainability transitions: Insights from two blockchain-based green experiments in China's electricity sector. *Environ. Innov. Soc. Transit.* **2024**, *50*, 100801. [CrossRef]
119. Banerjee, S.; Roy, S.; Khatua, S. Game Theory Based Energy-aware Virtual Machine Placement towards Improving Resource Efficiency in Homogeneous Cloud Data Center. In Proceedings of the 2022 IEEE Calcutta Conference, CALCON 2022-Proceedings, Kolkata, India, 10–11 December 2022; pp. 293–298.
120. Rentschler, J.; Flachenecker, F.; Kornejew, M. Assessing carbon emission savings from corporate resource efficiency investments: An estimation indicator in theory and practice, Environment. *Dev. Sustain.* **2020**, *22*, 835–861. [CrossRef]

energies

MDPI

Review

Business Models on the Energy Market in the Era of a Low-Emission Economy

Arkadiusz Sułek [1] and Piotr F. Borowski [2,*]

1 Prozap Sp. z o.o., al. Tysiąclecia Państwa Polskiego 13, 24-110 Puławy, Poland; prozap@prozap.com.pl
2 Faculty of Business and International Relations, Vistula University, 3 Stokłosy Str., 02-787 Warsaw, Poland
* Correspondence: p.borowski@vistula.edu.pl or pborowski@autograf.pl

Abstract: In the energy market, we observe a dynamic development of innovative business models that take into account various aspects related to the direction of zero-emission economic growth. Companies are intensifying their efforts in utilizing renewable energy sources, implementing significant photovoltaic projects, and advancing technologies related to wind and hydrodynamic energy. Within this trend, microgrids become a crucial element, enabling efficient management of local energy sources. Contemporary energy companies also focus on innovative digital technologies, harnessing the potential of the Internet of Things (IoT) and artificial intelligence (AI). These tools allow for precise monitoring and optimization of energy consumption, resulting in increased operational efficiency. The expansion of subscription-based energy services encompasses not only traditional energy deliveries but also new aspects, such as intelligent management of home energy installations or the provision of advisory services on energy conservation. This approach emphasizes the customer as a partner in sustainable energy usage. Hybrid energy models, integrating diverse energy sources, constitute a key element in the transformation of the sector. The combination of photovoltaic, wind, and traditional power plants allows for flexible adaptation to changing conditions and maintains stability in the energy supply. In the face of a changing energy landscape, companies consistently strive for sustainable practices, implementing strategies that not only reduce their carbon footprint but also contribute to improving efficiency, ecology, and the decentralization of the energy system. Adapting to these dynamic changes becomes not only a challenge but also an opportunity to create a more sustainable energy future. The objective of this research is to analyze key business models in the energy market and identify their impact on operational efficiency and market competitiveness. The main results indicate significant improvements in energy management and sustainability through the adoption of these models.

Keywords: business model; energy transition; energy market; ESCO; EaaS; business model classification

1. Introduction

In the face of the growing challenge of climate change and the need to reduce greenhouse gas emissions, the energy sector is undergoing fundamental transformations. The leading idea is the transition towards a zero-emission economy, resulting in a revolution in business models in the energy market. The shift to renewable energy sources (RES) and modern technologies poses challenges for businesses but, at the same time, opens up broad perspectives for innovative approaches to energy generation, distribution, and consumption.

The research problem is the issue of building and implementing business models that will enable effective management of enterprises in the energy sector and will allow for flexible energy management through energy storage, intelligent consumption control, and integration with modern technologies. The main question that the researchers faced was what business models should be introduced in connection with the development of new technologies to effectively manage the energy sector.

One of the key business models gaining significance in the era of zero emissions is the model of distributed energy. Instead of relying on central energy sources, companies increasingly invest in distributed generation systems, such as micro photovoltaic installations or microwind turbines. This not only allows for emission reduction but also creates local power sources, increasing the energy independence of communities.

An innovative approach is also the Energy-as-a-Service (EaaS) model [1]. Instead of the traditional model of charging for the amount of energy consumed, companies offer customers comprehensive energy services, including efficient consumption management, RES, and even energy storage. This approach promotes energy efficiency and encourages more sustainable practices.

The energy platform model is also gaining importance. Creating digital platforms that integrate producers, distributors, and consumers enables more flexible and efficient energy exchange. Blockchain, the Internet of Things (IoT), and artificial intelligence play a crucial role in this revolution, providing transparency, security, and process automation [2].

However, on the path to a zero-emission economy, changes in financing models are also essential. The "pay-as-you-save" model, paying for energy savings, is gaining popularity. Investments in energy efficiency or RES installations are funded from the savings they generate, eliminating traditional financial barriers.

It is worth emphasizing that companies in the energy market, to operate effectively in the era of a zero-emission economy, must consider not only economic aspects but also social and ecological factors.

Previous research has shown that new skills will be needed if new business models need to be implemented. New business models will require people to have sales and marketing skills and employ knowledgeable workers [3,4]. The literature has also highlighted the need to implement models that have great potential to support energy flexibility in a cost-effective and sustainable way [5].

Recent studies by Richter [6] and Wainstein and Bumpus [7] have extensively examined the role of business models in the energy sector's transition towards sustainability. Richter [6] discusses the impact of distributed generation and the shift from traditional utilities to prosumer-oriented models. Wainstein and Bumpus [7] analyze the potential of innovative business models like EaaS in enhancing energy efficiency and customer engagement.

Other significant contributions include the work of Boamah [8], who highlights the integration of digital technologies in business models for the energy market, and Lüdeke-Freund et al. [9], who provide a comprehensive framework for sustainable business models in the energy sector.

The business model should be integrated with the principles of sustainable development, simultaneously ensuring active engagement with local communities and support for pro-environment initiatives. The novelty and added value to the existing knowledge lie in a detailed analysis of two business models and indicating the possibility of utilizing additional models that will enable the management of the energy sector in an innovative, efficient manner, allowing for the achievement of a zero-emission economy.

In the face of climate change, business models in the energy market must evolve towards a zero-emission future. Distributed energy, energy as a service, energy platforms, and innovative financing models are key elements of this revolution. Companies that take on these challenges and invest in sustainable practices will not only survive but also succeed in a dynamic and sustainable energy sector.

2. Materials and Methods

Scientific research forms the cornerstone of knowledge development and understanding the world around us. When analyzing business models in the energy sector, it is valuable to scrutinize existing and already published sources. One of the crucial stages of this process is the selection of appropriate research methods. One such method gaining increasing popularity is known as desk research, involving archival research also referred to as secondary research, or the use of existing sources of information. Many researchers use this method [10,11], and it is widely used in social sciences, economic sciences, and technical sciences [12,13], which confirms the validity of using this method in the research conducted.

Desk research relies on the analysis of available data, documents, reports, publications, or other existing sources before commencing the study. In the context of hypothetical-deductive research, this method proves highly effective in seeking answers to posed questions.

The primary strength of the desk research method is its accessibility. Researchers leverage pre-existing information, saving time and resources, particularly in the initial stages of the research process. This is particularly significant in the context of hypothetical-deductive research, where there is a need to gather the appropriate theoretical context on which subsequent stages of the study will be based.

Another aspect worth emphasizing is the extensive availability of diverse sources of information, such as Internet resources, databases, scientific articles, and institution reports—these are just a fraction of the sources that can be utilized in desk research. The wealth of available data allows for a holistic perspective on the research topic, which can be crucial in developing a specific research question.

It is also worth noting that the desk research method facilitates the analysis of historical data, which can be essential in studying the evolution of phenomena, trends, or processes. This enables researchers to direct their research questions toward understanding changes over time, providing invaluable contributions in the context of hypothetical-deductive research, where the aim is to deduce general regularities.

The utilization of the desk research method provides a comprehensive framework for the in-depth analysis of various issues directly associated with the research question, particularly focusing on the exploration and development of a business model. This model is envisaged (assumed) to play a pivotal role in not only fostering but also facilitating innovation and efficiency within the operational dynamics of the energy market.

Desk research studies constitute an extremely important element of the research process, and their application has allowed for a deepening of knowledge in the field of business models. Through the review and analysis of previously published research in this area, it was possible to gather significant knowledge about various approaches and strategies employed by companies in the creation, development, and adaptation of business models. Thanks to desk research studies, trends, patterns, and proven practices in this area were identified, enabling an understanding of which strategies are most effective in different business contexts. Additionally, the research facilitated an analysis of factors influencing the effectiveness and durability of implemented models (including the prosumer model, ESCO model, network operator model, and virtual power plant model).

As a result, desk research studies not only allowed for a thorough understanding of existing business models but also contributed to proposing new and innovative solutions in the energy sector (such as virtual power plants and value-added activators). They enable better utilization of available knowledge and more effective adjustment of business strategies to changing market conditions and customer needs.

The main elements of desk research are listed below:

Defining research objectives—Determining the specific objectives and scope of the study to determine what information and data are needed. In the case of this scientific research, the main goal is to analyze and indicate the possibility of creating innovative market models among energy companies to achieve low emissions.

Setting search criteria—Selecting appropriate search criteria such as keywords, thematic categories, or specific sources of information related to business models, zero-emission or low-emission.

Data mining—Browsing available sources of information, such as databases, research articles, industry reports, or books, to gather relevant data and materials.

Selection and evaluation of materials—Analysis of the collected materials in terms of their usefulness, credibility, and connection with the research topic and their substantive value.

Data analysis and interpretation—Processing collected data to draw conclusions and identify trends, patterns, or relevant information.

Presentation of results—Preparation of a scientific article transparently presenting research results and documenting the sources and methods used.

Quality control—Checking the accuracy, completeness, and reliability of collected data and materials.

However, it is essential to remember that the desk research method has its limitations. Depending on the research topic, there is a risk that some information may not be available or may be incomplete. Furthermore, there is a risk of subjectivity, especially if the sources of information are not adequately verified.

The applied desk research method is a valuable tool in the analysis of hypothetical-deductive research. Its accessibility, wealth of sources, and ability to analyze historical data make it a technique that effectively supports the process of generating and testing scientific hypotheses. Of course, the application of this method requires a judicious selection of sources, critical analysis of information, and a focus on precisely defined research questions.

3. Results and Discussion

Energy companies aiming to function effectively in contemporary conditions, especially within the framework of the so-called energy transformation, must abandon the belief in their permanence and excellence and acquire the ability to adapt to market requirements. Due to the continuous and abrupt variability of conditions, they must create new characteristics, behaviors, and attitudes that will constitute an adequate response to the globally transforming reality. In this regard, the issue of business models and their utilization may become a priority for scientific exploration in the near future within the realm of management science for contemporary enterprises. Theoretical and empirical research on the business model in the energy industry is gaining importance. In the context of the energy market, the business model becomes a crucial tool describing how a company creates value, delivers products or services to customers, and generates revenues to sustain its operations. In this sector, the business model encompasses complex issues related to energy production, transmission, distribution, and sales.

The revenue structure in the energy business model may rely on various sources such as energy sales, distribution services, contracts for renewable energy sources, or innovative fee models like subscriptions for energy services. Cost sources encompass all expenses related to energy production, infrastructure maintenance, fuel purchases, or technologies. Contemporary business models in the energy market increasingly incorporate investments in renewable technologies and digital solutions, significantly influencing cost structures.

The value creation proposition in the energy business model may involve not only providing energy but also addressing issues related to sustainable development, energy efficiency technologies, or loyalty programs for customers.

The business model in the energy market must account for the unique challenges and opportunities of this sector, and its effectiveness depends on flexibility and the ability to adapt to the changing energy landscape and customer expectations.

Each business model defines the essence of a company's functioning, and in cases where it is not specified by intentional actions of the company's management, it can still be assumed that the company operates under a certain structure of mechanisms, a configuration of dependencies that determine a certain style of functioning. In the first and second decades of the 21st century, energy groups underwent consolidation and developed their business models. This is reflected in regulating internal and external relations, ways of organizing internal social relations, decision-making, planning actions, assuming risks, and approaches to fulfilling tasks and obligations towards the environment.

The process of energy transformation requires consistent implementation of energy efficiency and a fundamental shift in the source of energy supply towards renewable, CO_2-emission-neutral sources [14]. The decarbonization process of energy production favors the trend towards distributed production structures [15]. Furthermore, the digitization of processes and industries has an increasing impact on the energy sector and new players entering energy markets [16]. Traditional energy structures, which emerged after the liberalization of electricity and gas markets in Europe, including entrenched business models, are undergoing significant changes [17].

Strategic management in energy companies requires the selection of a business model that leverages the resources and competitive advantages available in the organization. A commercial approach to the market and the selection of a model that serves as a mechanism linking strategy to its economic outcomes are necessary. Business models are an art form because there is no true plan for their design, construction, or execution. Available business modeling tools facilitate this process. However, there is a research gap in the literature regarding tools for creating and describing business models in the energy industry. The business model determines and establishes the basis for how an entity will compete once it has defined the rules of the market game—where it will choose to participate and take its place in the game, the way the game is played, and the way to win.

There is a limited number of articles and studies in the scientific literature describing current or future business models related to the energy sector.

A critical review of existing literature on the subject allows for the identification of emerging trends. Firstly, paradoxically, research on business models often fails to define either the business model of energy companies or business models in a broader sense, as noted by J. Newcomb [18] and R. Lehr [19], among others. Secondly, many studies define and analyze individual business models or a small set of business models associated with a single technology without examining how these models may be competitive compared to others, for example, M. Bell, J. Creyts, V. Lacy, J. Sherwood [20], J. Vasconcelos, S. Ruester, He X., Chong E., Glachant [21], and Behrangrad [22].

Energy sector companies have little experience with non-traditional business models and often employ simple models, such as integrated and regulated or separate and unregulated, depending on the value chain segment. Until the end of the first decade of the 21st century, there was no need to refine these models to accommodate non-uniform markets or role separation. However, with energy becoming a commodity traded on the stock market, there arose the necessity to change the paradigm regarding energy supply. While energy companies are to ensure the country's energy security, this is to be achieved under market conditions.

In the second decade of the 21st century, energy companies developed and adopted various business models to stimulate value chain integration and integration within separate segments, such as energy trading, distribution, and generation. However, these business models, usually traditional in nature, were based on roles, such as asset owners—power plants, or broadly defined, such as distribution companies. The redesign of the business model accelerates with the increasing pace of technological development and the emergence of efficient startups in the sector. The scope of potential business models offers a broader picture of the nature of future roles of energy companies and questions conventional norms regarding their place in the value chain.

In such a delineated reality, management sciences, through their methodological potential, must identify, diagnose, and project emerging business models. The increased interest in the issue of business models in energy is associated with the following six elements [23]:

- Applicability of the business model as a transparent and understandable concept of creating value for customers and shareholders of energy companies at every stage of the value chain, i.e., extraction, generation, trading, and distribution;
- The potential use of the business model as a carrier of innovation in the areas of processes, products, marketing activities, and CSR. In the energy sector undergoing transformation, the use of business models in innovative initiatives will become an area of particular interest for company managers (emergence of new technologies, electromobility, relationships with customers—prosumers);
- Seeking methods to achieve competitive advantages. Business models focused on achieving competitive advantages in terms of process and cost optimization will be

replaced by business models focused on delivering value to customers and achieving greater profitability in operational activities through new revenue streams;
- Strategic management through the identification and application of the business model as a business activity architecture aimed at delivering lasting value to shareholders by generating income and transforming business activities into profitable areas while maintaining energy security in the face of increasing climate policy requirements;
- Application of the business model to build the company's development strategy. The possibility of examining the impact of the business model on the way of development, both internal and external, through mergers, acquisitions, and business partnerships;
- Benchmarking of the business models used in Poland to those used in Europe. Considering the Energy Union and policies related to the harmonization and liberalization of energy markets, business models in Poland must follow more developed foreign markets.

As a result of the review of research and analysis concerning energy, five main trends shaping business conduct and creating value for customers and shareholders can be presented [24,25] as follows:

- Trend 1—Introduction of a two-way market—capacity market, EU decarbonization policy, and subsequent regulations aimed at reducing emissions, the introduction of the qualitative regulation model in distribution, changes in the support system for RES installations, and EU actions towards a common electricity market;
- Trend 2—Changing forecasts for electricity prices, oversupply and coal prices, electricity demand, power demand, increasing competition in the retail market, increasing level of RES generation with simultaneous withdrawal of European players, and limited funding for conventional energy;
- Trend 3—Increasing customer awareness and demands in terms of meeting needs and the complexity of offerings, increasing expectations regarding quality and availability of customer service;
- Trend 4—Falling prices of renewable and distributed technologies, increasing competitiveness of these sources compared to conventional sources, changing role of distribution in relation to the development of distributed energy, development of smart technologies, microgeneration, and energy storage;
- Trend 5—Electromobility.

Trends 1 and 2 signify the introduction of a two-way market and a change in the generation structure, resulting in an increased share of renewable energy sources and the gradual phasing out (by 2040) of decommissioned coal power plants. This trend, visible to a much lesser extent in the Polish market, will require players in the generation market to adjust their strategies. To create investment incentives in the generation sector, a capacity market was introduced, allowing for new investments.

The situation in the energy market and climate policy leads to reduced profitability of electricity production in conventional sources and even significant financial losses. Consequently, energy producers in such sources are forced to take action to restore profitability in the medium and long term. Besides the capacity market, which stimulates investments in generation, companies are conducting development activities aimed at optimizing and reducing the level of fixed and variable production costs and effectively managing capital expenditures (e.g., for new investments, modernization, and repairs).

Trend 3 signifies the development of product offerings in the electricity sales market, which has ceased to be treated by end customers and energy companies as a public good and has become a market product. To compete in the market, sellers must build an enticing value proposition for diverse customer groups.

In the sales segment, changing customer behavior and intensifying competitive struggles indicate three main challenges for electricity sellers: loss of customer base, margin erosion, and acquiring new customers. A changing customer base and deteriorating profitability require companies to develop tools and take action to secure their current position on one hand and capitalize on emerging market opportunities on the other. Three strategies to win in the market are as follows:

- Understanding and managing the customer base;
- Intelligent partnership;
- Building an "energy" value proposition.

Trend 4 signifies assigning new roles to distribution system operators (DSOs)—the development of renewable energy sources (RES) and new obligations for DSOs necessitate significant investments in the distribution network and the creation of a platform for collaboration and information exchange with consumers. Due to the natural monopoly in the electricity distribution area, DSOs do not have as tangible a dimension as in competitive areas of the energy market. This allows for the identification of three groups of business actions/strategies that distinguish efficient and modern DSOs from distributors solely relying on their historical position. With the introduction of elements compensating for the effects of actions into the tariff system by the Energy Regulatory Office, higher efficiency should also lead to better results in the medium term. The three identified strategies are as follows:

- Smart DSOs;
- Efficient DSOs;
- DSOs as distributors of information, not just energy.

Trend 5 denotes the development of business around charging infrastructure, which can, in turn, contribute to the achievement of various strategic goals. Some companies may opt for limited involvement, including only single chargers or company cars. These actions aim to improve their image and support company promotion. More complex strategies may involve using funds from investments in this area to strengthen and modernize the power grid. This business can also improve customer retention by creating appropriate products. It can also strengthen defense against competition, which will have an attractive e-mobility offer combined with basic energy products, which in turn may increase customer churn rate. A well-crafted strategy can thus lead to an increase in company value through increased revenue and margin mass from product and service sales.

Creating value for customers, one of the key elements that must be taken into account when defining a business model, involves considering trends in customer behavior. Based on PwC's analysis of consumer trends shaping the service market, those implying the type of business of energy companies in the energy trading area have been identified [26] as follows:

- Digitization of life;
- Seeking convenience;
- Value for money;
- Supporting ecology;
- Increasing transparency.

From the perspective of the market situation in Poland, three of them have been prioritized as the most relevant as follows:

- Digitization of life—the growing penetration of the Internet translates into the development of e-commerce and other alternative channels and models of service and sales (so-called multichannel);
- Seeking "convenience"—for customers, this element becomes an increasingly important purchasing criterion, which affects the development of convenient products and sales channels;
- "Value for money"—an increasing number of customers prefer good quality at a reasonable (not the lowest) price—a visible trend in the development of retail networks and private labels.

The analysis of energy companies in foreign markets indicates that these trends are taken into account when constructing business models and product offerings. One example could be the widespread introduction of simple and inexpensive products for customers seeking "value for money"; the section may be divided by subheadings. It should provide a concise and precise description of the experimental results, their interpretation, as well as the experimental conclusions that can be drawn.

Business Model in the Energy Sector

"It would be foolish to ignore the potential for significant changes in the business models of energy companies."—Theodore Craver Jr., CEO of Edison International [27]. Craver, working for one of the largest energy companies in the USA and serving as the president of the Edison Electric Institute, an industry organization representing American energy companies, is not alone in his belief that the business model of energy companies is on the verge of radical changes. A study conducted in 2013 showed that 94% of senior energy executives surveyed predict "a complete transformation or significant changes in the business model of energy companies" by 2030 [28].

These changes are primarily driven by the influx of distributed energy resources, including photovoltaics and other distributed generation, thermal and electrical energy storage, as well as more flexible and price-responsive demand-side management of electricity. Many predict that the changes brought about by distributed generation assets will be highly disruptive to the electric power sector and that without adaptation, traditional energy companies may fall into a "death spiral," threatening their financial profitability [29].

In a series of studies on business models in the energy sector, analyses of specific technologies are conducted without examining the full range of services that this technology provides or could provide (e.g., Huijben [30] or Weiller [31]). Only a small fraction of business model researchers have analyzed the business models of energy companies using an ontological approach. Some of these studies focus on subsets of business models that utilize specific technology (e.g., Schoettl and Lehmann-Ortega [32] or Okkonen and Suhonen [33]). Richter [34], in his research, utilizes case studies and surveys combined with an ontological approach to understand the business models of energy companies that utilize various renewable energy technologies.

Energy companies are undergoing a cultural shift towards a data-driven digital economy where core processes are digitized, and market flexibility is achieved through technical optimization [35]. Overall, researchers focusing on energy have primarily concentrated on technological aspects because the energy trading system and energy as a commodity are still in the early stages from a business perspective [36]. Alvaro-Hermana et al. [37] proposed a new business model for energy trading between two sets of electric vehicles. Inam et al. [38] discussed the microgrid structure in a distributed energy system with the division of electrical energy. Much theoretical work has been performed on blockchain technology, which can be used in a wide range of applications, including using electric vehicles in grid services to assist with congestion management, recording energy consumption data, billing, changing suppliers, exchanging production capacity, and so on [39]. Burger and Weinmann studied the development of blockchain technology and its potential in the energy sector [40]. Kim et al. [41] presented a scheme of peer-to-peer (P2P) energy loans using a blockchain-based P2P energy loan procedure. Recently, more attention has been given to the social and economic consequences of electricity trading. Giotitsas et al. [42] reviewed the evolution procedure of energy trading technology and examined its potential impact on the global socio-economic structure. Roy et al. [43], on the other hand, discussed the benefits and feasibility of peer-to-peer electricity trading on the Australian domestic electricity market, analyzing its social, economic, and technical aspects.

Existing energy trading systems are generally based on traditional energy contracts, which are settled centrally, and transactions must be made with the help of network companies. Bilateral agreements [44], hourly (or daily, so-called day-ahead) markets, ancillary service markets, and capacity markets are commonly used forms of the traditional business model in the energy sector [45]. On the other hand, the advancing digitization in energy provides opportunities for decentralization and deepening of these markets through the creation of new business models. Digitization in energy provides corporations, entrepreneurs, and startups with new opportunities to provide solutions and create new businesses in this sector. At this point, new business models are emerging that will capture a share of this total amount of existing energy expenditure. According to Cuofano, "(...) business modeling is about finding a systematic way to unlock long-term value for an organization while delivering valuable products and services simultaneously" [46].

In Poland, research on new concepts of the functioning of innovative energy companies suitable for the liberalizing energy market has been undertaken by B. Matusiak. The author believes that such a model should include the following information [47]:

- What is the specific value proposition for the participants of the model?
- What customer relationship model has been applied?
- What is the model for returning value from the invested capital?
- What is the configuration of all material and immaterial values involved in obtaining this value?

B. Matusiak characterizes the following five business models suitable for the integrated energy market [48]:

- The prosumer business model (exchange of value between customer and supplier);
- The ESCO (Energy Saving Company) model, which provides comprehensive energy management services;
- The market aggregator business model (a company operating between distributed energy producers and consumers);
- The business model of electric vehicle users in the energy market;
- The producer business model (application of large-scale ecological, often renewable energy sources).

The authors of this research have detailed the first two business models, the prosumer business model and ESCO, which are currently very popular and used in Poland and many other countries. The case of Poland serves as a valuable example for other countries due to its significant reliance on coal and the resulting urgent need for energy transformation. Transitioning to renewable energy sources is not only crucial for achieving climate goals but also for reducing emissions and mitigating environmental impact. Poland's proactive approach to investing in renewable energy showcases a feasible path towards achieving sustainable development and resilience in the energy sector. Additionally, the emergence of innovative business models in response to this transformation underscores the flexibility and potential scalability of these solutions to fit various contexts worldwide. The case of the applied business models in Poland and the country's situation allows for the extrapolation of these experiences to other countries. By observing and learning from Poland's experiences, other countries can gain valuable insights and strategies for effectively and efficiently conducting their own energy transformations.

The term "prosumer" is a portmanteau of "producer" and "consumer", representing a paradigm shift in the energy market. In the prosument model, individuals or entities are both producers and consumers of energy, actively participating in the generation and consumption of electricity. This concept challenges the traditional one-way flow of energy from centralized power plants to end-users, promoting a more decentralized and participatory approach [49]. For instance, countries like Poland, which is renowned for its progressive energy policies, and companies such as Columbus Energy S.A., with their innovative approach to integrating prosumer technologies, have led the way in transforming the energy landscape [50].

The prosument model encourages the decentralization of energy production. Rather than relying solely on large, centralized power plants, individuals, businesses, and communities can generate their own energy through renewable sources like solar panels, wind turbines, or other distributed energy systems.

Prosument models often emphasize the integration of renewable energy sources. Prosumers commonly invest in technologies that harness sustainable energy, reducing reliance on fossil fuels and contributing to environmental sustainability.

By becoming both producers and consumers, prosumers aim to achieve a degree of energy independence. They can generate their own power, store excess energy, and potentially even sell surplus energy back to the grid, contributing to a more resilient and self-sufficient energy ecosystem.

Prosument models leverage smart technologies and automation to optimize energy consumption. Smart meters, home energy management systems, and other IoT devices enable prosumers to monitor, control, and optimize their energy usage in real time, enhancing efficiency.

The prosument model challenges the traditional power grid by enabling two-way communication. Prosumers can feed excess energy into the grid during periods of surplus and draw power when their own production is insufficient, fostering a more dynamic and responsive energy infrastructure.

Prosument models often lead to the formation of community energy initiatives. Communities may collaborate to install shared renewable energy systems, creating local energy resilience and fostering a sense of collective responsibility for sustainability.

Prosumers can enjoy economic benefits through the prosument model. They may receive financial incentives for generating excess energy, selling it back to the grid, or participating in demand response programs. Additionally, prosumers can potentially reduce their energy bills through increased efficiency and on-site generation.

The prosument model poses regulatory challenges related to grid management, pricing structures, and energy market dynamics. Policymakers need to adapt regulations to accommodate the evolving role of prosumers, providing fair compensation for the energy they contribute to the grid.

The shift towards the prosument model aligns with broader environmental and social goals. Increased use of renewable energy and community-driven initiatives contribute to mitigating climate change, reducing dependence on non-renewable resources, and fostering a sense of shared responsibility for environmental stewardship.

The prosument model reflects a transformative approach to the energy market, emphasizing sustainability, community engagement, and technological innovation. As it continues to evolve, the prosument model has the potential to reshape the energy landscape, making it more resilient, inclusive, and environmentally friendly.

Solar photovoltaic installations installed and used by consumers are currently one of the most rapidly developing markets according to the prosumer model in the energy sector worldwide. In Table 1 below, the characteristic features of such a business model are presented.

Another business model worth exploring and analyzing due to its market potential is the ESCO (Energy Service Company), business model. The energy transformation and the aim to reduce CO_2 emissions and energy consumption require companies to increase energy efficiency.

ESCO, which stands for Energy Service Company or Energy Saving Company, is a business model that operates in the energy market with a focus on providing energy efficiency and sustainable energy solutions. ESCOs play a crucial role in helping clients, often commercial and industrial entities, improve energy performance, reduce energy consumption, and lower overall operational costs [51].

ESCOs typically offer a range of energy efficiency services, including energy audits, retrofitting, and the implementation of energy-efficient technologies. They identify areas where energy consumption can be optimized and propose solutions to enhance efficiency.

One distinctive feature of the ESCO model is performance contracting. In this arrangement, the ESCO guarantees a certain level of energy savings for its clients. The payment for ESCO services is often linked to the actual energy savings achieved, creating an incentive for the ESCO to deliver tangible results.

ESCOs may provide upfront funding for energy efficiency projects. Instead of clients bearing the entire financial burden of upgrades and improvements, ESCOs invest in the necessary technologies and services. The costs are then recovered over time through the achieved energy savings.

ESCOs are typically equipped with technical expertise in energy systems, renewable energy, and energy-efficient technologies. They leverage this knowledge to recommend and implement solutions that align with their clients' sustainability goals.

In addition to energy efficiency measures, many ESCOs are increasingly involved in the integration of renewable energy sources. This includes the installation of solar panels, wind turbines, or other sustainable energy solutions to further reduce reliance on conventional energy sources.

ESCOs often provide ongoing monitoring and maintenance services to ensure that the implemented energy efficiency measures continue to deliver the expected results. Regular assessments help identify opportunities for further optimization.

ESCOs are familiar with energy regulations and incentives, helping clients navigate compliance requirements and take advantage of available financial incentives for energy-efficient projects.

ESCOs tailor their services to meet the specific needs and requirements of each client. This customization ensures that energy efficiency solutions align with the operational context and goals of the businesses they serve.

The ESCO model contributes to the broader goal of creating a more sustainable and energy-efficient future by addressing energy challenges in a holistic manner. It aligns economic interests with environmental responsibility, making it an attractive option for organizations seeking to improve their energy performance [52].

ESCO services are used for energy management (services related to reducing consumption and demand for energy for their clients). The characteristics of the ESCO business model are presented in Table 2.

An interesting approach enabling the analysis and creation of business models for innovative companies in the energy sector is the concept of the so-called New Era of Innovation, created by C.K. Prahalad and M.S. Krishnan [53]. The business model consists of three basic components. The first two are social architecture and technical architecture, representing specific resources. The third component consists of business processes. Based on conducted studies of business modeling theory and our own research, the business model of prosumer energy was defined as a configuration of business processes combining and developing resources, shaped in the form of social and technical architecture, creating value based on renewable energy sources.

Table 1. Characteristics of the prosumer business model parameters for investments in photovoltaic sources [54].

Business Model Elements	Market Participants	Brief Description
Basic Financing	NFOŚiGW (National Fund for Environmental Protection and Water Management)	Grants and incentives for solar investments. Non-repayable amounts reduce the value of construction credit—if PV investments are made.
Supporting Financing	BOŚ (Bank of Environmental Protection) and other banking institutions, as well as private funds	Preferential loans and support in various forms: municipal bonds, debt buyback, bank guarantees.
Advanced Financing (for the construction of source diversification and microgrids)	Private funds, venture capital, purpose-built foundations, special-purpose companies	Financing requires a long-term, stable state tax policy for the application of exemptions and supportive incentives. For example, this applies to issues such as wholesale gas prices freed for negotiation or freed prices for retail customers.
Consumer Leasing Lessors	Consumer leasing companies	Current market interest rates, averaging around 4.12%.
Technological Partner	Companies supplying panels and installations with assembly and service, additional necessary equipment, and on-site installation	The market offer is very wide and diverse in terms of price; the choice of panel technology is crucial for its efficiency and acquisition costs.

Table 1. *Cont.*

Business Model Elements	Market Participants	Brief Description
Technological Partner	OSD (Distribution System Operator)	Connection agreement and connection capacity, no connection fee, installation of meters and metering, for which OSD is responsible.
Purchase of Energy from the Grid (Energy Trading)	Selected electricity supplier under the TPA principle (or aggregator—assuming settlement services)	Prosumer settlement and balancing—on average, in the G11 tariff (tariff for individuals), the price of 1 kWh gross (including transmission fee and excise duty) is approximately 0.5 to 0.6 PLN (0.12 to 0.15 USD); there is also a need to free prices for users from the G tariff.
Selling Energy to the Grid—Support: Feed-in Tariff (FIT) Support System	State regulations. Implemented by the designated electricity supplier	Regulation resulting from the RES Act: installation with a total electrical power of up to 3 kW—the rate is 0.75 PLN for 1 kWh for 15 years, up to 10 kW is 0.65 PLN for 1 kWh, >10 kW is 0.16 PLN for 1 kWh.

Table 2. Characteristics of the ESCO Business Model Parameters [54].

Business Model Elements	Market Participants	Brief Description
Basic Financing	Third parties—all institutions financing investments in RES and internal infrastructure modernization	Banking products, leasing, credits. Grants and incentives for PV investments.
Supporting Financing	White certificates	Trading white certificates.
Advanced Financing (in the case of diversification of sources and microgrid construction)	Private funds, venture capital, target foundations, special-purpose companies	Financing requires a long-term, stable state tax policy, for applying exemptions and supporting incentives. For example, it concerns issues such as negotiated wholesale gas prices or released prices for retail customers.
Technological Partner	ESCO, and through it, companies providing panels and installations with assembly and service, additional necessary equipment, and on-site installations	The market offer is very wide and diverse in terms of price; the choice of panel technology itself is crucial for its efficiency and acquisition costs.
Technological Partner	OSD and/or market aggregator	Connection and connection power agreement, no connection fee, installation of meters and metering, for which OSD is responsible. Alternatively, cooperation with a market aggregator for balancing ESCO clients.
Purchase of Energy from the Grid for Own Needs	Selected electricity supplier according to the TPA principle (or aggregator—taking over settlement services)	Settlement and balancing according to agreements for a large business customer (ESCO), according to free-market prices. In special cases, selling energy to the grid if provided for in the contract, e.g., leasing the customer's roof—for energy production for the grid (producer model)

B. Nogalski conducted research on business models in energy companies [55]. The study characterized the elements and features of the business model in the following areas:

- General Characteristics of the Business Model, including:

  Dominant types of models
  Way of creation and legal forms
  Basic areas of activity
  Main products
  Ownership structure

- Competitive Advantage, covering:

  Values offered to customers
  Market segments served

Types of competitive advantage
Sources and methods of profit generation related to value capture

- Resources and their Utilization, addressing:

Material resources
Intangible resources
Strategic competencies

- Scope of Operations, analyzing:

Character of basic activities
Type of value chain enabling the identification of sources of value creation
Value chain structure indicating cost determinants and differentiation

From the conducted review of research on business models in energy companies, it emerges that researchers have identified at least eight business models that will be implemented in energy companies in the future. Apart from the current traditional business model based on energy generation in conventional power plants, sales, and distribution of energy, new business models will appear, complementing the offerings of so-called incumbents, i.e., traditional energy sector companies. Figure 1 presents a comparison of business models that can be applied in the energy industry.

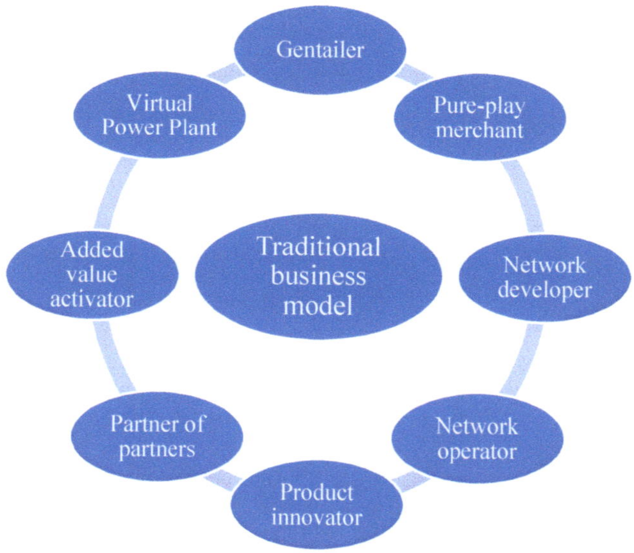

Figure 1. Business Models in the Energy Sector.

The results of the literature review in this area have identified the following business models in the energy sector that could be implemented in Polish energy companies in various areas of activity (generation, distribution, trading), as presented in Table 3.

In summarizing the table of business models in the energy sector, it is evident that a diversity of models is implemented by energy companies. The emergence of new actors, such as network operators and virtual power plants, who utilize technologies and platforms for integrating distributed energy resources, is also significant. This indicates that the future of the energy industry will be based on various business models that must be flexible and ready to adapt to the changing market and technological innovations with the aim of achieving sustainability for the economy and society [56]. Diversification and flexibility in approaching business models can enable energy companies to maintain competitiveness and effectively adapt to dynamically changing market conditions.

Table 3. Business Models in the Energy Sector.

Business Model in Energy	Description of Business Model
Gentailer	Gentailer operates at both ends of the value chain, owning generation assets and selling energy to end customers in the competitive market. The gentailer model provides a natural hedge for trading activities and can take either a long position (generates more energy than it sells) or a short position (sells more energy than it generates).
Pure-play merchant	The company owns and operates generation assets, selling energy on competitive wholesale markets at market prices or based on negotiated bilateral contracts. Companies generate Dark Spread or Clean Dark Spread margins (coal or gas-fired power plants). Generating energy from wind and solar power plants is becoming increasingly important.
Network developer	Companies in this model acquire, develop, build, own, and maintain transmission and distribution assets that connect decentralized generation units with local distribution system operators. Network developers continuously assess the system's capability to meet current and future needs and seek new assets and renewal opportunities.
Network operator	The network operator manages distribution assets and provides access to its networks for generators, connections, and retail service providers. A potential new role could be a platform optimizer providing data and services such as integrating distributed energy resources.
Product innovator	The product innovator offers electricity and gas, as well as over-the-counter products, expanding the role of the entrenched operator. Products beyond the meter could combine retail supply packages, such as green energy, with service and pricing packages for flexibility. This model may evolve towards smart devices, such as energy hubs.
Partner of partners	The partner program offers energy products and services, such as vehicle support throughout the life cycle of an electric vehicle and home comfort services. The energy company may be the exclusive provider of these services, but it is more likely that customers will take advantage of the offer when services are bundled with branded providers.
Value-added activator	A company positioning itself as a value-added provider uses its core capabilities, such as data collection and analysis, to enhance its role and value to the customer. The energy company becomes a value-added energy manager, considering its knowledge and the customer's lack of knowledge about basic energy management activities.
Virtual power plant	A virtual power plant aggregates generation from various distributed systems and acts as an intermediary in energy markets. The virtual power plant also serves as an integrator of non-traditional resources and services provided by third parties to customers, such as distributed energy resources beyond its traditional service territory.

4. Conclusions

This scientific paper provides a synthetic overview of definitions and concepts of business models, focusing on the diversity in methods of their construction and structure. Each case emphasizes that the primary goal of building a business model is to identify the value that the enterprise provides to both customers and stakeholders.

The article discusses the essence of selected business models, such as the prosumer model and the ESCO model, highlighting their roles as key elements in enterprise management. The prosumer model, where customers act as producers and consumers of energy, and the ESCO model, based on energy efficiency, exemplify significant approaches in contemporary enterprise strategies.

In the context of the energy industry, the paper underscores the challenge of introducing innovative business models that simultaneously deliver value to customers and ensure energy security. It concludes with an analysis of the essence and diversity of these models, stressing their crucial role in shaping enterprise development strategies, particularly within the dynamic energy landscape.

New business models in the energy sector are pivotal for future transformations in the field, facilitating the realization of goals related to a zero-emission economy. Through innovative management and technological solutions, these models enable flexible adaptation to evolving environmental requirements, addressing global challenges posed by climate change.

The article focuses on business models specific to the Polish energy sector, acknowledging their potential impact on global trends, albeit requiring further exploration of their transferability. Given the sector's continual evolution, ongoing scrutiny of these prototypes is necessary to accommodate new developments and empirical evidence.

The implementation of modern technologies, such as intelligent energy grids, energy storage, or distributed energy sources, allows for increasing energy efficiency and reducing losses in the energy supply system. New business models also focus on the integration of renewable energy sources, which promotes the sustainable development of the energy sector. The use of new technologies, such as artificial intelligence or data analysis, enables optimization of the management of energy networks, which leads to more efficient use of resources and minimization of negative impact on the environment.

In the long term, new business models will not only contribute to meeting zero emissions requirements but will also open the way to new growth opportunities for energy companies. Innovation in the area of management, technology, and operational strategy will become a key element of success in the energy industry while allowing for the harmonious integration of economic goals with the needs of sustainable development. Ultimately, new business models in the energy sector are a key step toward building a more sustainable, efficient, and eco-friendly energy infrastructure.

The current development of the energy sector, driven by innovative technologies and evolving market structures, poses numerous challenges to the industry while opening the field for the emergence of new, sustainable business models. Dynamic changes in the production and distribution of energy, the introduction of renewable sources, and the growing demand for effective management of energy resources encourage in-depth research on business strategies that will be adequate to contemporary challenges. Given the growing importance of renewable energy sources and the growing emphasis on reducing greenhouse gas emissions, there is a need to develop business models that favor ecology, energy efficiency, and sustainable development.

It is worth noting that changing market structures, including the growing role of energy microgrids and prosumer energy producers, require innovative approaches to energy management, distribution, and trading. This, in turn, generates the need to systematically conduct research on appropriate business models that will respond efficiently to these dynamic changes.

Looking ahead, these innovative business models not only aim to meet zero-emission goals but also create new growth opportunities for energy companies. Embracing innovation in management, technology, and operational strategies is pivotal for achieving a harmonious integration of economic objectives with sustainable development goals.

In conclusion, the ongoing evolution of the energy sector, driven by technological innovation and evolving market structures, presents challenges and opportunities for developing sustainable business models. Research into adaptive strategies is imperative to effectively respond to dynamic changes and contribute to a resilient, eco-friendly energy infrastructure.

Author Contributions: Conceptualization, A.S. and P.F.B.; methodology, A.S. validation, A.S. and P.F.B.; formal analysis, A.S.; resources, A.S. and P.F.B.; writing—original draft preparation, A.S.; writing—review and editing, A.S. and P.F.B.; supervision, P.F.B.; funding acquisition, A.S. All authors have read and agreed to the published version of the manuscript.

Funding: The APC was funded partially by Prozap Sp. z o.o.

Data Availability Statement: No new data were created or analyzed in this study.

Acknowledgments: The authors would like to thank the innovative solutions for using chatGPT in the final process to proofread the language and improve and smooth the text of the paper.

Conflicts of Interest: The authors declare that this study received funding from Prozap Sp. z o.o. The funder had the following involvement with the study: made a decision to submit it for publication. The remaining authors declare that the research was conducted in the absence of any commercial or financial relationships that could be construed as a potential conflict of interest.

References

1. Iria, J.; Soares, F. An energy-as-a-service business model for aggregators of prosumers. *Appl. Energy* **2023**, *347*, 121487. [CrossRef]
2. Borowski, P.F. Digitization, digital twins, blockchain, and industry 4.0 as elements of management process in enterprises in the energy sector. *Energies* **2021**, *14*, 1885. [CrossRef]
3. Herbes, C.; Rilling, B.; Holstenkamp, L. Ready for new business models? Human and social capital in the management of renewable energy cooperatives in Germany. *Energy Policy* **2021**, *156*, 112417. [CrossRef]
4. Paiola, M.; Schiavone, F.; Grandinetti, R.; Chen, J. Digital servitization and sustainability through networking: Some evidences from IoT-based business models. *J. Bus. Res.* **2021**, *132*, 507–516. [CrossRef]
5. Hamwi, M.; Lizarralde, I.; Legardeur, J. Demand response business model canvas: A tool for flexibility creation in the electricity markets. *J. Clean. Prod.* **2021**, *282*, 124539. [CrossRef]
6. Richter, M. Business model innovation for sustainable energy: German utilities and renewable energy. *Energy Policy* **2013**, *62*, 1226–1237. [CrossRef]
7. Wainstein, M.E.; Bumpus, A.G. Business models as drivers of the low carbon power system transition: A multi-level perspective. *J. Clean. Prod.* **2016**, *126*, 572–585. [CrossRef]
8. Boamah, F. Digital technologies in the energy sector: Opportunities for innovation in business models and policy frameworks. *Renew. Sustain. Energy Rev.* **2020**, *128*, 109878. [CrossRef]
9. Lüdeke-Freund, F.; Carroux, S.; Joyce, A.; Massa, L.; Breuer, H. The sustainable business model pattern taxonomy—45 patterns to support sustainability-oriented business model innovation. *Sustain. Prod. Consum.* **2019**, *18*, 103–119. [CrossRef]
10. Largan, C.; Morris, T. *Qualitative Secondary Research: A Step-by-Step Guide*; Sage: Thousand Oaks, CA, USA, 2019.
11. Walliman, N. *Research Methods: The Basics*; Routledge: London, UK, 2021.
12. Czepło, F.; Borowski, P.F. Innovation Solution in Photovoltaic Sector. *Energies* **2024**, *17*, 265. [CrossRef]
13. Ajibade, P. Technology acceptance model limitations and criticisms: Exploring the practical applications and use in technology-related studies, mixed-method, and qualitative researches. *Libr. Philos. Pract.* **2018**, *9*, 1–13.
14. Mühlemeier, S.; Wyss, R.; Binder, C.R. Und Aktion!—Konzeptualisierung der Rolle individuellen, Akteurshandelns in sozio-technischen Transitionen am Beispiel der regionalen Energiewende im bayerischen, Allgäu. *Z. Für Energiewirtsch* **2017**, *41*, 187–202. [CrossRef]
15. Quitzow, L.; Canzler, W.; Grundmann, P.; Leibenath, M.; Moss, T.; Rave, T. The German Energiewende—What's happening? Introducing the special issue. *Util. Policy* **2016**, *41*, 163–171. [CrossRef]
16. Shaffer, B.; Flores, R.; Samuelsen, S.; Anderson, M.; Mizzi, R.; Kuitunen, E. Urban energy systems and the transition to zero carbon -Research and case studies from the USA and Europe. *Energy Procedia* **2018**, *149*, 25–38. [CrossRef]
17. Löbbe, S.; Hackbarth, A. The Transformation of the German electricity sector and the emergence of new business models in distributed energy systems. In *Innovation and Disruption at the Grid's Edge*, 1st ed.; Sioshansi, F.P., Ed.; Academic Press: Cambridge, MA, USA, 2017; pp. 287–318.
18. Newcomb, J.; Lacy, V.; Hansen, L.; Bell, M. *Policy Implications of Decentralization*; Rocky Mountain Institute: Boulder, CO, USA, 2013; p. 66.
19. Lehr, R. New Utility Business Models: Utility and Regulatory Models for the Modern Era. *Electr. J.* **2013**, *56*, 35–53. [CrossRef]
20. Bell, M.; Creyts, J.; Lacy, V.; Sherwood, J. *Bridges to New Solar Business Models: Opportunities to Increase and Capture the Value of Distributed Solar Photovoltaics*; Rocky Mountain Institute: Boulder, CO, USA, 2014; p. 46.
21. Vasconcelos, J.; Ruester, S.; He, X.; Chong, E.; Glachant, J.-M. *Electricity Storage: How to Facilitate Its Deployment and Operation in the EU*; Publications Office of the EU: Luxembourg, 2012. [CrossRef]
22. Behrangrad, M. A review of demand side management business models in the electricity market. *Renew. Sustain. Energy Rev.* **2015**, *47*, 270–283. [CrossRef]
23. PwC and ING Bank. *New Business Models for the Distribution Edge: The Transition from Value Chain to Value Constellation*; Rocky Mountain Institute: Boulder, CO, USA, 2013.
24. PwC and ING Bank. *Koniec Tradycyjnej Energetyki?—Jak Wygrać w Dobie Zmian*; PwC and ING Bank: Warsaw, Poland, 2015.
25. Provance, M.; Donnelly, R.G.; Carayannis, E.G. Institutional influences on business model choice by new ventures in the micro generated energy industry. *Energy Policy* **2011**, *39*, 5630–5637. [CrossRef]
26. PwC. *A Different Energy Future: Where Energy Transformation Is Leading Us*; PwC: Warsaw, Poland, 2015.
27. Amusa, M. How the Biggest U.S. Utilities Are Planning for the Future. Util Dive 2013. Available online: http://www.utilitydive.com/news/how-the-biggest-us-utilities-are-planning-for-the-future/130302/ (accessed on 13 June 2019).
28. PwC. *Energy Transformation: The Impact on the Power Sector Business Model*; PwC: Warsaw, Poland, 2013.
29. Kind, P. *Disruptive Challenges: Financial Implications and Strategic Responses to a Changing Retail Electric Business*; Edison Electric Institute: Washington, DC, USA, 2013.

30. Huijben, J.C.C.M.; Verbong, G.P.J. Breakthrough without subsidies? PV business model experiments in the Netherlands. *Energy Policy* **2013**, *56*, 362–370. [CrossRef]
31. Weiller, C.; Shang, T.; Neely, A.; Shi, Y. Competing and co-existing business models for EV: Lessons from international case studies. *Int. J. Automot. Technol. Manag.* **2015**, *15*, 126–148. [CrossRef]
32. Schoett, J.-M.; Lehmann-Ortega, L. Photovoltaic business models: Threat or opportunity for utilities? In *The Handbook of Research on Energy Entrepreneurship*; Wustenhagen, R., Wuebker, R., Eds.; Edward Elgar Publishing: Cheltenham, UK, 2011; pp. 145–171.
33. Okkonen, L.; Suhonen, N. Business models of heat entrepreneurship in Finland. *Energy Policy* **2010**, *38*, 3443–3452. [CrossRef]
34. Richter, M. Utilities' business models for renewable energy: A review. *Renew. Sustain. Energy Rev.* **2012**, *16*, 2483–2493. [CrossRef]
35. Schoklitsch, H. Digitalization Is Revolutionizing the Renewable Energy Sector—Renewable Energy World. 2018. Available online: https://www.renewableenergyworld.com/articles/2018/02/digitalisation-is-revolutionizing-the-renewable-energy-sector.html (accessed on 8 June 2018).
36. Pouttu, A.; Haapola, J.; Ahokangas, P.; Xu, Y.; Kopsakangas-Savolainen, M.; Porras, E.; Matamoros, J.; Kalalas, C.; Alonso-Zarate, J.; Gallego, F.D. P2P model for distributed energy trading, grid control and ICT for local smart grids. In Proceedings of the European Conference on Networks and Communications (EuCNC), Oulu, Finland, 12–15 June 2017; IEEE: Piscataway, NJ, USA, 2017; pp. 1–6.
37. Alvaro-Hermana, R.; Fraile-Ardanuy, J.; Zufiria, P.J.; Knapen, L.; Janssens, D. Peer to peer energy trading with electric vehicles. *IEEE Intell. Transp. Syst. Mag.* **2016**, *8*, 33–44. [CrossRef]
38. Inam, W.; Strawser, D.; Afridi, K.K.; Ram, R.J.; Perreault, D.J. Architecture and system analysis of microgrids with peer-to-peer electricity sharing to create a marketplace which enables energy access. In Proceedings of the 2015 9th International Conference on Power Electronics and ECCE Asia (ICPE-ECCE Asia), Seoul, Republic of Korea, 1–5 June 2015; IEEE: Piscataway, NJ, USA, 2015; pp. 464–469.
39. Dena, Deutsche Energie-Agentur GmbH, German Energy Agency. *Dena Multi-Stakeholder Study, Blockchain in the Integrated Energy Transition, Study Findings*; Dena: Berlin, Germany, 2019.
40. Burger, C.; Weinmann, J. Innovation performance of the US American and European electricity supply industry. *Energy Policy* **2015**, *86*, 351–359. [CrossRef]
41. Kim, M.; Song, S.; Jun, M.S. A Study of Block Chain-Based Peer-to-Peer Energy Loan Service in Smart Grid Environments. *J. Comput. Theor. Nanosci.* **2016**, *22*, 2543–2546. [CrossRef]
42. Giotitsas, C.; Pazaitis, A.; Kostakis, V. A peer-to-peer approach to energy production. *Technol. Soc.* **2015**, *42*, 28–38. [CrossRef]
43. Roy, A.; Bruce, A.; MacGill, I.; The Potential Value of Peer-to-Peer Energy Trading in the Australian National Electricity Market. Asia-Pacific Solar Research Conference. Available online: https://www.ceem.unsw.edu.au/sites/default/files/documents/A%20Roy,%20A%20Bruce,%20I%20MacGill-Potential%20Value%20of%20Peer-To-Peer%20Energy%20Trading%20in%20the%20Australian%20National%20Electricity%20Market.pdf (accessed on 15 May 2019).
44. Gao, G.; Lo, K.; Lu, J. Risk assessment due to electricity price forecast uncertainty in UK electricity market. In Proceedings of the 2017 52nd International Universities Power Engineering Conference (UPEC), Heraklion, Greece, 28–31 August 2017; IEEE: Piscataway, NJ, USA, 2017; pp. 1–6.
45. Khan, A.S.M.; Verzijlbergh, R.A.; Sakinci, O.C.; De Vries, L.J. How do demand response and electrical energy storage affect (the need for) a capacity market? *Appl. Energy* **2018**, *214*, 39–62. [CrossRef]
46. Cuofano, G. What Is a Business Model? 30 Successful Types of Business Models You Need to Know. Available online: https://fourweekmba.com/what-is-a-business-model/ (accessed on 12 May 2019).
47. Matusiak, B. *Modele Biznesowe na Nowym Zintegrowanym Rynku Energii*; Uniwersytet Łódzki: Łódź, Poland, 2013; p. 120.
48. Matusiak, B. Usługa bilansowania—Business case dla platformy e-commerce. *Rynek Energii* **2017**, *1*, 62–66.
49. Borowski, P.F. Zonal and Nodal Models of energy market in European Union. *Energies* **2020**, *13*, 4182. [CrossRef]
50. Instytut Energetyki Odnawialnej. Rynek Fotowoltaiki w Polsce 2023. 2023. Available online: https://ieo.pl/raport-rynek-fotowoltaiki-w-polsce-2023 (accessed on 15 March 2024).
51. Segers, J.P.; Franco, D.V.; Van Caillie, D.; Gaile-Sarkane, E.; Macke, J. A Holistic Model for Measuring Sustainable Performance Generated by Innovative Projects: The ESCO Energy Transition Case. In *European Perspectives on Innovation Management*; Springer International Publishing: Cham, Switzerland, 2024; pp. 435–455.
52. Künar, A.; Uyar, T.S.; Bilto, M. ESCO and EPC Models for Energy Efficiency Transformation. In *Renewable Energy Based Solutions*; Springer International Publishing: Cham, Switzerland, 2022; pp. 241–272.
53. Prahalad, C.K.; Krishnan, M.S. *New Age of Innovation*; McGraw Hill: New York, NY, USA, 2008; pp. 38–47.
54. Matusiak, B. *Business Models in the New Integrated Energy Market*; University of Lodz: Lodz, Poland, 2013.
55. Nogalski, B. *Modele i Strategie Biznesu w Obszarze Dystrybucji Energii Elektrycznej w Polsce*; Wydawnictwo Uniwersytetu Gdańskiego: Sopot, Poland, 2017.
56. Stankovic, S.; Ilic, B.; Rabrenovic, M. Using the Composite EEPSE Green Economy Index to Assess the Progress of Emerging Economies in Achieving the Sustainable Development Goals. *Probl. Ekorozwoju* **2024**, *19*, 78–88. [CrossRef]

energies

MDPI

Review

Renewable Energy Source Utilization Progress in South Africa: A Review

Abayomi A. Adebiyi * and Katleho Moloi

Department of Electrical Power Engineering, Faculty of Engineering and the Built Environment, Durban University of Technology, Durban 4001, South Africa; katlehom@dut.ac.za
* Correspondence: abayomia@dut.ac.za

Abstract: Renewable energy has emerged as a promising solution to address the challenges of climate change, energy security, and socio-economic development. South Africa, with its abundant renewable energy resources, has made significant strides in the utilization of renewable energy over the past decade. This paper provides a comprehensive review of the progress of renewable energy advancement in South Africa, examining the policies, initiatives, and achievements in various renewable energy sectors. This study explores the country's transition from a heavily coal-dependent energy system to a diversified and sustainable energy mix. It analyses the growth of renewable energy technologies, such as wind power and solar photovoltaic (PV), highlighting the key milestones, challenges, and opportunities. Furthermore, this paper discusses the role of government support, regulatory frameworks, and private sector investments in driving renewable energy deployment in South Africa. Finally, it identifies the prospects and potential areas for further advancement in the renewable energy sector. This review aims to contribute to the understanding of South Africa's renewable energy journey and provides valuable insights for policy-makers, researchers, and stakeholders involved in the sustainable energy transition.

Keywords: South Africa; renewable energy utilization; policy; challenges; energy mix; environmental sustainability

1. Introduction

A developing country's ability to generate, transmit, and consume appropriate electric power is significantly linked to its socioeconomic and infrastructure expansion [1–7]. Being Africa's largest electric power consumer, South Africa has established itself as one of the most advanced countries on the continent [1,8]. Although South Africa possesses limited proven oil and gas reserves, it compensates for this shortfall by leveraging its extensive coal deposits to meet its energy requirements. As of 2022, coal has dominated the South African energy mix, contributing to 80% of South Africa's overall primary energy consumption. Renewable energy technologies' (wind, solar PV, and CSP) contribution increased in 2022 to a total of 6.2 GW installed capacity and provided 7.3% of the total energy mix, as depicted in Figures 1 and 2 [9]. Despite the effectiveness of coal in power generation, its use has contributed to South Africa being ranked amongst the top ten greenhouse gas emitters globally, emitting 680 kg of CO_2 per USD 1000 of GDP [10–13].

In response, South Africa has embarked on efforts to reduce CO_2 emissions whilst simultaneously developing renewable energy sources (RESs) [14–16]. This review discusses the development of solar photovoltaic (SPV), wind energy, biomass, and geothermal energy as prominent RESs in South Africa, highlighting the progress made and challenges encountered in the process. This article delves into the overview of non-renewable energy in South Africa and the historical background of South Africa's power generation in order to provide an in-depth comprehension of the expansion of renewable energy sources (RESs) in the country. Moreover, it offers a thorough analysis of the development of RESs in South

Africa, building on the extant literature in [1,16], which only provides an overview of the issues and advancements in the South Africa RES sector. The use of solar PV, wind energy, biomass, and geothermal energy is prioritized for the generation of both power and heat. Reliable databases (SCOPUS, Web of Science, and Google Scholar) provided the extant literature for this study.

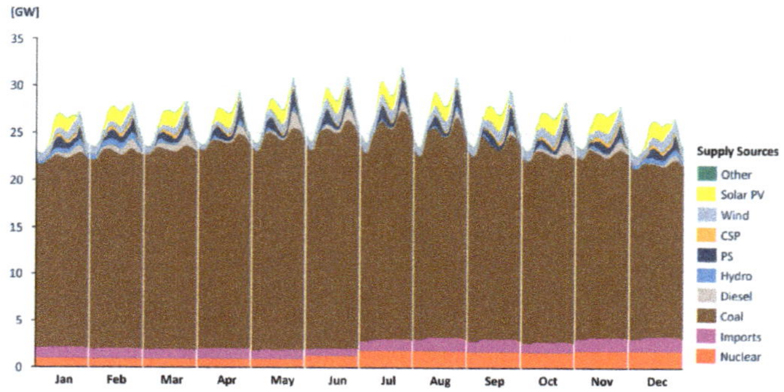

Figure 1. South Africa's energy mix composition [9].

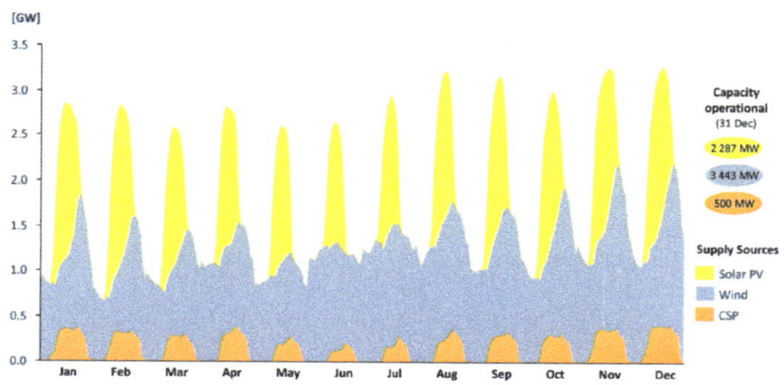

Figure 2. The renewable energy technology energy mix in South Africa [9].

2. Non-Renewable Energy in South Africa

Non-renewable energy typically refers to energy sources such as fossil fuels (coal, oil, and natural gas) and nuclear power, which are finite and cannot be naturally replenished on a human time-scale [17]. In the case of South Africa, non-renewable energy plays a significant role in the country's energy mix, particularly with coal being the dominant source of electricity generation. Despite the abundant accessibility of this fossil fuel, South Africa and many other countries persist in seeking alternative energy sources to mitigate the detrimental impacts of non-renewable energy on environmental well-being.

These impacts are far-reaching, environmentally destructive, and pose a threat to the economy [18–21]. The drive for sustainable development on a global scale has altered the outlook on the energy sector. According to [22–25], a significant element influencing economic growth has been found to be the increase in carbon dioxide emission levels. The creation of programs like the Kyoto Protocol and Sustainable Energy for All, a United Nations initiative aimed at mitigating the gradual degradation of the world and reducing climate change, is due to the global recognition of greenhouse gas emissions resulting from the combustion of fossil fuels.

3. The Historical Context of Power Generation in South Africa

The unveiling of an arc light in Cape Town in 1860 is considered to be the first recorded incidence of electricity in South Africa [1]. The first electric street lights in the country were then installed in the mining town of Kimberly, which helped mining activities at night [26,27]. The enhanced throughput made possible by the Cape Town mines' electrification after the discovery of gold in Johannesburg in 1886 led to the construction of Johannesburg's first electricity plant in 1889 [28]. Subsequently, municipality lightning started from Rondebosch in 1892 and rapidly spread to numerous South African cities in the ensuing years [29]. South Africa experienced a remarkable surge in electricity generation and distribution during the period between the 1890s and the 1900s, primarily to support mining operations and illuminate urban areas.

However, this growth was accompanied by exorbitant electricity prices due to the limited capacity and inefficiency of the power utilities, along with inadequate connectivity amongst consumers. As a solution to these challenges, the idea of establishing a centralized power enterprise began to emerge, leading to the eventual founding of the Victoria Falls Power Company (VFP) in 1906 [1,30]. In 1915, VFP established a central control centre at Simmerpan, which had a combined installed capacity exceeding 160 megawatts (MW) across its thermal power plants [27,31]. Recognizing the significance of electricity, the Transvaal colonial administration introduced The Power Act, granting the government the authority to acquire privately owned electric enterprises after a span of thirty-five years. Subsequently, the Electricity Act of 1922 was enacted following a study on electrification in South Africa, which ultimately led to the establishment of the Electricity Supply Commission (ESCOM) in March 1923 [31].

ESCOM played a crucial role in electrifying the railway network and establishing additional power stations throughout South Africa. In 1925, ESCOM constructed the Malieveld Spruit hydropower station as a temporary solution while the Sabie River Gorge plant, which became operational in 1927, was being developed [32]. These two plants marked ESCOM's initial ventures in power generation, and they served as important milestones. To meet the growing electricity demand resulting from the gold rush, ESCOM proceeded to commission additional power plants. Electricity sales for ESCOM reached an impressive 800 million units by 1929 [31]. To meet the increasing power demands, ESCOM took significant steps such as acquiring VFP in 1948 and Kimberley's central power station in 1950. In 1951, ESCOM established the Rural Electrification Department to deliver electricity to rural communities in South Africa.

Alongside the establishment of new power plants, ESCOM pursued a strategy of acquiring independent plants, bolstering its existing accomplishments and responding to the need for a more integrated power generation and transmission infrastructure. This expansion continued until South Africa faced a severe drought in early 1983 [1]. In conjunction with other contributing factors, the drought led to a decline in ESCOM's power output, resulting in a subsequent increase in electricity tariffs. In response, the government established the Electricity Council, comprising fifteen individuals who were entrusted with the task of overseeing ESCOM's operations [33].

To increase power generation further, ESCOM underwent a re-organization, and a revised Electricity Act was introduced in 1987. This led to an expansion of the company's customer base and a name change from ESCOM to ESKOM, derived from the former name, ESCOM/EVKOM, by combining the acronyms [33]. Considering ESKOM's concerns regarding deregulated electricity pricing and non-compliant consumers, the government amended the Electricity Act and appointed a National Electricity Regulator to assume the responsibilities of the Electricity Council board. This modification helped to ameliorate ESKOM's financial issues to a degree. However, ESKOM advised the government of an impending power supply shortfall in the years ahead if an urgent revamping of the power stations was not commenced. The government seemed to dismiss the need for capital investment expansion in ESKOM because it was considering privatizing or corporatizing ESKOM and because it wanted to distribute power more widely rather than generate it. As a result, in 2007, ESKOM started to be inconsistent and less efficient in power delivery. After the government realized its error, it issued a public apology.

As a result of increasing energy demands in the years that followed, power disruptions increased in frequency [34,35]. Moreover, because of the ageing of several coal plants, there was a growing need for frequent maintenance, often requiring plants to temporarily go off-grid for repairs. This has been exemplified by incidents such as the turbine unit failure at the Duhva power plant in 2011, which caused a blast resulting in a loss of 600 MW of power from the grid. Consequently, the incident significantly escalated both the plant's maintenance expenses and its overall productivity costs [36]. Similarly, in 2018, a tragic accident occurred at Lethabo Power Station Unit 5 when the steam pressure from the boiler ruptured [37]. Furthermore, it would take at least another two years to repair the catastrophic hydrogen explosion that occurred in Unit 4 of the Medupi coal-fired power plant in 2022 [38].

In 2022, South African power plants experienced a significant surge in breakdowns, resulting in a staggering 19,052 megawatts of generating capacity being rendered inactive because of malfunctions. This amount accounts for approximately 40% of the total generating capacity in the country [39]. Despite possessing a total installed capacity exceeding 58 gigawatts (GW) to serve a population of 60.5 million people, ESKOM has not achieved its objective of delivering uninterrupted and affordable energy to all South Africans. Furthermore, the country is grappling with mounting debt, which has surpassed USD 263.9 billion [40–42]. While ESKOM continues to face developmental challenges, it is important to acknowledge that the state-owned electric utility company remains one of the most reliable and efficient power providers in sub-Saharan Africa. In fact, ESKOM surpasses the power generation capabilities of the continent's most populous nation, Nigeria, which has a total installed capacity of less than 5 GW [43,44] and Egypt's 595 GW installed power generation capacity to a population of 109.3 million [45,46].

4. South Africa's Carbon Footprint

The total amount of greenhouse gas emissions, mostly carbon dioxide (CO_2), brought about by human activity is referred to as a country's "carbon footprint". These emissions have an impact on global warming and climate change. Several industries, including the generation of power, transportation, industrial operations, and agriculture, have an impact on South Africa's carbon footprint. Moreover, South Africa's high reliance on coal for the generation of power has significantly increased its carbon footprint. The country ranks 14th globally in terms of carbon emissions, making it the highest emitter in Africa. Additionally, South Africa accounts for 33% of total air pollution in Africa. In 2020, the nation experienced a 1.5% increase in carbon emissions [47–49].

Figure 3 depicts the global carbon project analysis of South Africa's per capita CO_2 emissions. Promoting renewable energy sources, enhancing energy efficiency, putting emission reduction plans into practice, and taking part in global accords and actions to address climate change are all ways that South Africa is engaging to minimize its carbon footprint.

Figure 3. South Africa per capital CO_2 emissions [50].

A national advancement renewable energy source technology policy statement outlining South Africa's plan of generating 10 terawatt-hours (TWh) of power from RES was issued in 2003. In 2011, an integrated resource plan was unveiled, setting a new objective to increase the share of RESs in the energy mix to 17,800 MW by 2030. In addition, the Renewable Energy Independent Power Producer's Program (REIPPP) was launched in 2011 to promote private investments and expedite the nation's energy transformation. The program focuses on lowering CO_2 emissions, creating economic growth possibilities, and increasing generating capacity [51,52]. ESKOM, the predominant electricity producer in South Africa, currently supplies 96% of the country's total electricity production. The remaining 4% is generated by municipalities and independent power providers. ESKOM's power generation portfolio comprises 14 coal-fired power plants, one nuclear power plant, and a limited number of gas-fired stations. Furthermore, ESKOM operates a small number of hydroelectric power units alongside its other facilities [53]. The burning of coal, which is a finite and environmentally hostile resource for South Africa's power generation, can be substantially mitigated by shifting towards RESs as a viable alternative. Embracing renewable energy has the potential to not only reduce the country's carbon footprint but also create new employment opportunities, thereby bolstering the economy of the nation. The next sections will explore the prospects for renewable energy sources (RESs) in South Africa. However, before examining each energy source in detail, it is essential to compare the different technologies in terms of their life-spans. Notably, research conducted by the National Renewable Energy Laboratory (NREL) reveals that solar photovoltaic (PV) technology boasts the longest life-span amongst renewable energy sources [54]. Table 1 offers a comprehensive overview of the life-cycles of major technologies.

Table 1. A comparison of the life-cycles of RES technologies [54].

RES Type	System Useful Life
Wind energy	20 years
Biomass energy	20 to 30 years
Geothermal energy	24 years
Photovoltaics energy	25 to 40 years

These RES technologies offer additional social benefits, such as improved health outcomes and technological advancements, contingent on a few fundamental considerations. The fundamental factors in each nation's development are its social features. The social implications of each resource with its proportion are mentioned in Table 2.

Table 2. Social impact of RES technologies.

RES Technology	Social Impact	Proportion
Wind energy	Aesthetics	Minimal
Biomass energy	Impaired biodiversity	Minimal
Geothermal energy	Noise	Minimal
Photovoltaics energy	Contaminants	Minimal

5. Wind Energy

The application of wind energy in South Africa dates back several decades. Early settlers in the country, who were predominantly farmers, relied on wind energy for many years to pump water for irrigation and domestic purposes. These farmers' exceptional agricultural yields were made possible by this method [55]. The country's energy mix has, however, largely relied on coal because of its low cost and wide availability, making the integration of wind energy difficult. To supply home and agricultural water demands, some 30,000 windmill installations had been installed by the early 1990s throughout South Africa's rural and agrarian regions. However, the use of wind energy to generate power for commercial use has not yet attracted much attention [56].

The capacity of the wind power sector in South Africa has gradually evolved, enabling the nation to achieve its targets for renewable energy. With its ability to cut carbon emissions and diversify the mix of energy sources, wind power has come to be recognized as a clean and sustainable energy source. The 1995 Diab's wind atlas was the first attempt to quantity the wind energy potential in South Africa. It offers detailed estimates and analyses of the country's wind resources, helping to identify suitable locations for wind energy projects and furthering the development of the wind energy sector in South Africa [57]. In 2008, a second wind atlas mapping effort was conducted in South Africa, focusing on analysing wind speeds at heights above 10 m to identify prime locations for wind energy exploitation. The resulting map highlighted the most favourable areas for wind energy generation throughout the country.

With optimistic assumptions, the analysis suggested a potential of generating 50 GW of wind energy. However, taking a more conservative approach, it was estimated that South Africa could potentially harness approximately 6 GW of wind energy. These findings provided valuable insights into the country's wind energy potential and helped guide decision-making and planning in the renewable energy sector [58]. In response to the discrepancies observed in previous wind energy atlas mapping efforts, the South African government, through the South African Wind Energy Project (SAWEP), undertook the development of a comprehensive Wind Atlas for South Africa (WASA) in 2018. This updated wind atlas provides detailed information on the mean wind speed throughout the country and the mean power density across different provinces. The data presented in the WASA is particularly valuable for investors interested in establishing large-scale wind farms in South Africa as it assists in identifying optimal locations with favourable wind resources.

The WASA serves as a crucial tool in supporting informed decision-making and facilitating the development of the wind energy sector in the country [59]. Figure 4 depicts the WASA high-resolution map of South Africa's energy potential. In 2002, ESKOM commissioned the 27 MW Klipheuwel wind farm, the first wind farm in South Africa, on the country's west coast as part of the government's commitment to fostering the application of RESs. The wind farm served as an experimental site to explore the potential of wind energy for power generation. The wind farm supplies about 20,000 South Africans with 86,000 MWh per annum and has successfully achieved a reduction of 24,080 tonnes in annual carbon emissions and a total reduction of 481,600 tonnes in lifetime carbon emissions [60]. Multiple wind farms have since been built all around South Africa over the years. Because of favourable wind resources, provinces including the Western Cape, Eastern Cape, and Northern Cape have experienced major wind farm installations [61].

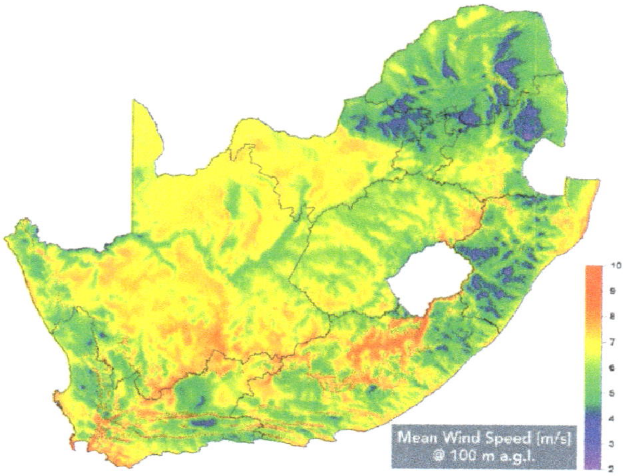

Figure 4. Wind Atlas South Africa (WASA) wind energy map [62].

Since the 27 MW Klipheuwel wind farm came into service, the installed wind power capacity in South Africa has grown rapidly, reaching 3357 MW in 2022, and it is projected to reach 17,742 MW by 2030 [63,64]. While South Africa's wind energy potential is being harnessed, there are still technical and environmental issues that must be resolved if the industry is to continue expanding.

5.1. Technical Difficulties

One of the issues hindering the widespread deployment of wind energy is the problem of situating wind farms. These power plants are often situated in remote areas, posing difficulties in connecting them to the power grid since wind farms are typically required to be in specified zones. Consequently, a significant portion of the generated power is lost because of inadequate grid infrastructure. To overcome this issue, expensive transmission lines are typically required to be erected, linking the wind farms to urban centres, thereby increasing the overall installation costs of the wind farm [65].

The variability in wind speed, upon which wind power relies, introduces potential technical issues such as voltage fluctuations, even in the presence of a stable grid infrastructure. The power output of wind power plants is directly proportional to changes in wind speed, and managing these fluctuations is costly [65]. This variability results in an excess or insufficient supply of power, ultimately causing disruptions in the power grid [66]. Despite the undeniable benefits of widespread wind energy adoption in South Africa, it is crucial to address the existing technical difficulties before the sector can thrive effectively.

5.2. Environmental Difficulties

Wind energy in South Africa faces various environmental challenges that need to be considered and addressed for the sustainable development of the technology. Some of these challenges include avian and bat mortality, habitat fragmentation, noise and visual impact, and land and water use. Wind turbines pose a risk to birds and bats, particularly if they are located in migration corridors or areas with high biodiversity. Collisions with rotating turbine blades can result in bird and bat fatalities [1,67,68]. Additionally, wind farm construction can lead to habitat fragmentation, disrupting natural ecosystems and potentially affecting wildlife populations.

It is important to consider the potential impacts on habitats and implement mitigation measures, such as establishing buffer zones and corridors, to preserve ecological connectivity and minimize habitat fragmentation [69]. Wind turbines also produce noise during operation, which can have adverse effects on local communities, wildlife, and even marine ecosystems if located near coastlines. Visual impact is also a concern, as large-scale wind farms can alter the aesthetic value of landscapes. The visual impact of wind energy infrastructure can be influenced by various factors, including colour contrast, size, proximity to residential areas, presence of flashing shadows, and duration of operation. These factors collectively define the "visual impact zone", according to contemporary planning standards.

By considering the distance and carefully selecting a wind farm's placement, it is possible to assess and plot the degree of visual impact. Proper planning, stakeholder engagement, and adherence to noise regulations can help mitigate these challenges [70,71]. Wind farms require significant land area for installation, which can potentially compete with other land uses, such as agriculture or conservation. Researchers have long been interested in understanding the competition that arises from land utilization for agriculture and energy purposes. Different studies have focused on examining the suitability of different types of land, including arable and marginal land, for both agricultural and energy-related activities [72,73].

To effectively tackle these environmental challenges, a comprehensive approach is necessary, encompassing rigorous environmental impact assessments, active stakeholder engagement, strategic site selection based on ecological considerations, and the implementation of efficient mitigation measures. By prioritizing environmental sustainability, South

Africa can optimize the benefits of wind energy whilst minimizing any potential adverse impacts on the environment. This can be achieved through the careful execution of thorough environmental impact assessments and the implementation of policies that address these challenges, such as appropriate siting practices and robust monitoring protocols [73]. Table 3 presents the net wind energy output in South Africa between 2013 and 2023. This indicates the growth in generation in line with government policy.

Table 3. South Africa's wind energy output from 2013 to 2023 [74].

Year	Generated Power (MW)
2013	257
2014	569
2015	1079
2016	1473
2017	2094
2018	2094
2019	2094
2020	2516
2021	2495
2022	3163
2023	3442

6. Solar Energy

South Africa, like many other African countries, possesses abundant solar energy resources. The Northern Cape region stands out with its remarkable Direct Normal Irradiation (DNI) reaching over 3200 kWh/m^2, while the KwaZulu-Natal region experiences a more moderate yearly DNI of around 1400 kWh/m^2. In recent years, there have been numerous applications of solar energy, both direct and indirect, in South Africa. This section explores the utilization of solar energy in the country, focusing on solar photovoltaics and solar water heating.

6.1. Solar Photovoltaic (SPV)

Solar photovoltaic (SPV) utilize the photo-electric effect to directly convert solar irradiation into electrical energy [75,76]. The photo-electric effect involves the emission of electrons from a material's surface when it is exposed to light of the appropriate wavelength, with these materials typically being referred to as semi-conductors [77,78]. SPV cells are devices designed to harness the energy of sunlight and convert it into usable electricity. SPV technology is a widely adopted and rapidly growing renewable energy source worldwide, including in South Africa. SPV systems consist of solar panels, which are composed of multiple PV cells that contain semi-conductors, typically made of silicon. SPV offers several advantages, making it an attractive option for generating electricity, including the following:

(a) Renewable and Sustainable: Solar energy is abundant and inexhaustible, making it a renewable and sustainable source of power. The sun's energy is freely available, and the technology to capture and convert it into electricity is constantly improving [79–81].
(b) Low Environmental Impact: Solar PV produces electricity without emitting greenhouse gases or other harmful pollutants, resulting in significantly reduced environmental impacts compared with fossil fuel-based power generation. This helps to combat climate change and contributes to cleaner air and water [82–84].
(c) Versatility and Scalability: SPV systems can be installed on a variety of scales, ranging from small rooftop installations for residential homes to large-scale utility solar farms, generating outputs ranging from microwatts to megawatts. This versatility enables SPV to be deployed in a wide range of applications, from powering individual households to supplying electricity to remote buildings, communication systems, satellites, and spacecraft [85].

(d) Economic Benefits: SPV systems provide opportunities for job creation, both in the manufacturing and installation sectors. Direct, indirect, and influenced employment opportunities are amongst the employment possibilities provided by the SPV value-chain. These employment opportunities not only support the expansion of the renewable energy industry but also have favourable effects on job creation and economic growth. South Africa could possibly add to employment via the solar-powered economy, which would also lower greenhouse gas emissions and promote RESs in the energy mix. Furthermore, solar energy has the potential to reduce reliance on expensive imported fossil fuels, thereby enhancing energy security and promoting local economic development [86,87].

In South Africa, SPV has experienced significant growth in recent years, driven by factors such as declining costs, government support, and increasing awareness of the benefits of RESs. The country benefits from abundant solar resources, particularly in regions such as the Northern Cape, which has excellent solar irradiation levels. The South African government has implemented various initiatives and incentives, such as the Renewable Energy Independent Power Producer Procurement Program (REIPPPP), to promote the development of solar PV projects. SPV deployment in South Africa can offset electricity consumption, reduce reliance on the grid, and potentially enable the selling of excess electricity back to the grid through net metering or feed-in tariff schemes.

While solar PV offers numerous benefits, challenges do exist, including the intermittent nature of sunlight, the need for sufficient space for installations, and the initial investment costs. However, ongoing technological advancements and supportive policies continue to drive the growth and adoption of SPV in South Africa, contributing to a more sustainable and cleaner energy future [88,89]. According to the most recent statistics available from South Africa's Department of Energy (DoE), as shown in Table 4, the total installed capacity of SPV increased from 1350.6 MW in 2016 to 1961.4 MW in 2021 [90].

Table 4. South Africa's SPV output from 2016 to 2021 [90,91].

Year	Generated Power (MW)
2013	262
2014	1164
2015	1353
2016	2176
2017	3451
2018	4805
2019	4908
2020	5994
2021	6316
2022	6326
2023	6164

In recent times, South Africa has successfully installed several large-scale solar power plants, which has bolstered the country's rapidly expanding renewable energy industry. In the Northern Cape province alone, there are more than one hundred independent power producers, with a total capacity of 900 MW and over twenty projects connected to the grid. At the end of the first quarter of 2023, there was 6164 MW of installed solar PV capacity in South Africa, and it is expected to surpass 12 GW in the next ten years [92,93]. Of the current installed capacity, 2200 MW came from governmental procurement, mostly through the Renewable Energy Independent Power Producer Program (REIPPP), and 3964 MW is from private installation.

The residential sector utilizes 11% of the installed capacity, with 30% of the capacity consisting of systems smaller than 1 MVA, encompassing residential, commercial, and industrial applications, which constitutes the small-scale embedded generation. Utility-scale systems exceeding 50 MW make up 27% of the installed capacity, whereas systems in the 1 MW to 50 MW range contribute 32%. With 586 MW of installed capacity, the City of Johannesburg in the province of Gauteng ranks highest on the list of installations. The Northern Cape province's Pixley ka Seme district municipality is next, with 583 MW.

Taken as a whole, they both account for more than 20% of the country's installed SPV capacity. The City of Tshwane in Gauteng leads the list in terms of residential installations with 22,956 installations, closely followed by the City of Cape Town with 21,342 installations. With an average of 10.8 kWp per system, eThekwini in KwaZulu-Natal has the largest average residential systems amongst them, followed by the City of Cape Town at 7 kWp [94]. Table 5 comprises a compilation of several solar photovoltaic projects across the country.

Table 5. South Africa's major SPV installations [95].

Project Name	Installed Capacity (MW)	Project Location	Energy Production (GWh)	Averted CO_2 Emissions (Tons/Year)	Energy Demand Supplied (Household)
Adams Solar Plant	82.5	Northern Cape province	167 GWh	171,700	100,000
Aggeneys Solar PV Park	46	Northern Cape province	117 GWh	90,600	20,000
Aries Solar Park	11	Northern Cape province	22 GWh	14,053	4900
Boshof Solar Farm	66	Free State province	132 GWh	98,700	30,000
De Aar project	175	Northern Cape province	195 GWh	84,554	76,000
Dreunberg Solar PV Park	75	Eastern Cape province	156 GWh	130,000	38,000
KaXu Solar One	100	Northern Cape province	400 GWh	315,000	65,000
Kalkbult Solar Station	75	Northern Cape province	150 GWh	130,000	35,000
Kotulo Tsatsi PV Farm	150	Northern Cape province	1800 GWh	750,000	320,000
Jasper Solar	96	Northern Cape province	180 GWh	145,000	80,000
Mulilo Prieska PV	180	Northern Cape province	220 GWh	160,000	80,000
Xina Solar One	100	Northern Cape province	400 GWh	205,763	65,000
Ilanga-1 CSP Plant	100	Northern Cape province	400 GWh	100,000	90,000
Lesedi Solar PV Plant	75	Northern Cape province	150 GWh	134,690	65,000
Letsatsi Solar Plant	75	Northern Cape province	150 GWh	134,690	65,000
Tom Burke PV Plant	66	Limpopo province	122 GWh	111,000	38,000
Greefspan II PV Plant	63	Northern Cape province	0.15 GWh	130,000	20,000
Paleisheuwel Solar Park	82.5	Western Cape province	153 GWh	140,000	48,000
Sishen PV Plant	94.3	Western Cape province	216 GWh	229,280	100,000

There has been impressive growth in South Africa's installed rooftop solar PV capacity. The increase from 983 MW in March 2022 to 4412 MW in June 2023 represents a remarkable 349% growth in just over a year. This surge in rooftop solar capacity suggests a growing interest in and adoption of renewable energy sources, likely driven by factors such as environmental concerns, cost savings, and advancements in solar technology. It is noteworthy that ESKOM, the state-owned electricity supplier in South Africa, has supplied these data, indicating a level of official recognition and monitoring of the country's renewable energy landscape. This trend aligns with global efforts to transition towards cleaner and more sustainable energy sources to mitigate the impacts of climate change.

The continued proliferation of rooftop solar PV installations in South African provinces, particularly in the Western Cape, Gauteng, and Kwa Zulu-Natal, is a noteworthy development. The increased affordability of solar technology is a key driver of this expansion, making it more accessible to a broader range of energy consumers. Eskom, South Africa's power utility, monitored and reported on the growth of rooftop solar PV capacity, indicating a growing recognition of the importance of renewable energy sources in the country's energy landscape.

The substantial growth from 983 MW in March 2022 to 4412 MW in June 2023 demonstrates a significant shift towards decentralized energy generation and a desire to harness clean and sustainable energy options [96]. This surge in rooftop solar adoption can have positive implications for energy resilience, reducing dependence on traditional grid infrastructure, and contributing to national efforts to address climate change by decreasing reliance on fossil fuels.

6.2. Solar Water Heating

Energy is used up significantly during the heating of water, and up to 40% of energy consumption in South Africa goes towards heating water. Unfortunately, most South Africans pay a high price for hot water, which is necessary to maintain personal and public health. Solar water heaters (SWHs) that use solar energy can assist in alleviating the South African energy issue. In South Africa, solar water heating offers an effective way to cut costs, conserve energy, and advance sustainable development. The usage of SWH systems is anticipated to increase, benefiting both homes and the environment, because of continued government initiatives and improved awareness [97,98]. Figure 5 depicts a map of South Africa's solar irradiation spread.

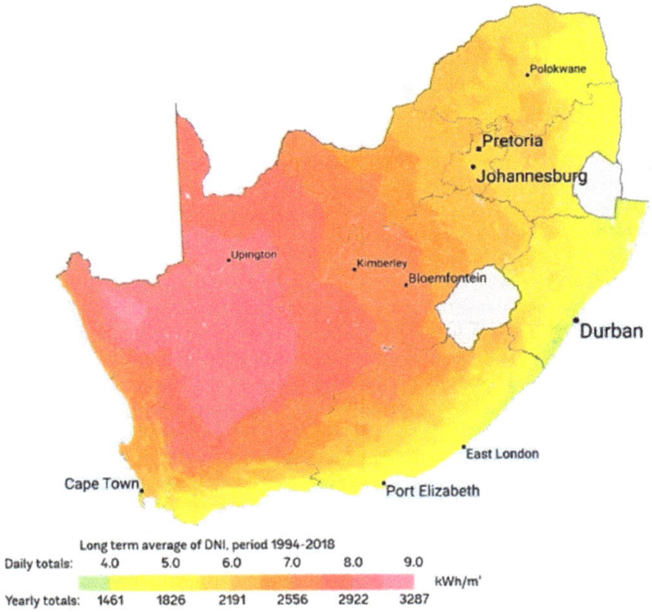

Figure 5. Map of South Africa's global horizontal irradiation [99].

6.3. Concentrated Solar Power

Globally, there is a growing interest in concentrated solar power (CSP) plants, with South Africa being a particularly notable adopter. South Africa, through the Renewable Energy Independent Power Producer Procurement (REIPPP) program, has played a crucial role in facilitating private investment in the implementation of large-scale CSP projects. The Redstone Solar Thermal Power Project, a 100 MW facility in the Northern Cape, is one of the major CSP projects in South Africa. This project incorporates molten salt energy storage, allowing it to generate 480,000 MW hours of electricity annually, even when the sun is not shining [100].

CSP systems use mirrors or lenses to concentrate sunlight, generating high-temperature thermal energy that is applicable in various ways. This thermal energy is versatile, allowing for conversion into power generation, material processing, process heat, and heating and cooling applications. The dual capability of CSP technology, producing both heat and electricity, makes it a flexible solution for diverse energy needs across various industries. Moreover, the incorporation of thermal energy storage technologies in CSP plants ensures continuous electricity generation even when solar irradiation is unavailable, thereby enhancing grid stability and reliability. CSP's adaptability positions it as a complementary element to other renewable energy sources like wind and PV power, contributing to the ongoing transition towards a more sustainable and renewable energy future.

Solar power towers (SPTs), parabolic dish collector systems (PDCSs), parabolic trough collector systems (PTCSs), and linear Fresnel reflectors (LFRs) are some of the technologies used in CSP. Furthermore, because of its benefits—which include increased efficiency, lower operating costs, and significant scale-up potential—the SPT has become a more significant CSP technology. Despite the advantages of CSP, a major problem facing all CSP technologies is the fluctuation in solar irradiation throughout the day and year [101–103]. This section compares the technologies and provides a thorough explanation of each CSP in relation to South Africa's potential for solar energy.

6.3.1. Parabolic Dish Collector Systems

The parabolic dish collector system (PDCS) comprises an energy generator, a Stirling engine at the centre of the dish, and parabolic reflectors in the form of dishes. These solar dishes automatically face the sun throughout the day to direct solar irradiation onto the Stirling engine. The system uses parabolic dishes with a concentration ratio of about 2000 to reach 9000 °C at their focal point and produce 200 bar working fluid pressure [103]. A parabolic dish collector typically has 50–120 square meters of surface area and a diameter of 6–10 m. These dish-reflecting surfaces are typically made of glass or plastic with aluminium or silver overlays. Mirrors made of silver on glass that have a thickness of around 1 mm exhibit the best performance. A certain quantity of iron is included in the glass to improve reflection. Mirrors composed of silver on iron-containing glass have a combined solar reflectivity and emittance value of between 90% and 94%. Figure 6 below presents a graphic illustration of a parabolic dish collector with a Stirling motor.

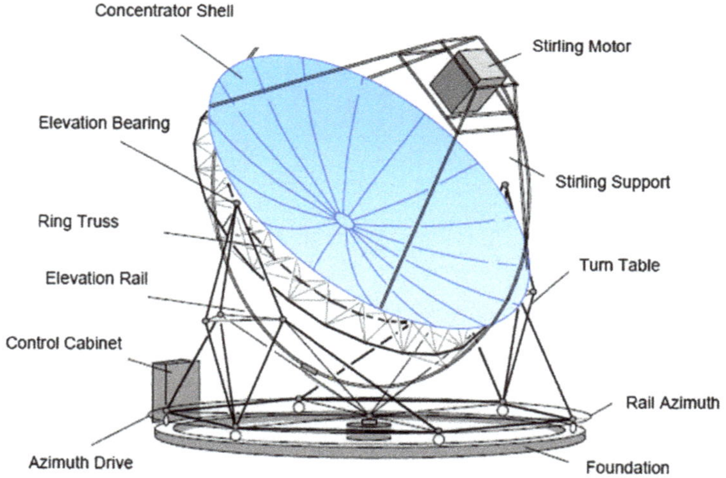

Figure 6. Parabolic dish schematic with a Stirling motor [104].

The technology of the parabolic dish collector system (PDCS) is advantageous for its ability to be deployed independently, making it suitable for areas that are not connected to the power utility grid, or regions with uneven and rough terrain. However, it comes with the following drawback: the inability to seamlessly integrate with thermal storage systems [1,105]. Moreover, the PDCS with a Stirling motor has an efficiency rate of 30% [103,106].

6.3.2. Parabolic Trough Collector Systems (PTCSs)

Parabolic trough collector systems (PTCSs) utilize a one-axis tracking mechanism, where parabolic mirrors aligned along a north–south or east–west axis track the Direct Normal Irradiance of the solar energy. Subsequently, these mirrors reflect the irradiation onto a central receiver system housing pressurized water, molten salt, or thermal oil. The produced heat is employed to generate steam, which in turn operates a steam engine for electricity generation. Subsequently, the utilized steam is condensed back into water and re-introduced into the system.

The operating temperature of the heat transfer fluid typically falls within the range of 400–600 °C, depending on the working fluid, while metal liquids have the capacity to reach temperatures of up to 900 °C [1,103,107]. The PTCS stands out as the most extensively employed system amongst CSP technologies because of its well-established technical advancement and stable performance. As of the conclusion of 2021, global CSP installations have reached a total capacity of 6.8 GW, with parabolic trough collector systems (PTCSs) representing over 80% of this figure, according to statistics [108]. Figure 7 below illustrates a parabolic trough collector system.

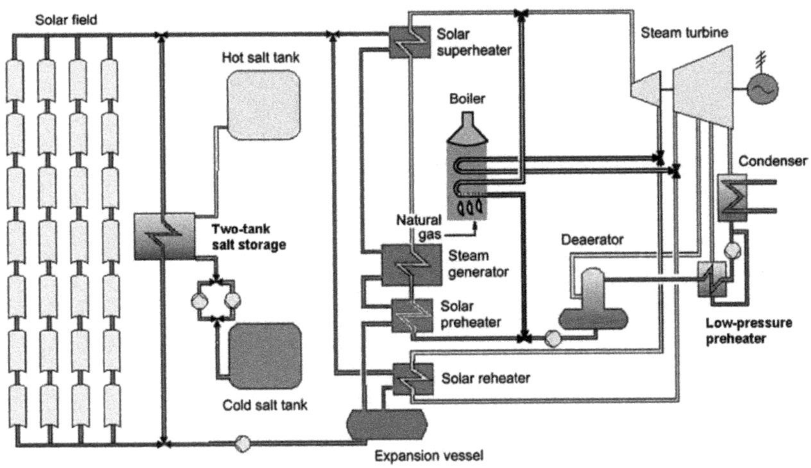

Figure 7. PTCS process flow schematic [109].

The South Africa Electricity Regulation Act of 2006 outlines the national electricity policy, providing a framework for power demand projections. The Act envisages an increase in generation capacity by 46 GW, with 24 GW coming from renewable energy sources, including CSP, aiming to achieve a preferred energy mix [110] and an augmented national installed capacity of 90 GW by 2030 [111,112]. The inclusion of CSP technology is motivated by the substantial solar energy potential in the country. Although there are currently a few PTCS-based CSP installations in existence, as indicated in Table 6, the full potential of CSP technologies has not been fully maximized at present.

Studies on the possibility of CSP generation in all of South Africa's provinces were conducted, as described in [113]. The research considered several variables, such as the amount of solar irradiation, the distance from current transmission lines, the topography of the province, and the environmental impact assessment. The results showed that the Northern Cape province was best suited for the widespread application of CSP technologies, with the Eastern Cape and the Free State provinces showing some degree of potential as well. The study avers that South Africa has a 500 GW potential CSP total generation capacity, especially if the Northern Cape province's solar resource is used to its full potential [110,113].

Table 6. CSP projects in South Africa.

CSP Projects	CSP Technology	Capacity (MW)	Location
Ilanga I	Parabolic trough	100	Northern Cape
ISCC Green Duba 1	Parabolic trough	43	Northern Cape
Kathu Solar Park	Parabolic trough	100	Northern Cape
KaXu Solar One	Parabolic trough	100	Northern Cape
Khi Solar One	Parabolic trough	50	Northern Cape
Xina Solar One	Parabolic trough	100	Northern Cape
Bokpoort CSP	Parabolic trough	50	Northern Cape

6.3.3. Linear Fresnel Reflector (LFR)

Linear Fresnel Reflector (LFR) solar power generation, as depicted in Figure 8, is a system designed to concentrate solar beam radiation onto a receiver tube positioned at the focal point of the Fresnel mirror. This concentration is achieved through the tracking movement of the FLR mirror to follow the sun's trajectory, generating a high-temperature working medium for thermal cycle power generation.

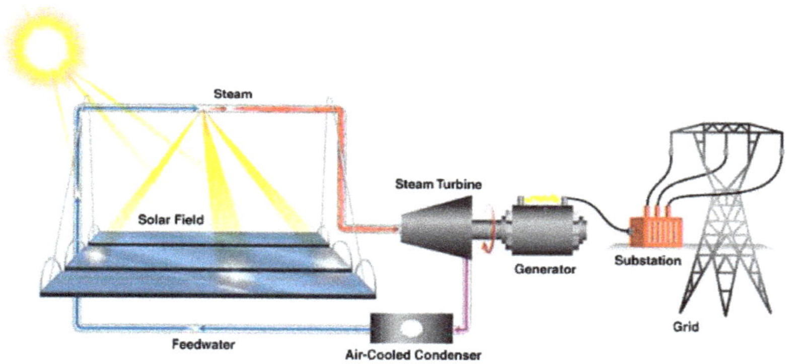

Figure 8. Fresnel reflector-integrated CSP plant [103].

Key components of the LFR power generation system include the linear reflective mirror, receiver tube, and transmission system. This system is essentially a condensed variant of the parabolic trough collector system (PTCS), in which a surface mirror takes the place of the conventional parabolic trough concentrator [101]. The mirror's characteristics include its proximity to the ground, low wind load, simple structure, intensive layout, and higher land-use efficiency. Additionally, the LFR system eliminates the need for vacuum treatment for the receiver tube, thus reducing both technical challenges and associated costs.

This results in a comparatively lower total system cost. However, it is worth noting that the system's low concentration ratio leads to a lower operational temperature, consequently affecting the overall system efficiency. A significant drawback of the Linear Fresnel Reflector (LFR) system lies in its substantial space requirement to avoid shading effects on the reflectors. While this issue can be mitigated by elevating the height of the absorber tower, such a modification would inevitably raise the overall cost of the system [114].

6.3.4. Solar Power Towers (SPTs)

Solar power towers (SPTs) utilize a heliostat field collector (HFC), which consists of solar tracking reflectors known as heliostats. Heliostats are mirrors, either flat or slightly concave, that track the sun in a two-axis movement. These heliostats reflect and concentrate solar irradiation onto a central receiver positioned at the top of a fixed tower [101,115]. In the central receiver, heat is absorbed by a heat transfer fluid (HTF), which then conveys the heat to heat exchangers driving a steam Rankine power cycle. Some operational commercial tower plants utilize Direct Steam Generation (DSG), while others employ various fluids, including molten salts, as both the HTF and storage medium [101].

The tower concept's concentrating power achieves exceptionally high temperatures, enhancing the efficiency of converting heat into electricity and reducing thermal storage costs. Moreover, this concept offers considerable flexibility, allowing designers to choose from a diverse range of heliostats, receivers, and transfer fluids. Some plants may feature multiple towers feeding into a single power block. Figure 9 presents a schematic layout of the solar tower power plant.

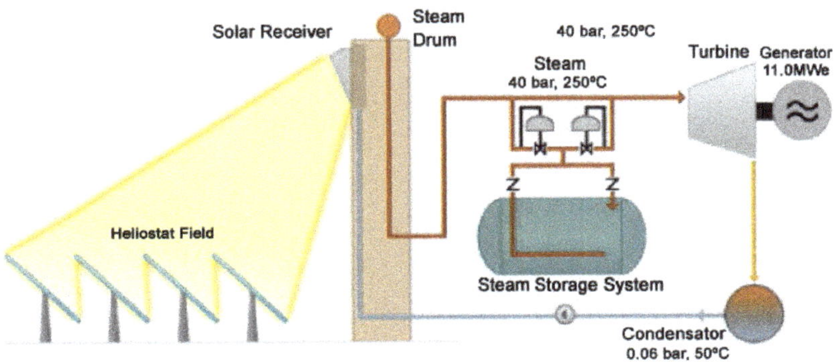

Figure 9. A schematic layout of a solar tower power plant [103].

6.3.5. Comparison

Table 7 provides a comparison of different CSP technologies based on basic criteria. With the potential to contribute at least 7% of global electricity needs by 2030 and 25% by 2050, under the projection of a high-energy-saving condition [116], CSP technologies possess substantial potential to address a portion of South Africa's future energy demand [117]. The anticipation is that the CSP market will experience rapid and continued expansion [103].

Table 7. CSP technology comparison.

Type of CSP	Capital Cost (USD/W)	Land Usage	Water for Cooling (L/MWh)	Heat Dynamic Efficiency (%)	Solar Concentration Ratio (C)
LFR	1.57	Medium	3000	Low	40
PDCS	2.91	Small	3000	High	45
PTCS	3.22	Large	Dry-cooled	Low	1000
SPT	4.62	Medium	1500	High	1500

The solar power tower (SPT) demonstrates the highest solar concentration ratio, as indicated in Table 7. The solar concentration ratio (C) is defined as the ratio of the concentrated flux on the receiver to the ambient flux from the sun. This ratio is calculated by dividing the receiver's area by the total area of the reflectors, assuming that the receiver is fully illuminated. The concentration ratio is a pivotal parameter in CSP plant performance, influencing power plant efficiency. Elevating the concentration ratio directs more sunlight into the collecting area, consequently leading to increased power output [118].

6.4. Technical Difficulties

Solar energy is a promising source of renewable energy for South Africa, given the abundant available sunlight. However, several technical difficulties need to be addressed in order to optimize its potential. This study found the following key challenges as an area of interest towards achieving South Africa's energy mix agenda:

(a) Grid Integration and Stability: Solar energy production is intermittent and variable, depending on weather conditions and time of day. This creates challenges for grid stability and reliability. The existing grid infrastructure may not be fully equipped to handle the large-scale integration of solar power, whilst upgrading and maintaining the grid can be costly and technically demanding. Effective energy storage systems, and storage technologies such as batteries, which are required to store excess solar energy and supply power during non-sunny periods are currently expensive and still developing [119,120].

(b) Geographical and Climatic Challenges: South Africa has abundant sunlight, but solar irradiation varies significantly across different regions. Optimal solar plant locations are often far from major urban centres, necessitating efficient transmission systems. Moreover, dust accumulation on solar panels can reduce their efficiency. Regular cleaning and maintenance are required, particularly in arid regions with high dust levels. High temperatures can also affect the efficiency of photovoltaic (PV) panels. In extremely hot climates, the efficiency of solar panels decreases [121].

Addressing these technical challenges requires a coordinated effort from the government, private sector, and research institutions to create a sustainable and efficient solar energy infrastructure in South Africa.

7. Conclusions

Despite South Africa's significant potential for renewable energy sources, coal remains its primary fuel source because of its low cost, although it is detrimental to the environment. The improved application of RESs is crucial for the nation, as the heavy reliance on coal is straining the existing coal-fired power plants. While wind and solar energy are viable options because of favourable wind speeds and high levels of annual solar radiation, the growth in renewable energy faces several challenges. These include environmental-, technological-, financial-, and policy-related issues. To address these issues and improve the integration of renewable energy into South Africa's energy mix, several recommendations are proposed. These recommendations include the following:

I. Enhance Policy and Regulatory Frameworks: The government needs to implement clear policies that establish clear long-term regulatory frameworks that support renewable energy development, providing stability and confidence for investors. The government needs to introduce financial incentives, such as tax breaks, grants, and subsidies, to encourage more investment in renewable energy projects, as well as to reduce bureaucratic delays in permitting and granting approval for renewable energy projects.

II. Upgrade Grid Infrastructure: The national grid could be upgraded by intentionally investing in modernizing and expanding the grid to better accommodate renewable energy inputs, thus ensuring stability and reliability. The country needs to develop and implement advanced energy storage systems to address the intermittency of renewable energy sources and provide a stable supply.

III. Explore Other Potentials: Further utilization of solar energy resources in South Africa can provide numerous benefits beyond the current major applications in solar PV, solar water heating (SWH), and concentrated solar power (CSP). Another potential application of solar energy, which remains under-utilized but could be highly beneficial for the country, is solar cooking using a solar cooker to harness sunlight for food processing. Adopting solar cooking technology could significantly reduce reliance on coal, thereby contributing to CO_2 mitigation and enhancing sustainability and energy efficiency in the country.

By implementing these recommendations, South Africa can overcome the challenges associated with renewable energy development and significantly enhance the integration of renewable sources into its energy mix, contributing to a more sustainable and resilient energy future. Future studies will carry out social and environmental impact analyses for various RES projects in particular provinces of South Africa. Focusing on these future research issues will support South Africa in its transition to a robust, sustainable, low-carbon energy system.

Author Contributions: Conceptualization, A.A.A.; methodology, A.A.A. and K.M.; software, A.A.A.; validation, A.A.A. and K.M.; formal analysis, A.A.A.; investigation, A.A.A.; resources, A.A.A. and K.M.; data curation, A.A.A.; writing—original draft preparation, A.A.A.; writing—review and editing, A.A.A. and K.M. All authors have read and agreed to the published version of the manuscript.

Funding: This research received no external funding.

Data Availability Statement: Upon request, the corresponding author will provide the data used in this work. Restrictions on the disclosure of the tools used to process the data prevent the data from being made publicly available.

Acknowledgments: The responsibility for the content is solely on the authors.

Conflicts of Interest: The authors declare no conflicts of interest. The funders had no role in the design of this study; in the collection, analysis, or interpretation of data; in the writing of this manuscript; or in the decision to publish the results.

References

1. Akinbami, O.M.; Oke, S.R.; Bodunrin, M.O. The state of renewable energy development in South Africa: An overview. *Alex. Eng. J.* **2021**, *60*, 5077–5093. [CrossRef]
2. Sehlapelo, T.; Inglesi-Lotz, R. Examining the determinants of electricity consumption in the nine South African provinces: A panel data application. *Energy Sci. Eng.* **2022**, *10*, 2487–2496. [CrossRef]
3. Bah, M.M.; Azam, M. Investigating the relationship between electricity consumption and economic growth: Evidence from South Africa. *Renew. Sustain. Energy Rev.* **2017**, *80*, 531–537. [CrossRef]
4. Azam, M. Relationship between energy; investment; human capital; environment, and economic growth in four BRICS countries. *Environ. Sci. Pollut. Res.* **2019**, *26*, 34388–34400. [CrossRef] [PubMed]
5. Mutumba, G.S.; Odongo, T.; Okurut, N.F.; Bagire, V. A survey of literature on energy consumption and economic growth. *Energy Rep.* **2021**, *7*, 9150–9239.
6. Azimoh, C.L.; Klintenberg, P.; Wallin, F.; Karlsson, B.; Mbohwa, C. Electricity for development: Mini-grid solution for rural electrification in South Africa. *Energy Convers. Manag.* **2016**, *110*, 268–277. [CrossRef]
7. Zebra, E.I.C.; van der Windt, H.J.; Nhumaio, G.; Faaij, A.P. A review of hybrid renewable energy systems in mini-grids for off-grid electrification in developing countries. *Renew. Sustain. Energy Rev.* **2021**, *144*, 111036. [CrossRef]
8. Apergis, N.; Polemis, M.; Soursou, S.-E. Energy poverty and education: Fresh evidence from a panel of developing countries. *Energy Econ.* **2022**, *106*, 105430–105441. [CrossRef]
9. CSIR. Statistics on Power Generation in South Africa for 2022. CSIR Strategic Communications. Available online: https://www.csir.co.za/csir-releases-statistics-on-power-generation-south-africa-2022#:~:text=Coal%20still%20dominates%20the%20South,of%20the%20total%20energy%20mix (accessed on 16 June 2023).
10. Zheng, X.; Streimikiene, D.; Balezentis, T.; Mardani, A.; Cavallaro, F.; Liao, H. A review of greenhouse gas emission profiles, dynamics, and climate change mitigation efforts across the key climate change players. *J. Clean. Prod.* **2019**, *234*, 1113–1133. [CrossRef]
11. Friedrich, J.; Ge, M.; Pickens, A.; Vigna, L. This interactive chart shows changes in the world's top 10 emitters. *World Resour. Inst.* **2020**, *10*, 1–5.
12. Olivier, J.G.; Schure, K.; Peters, J. Trends in global CO_2 and total greenhouse gas emissions. *PBL Neth. Environ. Assess. Agency* **2017**, *5*, 1–11.
13. Kumar, A.; Singh, P.; Raizada, P.; Hussain, C.M. Impact of COVID-19 on greenhouse gases emissions: A critical review. *Sci. Total Environ.* **2022**, *806*, 150349–150359. [CrossRef] [PubMed]
14. Ibrahim, I.D.; Hamam, Y.; Alayli, Y.; Jamiru, T.; Sadiku, E.R.; Kupolati, W.K.; Ndambuki, J.M.; Eze, A.A. A review on Africa energy supply through renewable energy production: Nigeria, Cameroon, Ghana and South Africa as a case study. *Energy Strategy Rev.* **2021**, *38*, 100740. [CrossRef]
15. Longe, O.M.; Myeni, L.; Ouahada, K. Renewable energy solution for electricity access in rural South Africa. In Proceedings of the 2019 IEEE International Smart Cities Conference (ISC2), Casablanca, Morocco, 14–17 October 2019; IEEE: Piscataway, NJ, USA, 2019; pp. 772–776.
16. Aliyu, A.K.; Modu, B.; Tan, C.W. A review of renewable energy development in Africa: A focus in South Africa, Egypt and Nigeria. *Renew. Sustain. Energy Rev.* **2018**, *81*, 2502–2518. [CrossRef]

17. Udeagha, M.C.; Ngepah, N. Disaggregating the environmental effects of renewable and non-renewable energy consumption in South Africa: Fresh evidence from the novel dynamic ARDL simulations approach. *Econ. Chang. Restruct.* **2022**, *55*, 1767–1814. [CrossRef]
18. Adedoyin, F.F.; Ozturk, I.; Agboola, M.O.; Agboola, P.O.; Bekun, F.V. The implications of renewable and non-renewable energy generating in Sub-Saharan Africa: The role of economic policy uncertainties. *Energy Policy* **2021**, *150*, 112115–112127. [CrossRef]
19. Ozcan, B.; Ozturk, I. Renewable energy consumption-economic growth nexus in emerging countries: A bootstrap panel causality test. *Renew. Sustain. Energy Rev.* **2019**, *104*, 30–37. [CrossRef]
20. Armeanu, D.S.; Joldes, C.C.; Gherghina, S.C.; Andrei, J.V. Understanding the multidimensional linkages among renewable energy, pollution, economic growth and urbanization in contemporary economies: Quantitative assessments across different income countries' groups. *Renew. Sustain. Energy Rev.* **2021**, *142*, 110818–110839. [CrossRef]
21. Nawaz, M.A.; Hussain, M.S.; Kamran, H.W.; Ehsanullah, S.; Maheen, R.; Shair, F. Trilemma association of energy consumption, carbon emission, and economic growth of BRICS and OECD regions: Quantile regression estimation. *Environ. Sci. Pollut. Res.* **2021**, *28*, 16014–16028. [CrossRef]
22. Agyeman, F.O.; Ma, Z.; Li, M.; Sampene, A.K.; Dapaah, M.F.; Kedjanyi, E.A.G.; Buabeng, P.; Li, Y.; Hakro, S.; Heydari, M. Probing the effect of governance of tourism development, economic growth, and foreign direct investment on carbon dioxide emissions in Africa: The African experience. *Energies* **2022**, *15*, 4530. [CrossRef]
23. Du, L.; Wei, C.; Cai, S. Economic development and carbon dioxide emissions in China: Provincial panel data analysis. *China Econ. Rev.* **2012**, *23*, 371–384. [CrossRef]
24. Salahuddin, M.; Gow, J.; Ozturk, I. Is the long-run relationship between economic growth, electricity consumption, carbon dioxide emissions and financial development in Gulf Cooperation Council Countries robust? *Renew. Sustain. Energy Rev.* **2015**, *51*, 317–326. [CrossRef]
25. Sufyanullah, K.; Ahmad, K.A.; Ali, M.A.S. Does emission of carbon dioxide is impacted by urbanization? An empirical study of urbanization, energy consumption, economic growth and carbon emissions-Using ARDL bound testing approach. *Energy Policy* **2022**, *164*, 112908–112925. [CrossRef]
26. Info, K.C.; Today in Kimberly's History. Kimberly City Info, Kimberly. Available online: https://www.kimberley.org.za/2-september-today-in-kimberleys-history/ (accessed on 5 July 2023).
27. Buraimoh, E.; Adebiyi, A.A.; Ayamolowo, O.J.; Davidson, I.E. South Africa electricity supply system: The past, present and the future. In Proceedings of the 2020 IEEE PES/IAS PowerAfrica, Nairobi, Kenya, 25–28 August 2020; pp. 1–5.
28. Marais, L.; Cloete, J.; Lenka, M. The plight of mining cities in South Africa: Planning for growth and closure. *Cities* **2022**, *130*, 103965. [CrossRef]
29. Todes, A.; Turok, I. Spatial inequalities and policies in South Africa: Place-based or people-centred? *Prog. Plan.* **2018**, *123*, 1–31. [CrossRef]
30. Christie, R. The Creation of a Monopoly: The Victoria Falls and Transvaal Power Company Limited, 1905–1914. In *Electricity, Industry and Class in South Africa*; Springer: Berlin/Heidelberg, Germany, 1984; pp. 27–49.
31. ESKOM. Empowering the Future Together. Available online: https://www.eskom.co.za/heritage/ (accessed on 5 July 2023).
32. Eskom. The History of the Sabie Gorge and Malieveld Hydro-Electric Power Stations. Available online: https://sabie.co.za/blog/?p=1849 (accessed on 5 July 2023).
33. Eskom. The Years of Expansion and Change. Available online: https://www.eskom.co.za/heritage/history-in-decades/eskom-1983-1992/ (accessed on 5 July 2023).
34. Molepo, J. Blackout: The Eskom Crisis, James-Brent Styan. *J. Public Adm.* **2018**, *53*, 795–798.
35. McGregor, L.; Nuttal, S. *Load Shedding: Writing on and over the Edge of South Africa*; Jonathan Ball Publishers: Johannesburg, South Africa, 2013.
36. Guardian. Power Plant Blast Threatens Supply. Available online: https://mg.co.za/article/2011-03-18-power-plant-blast-threatens-supply/ (accessed on 5 July 2023).
37. Energy. Lethabo Power Station Out of Action. Available online: https://www.bizcommunity.com/Article/196/693/182965.html (accessed on 5 July 2023).
38. Daily Investor. Medupi Unit 4 Explosion Repair Update. Available online: https://dailyinvestor.com/south-africa/10957/medupi-unit-4-explosion-repair-update/#:~:text=Unit%204%20of%20Eskom%E2%80%99s%20Medupi,service%20on%2031%20August%202024 (accessed on 5 July 2023).
39. Bloomberg. South African Power-Plant Breakdowns Hit a 2022 High. Bloomberg. Available online: https://www.bloomberg.com/news/articles/2022-12-07/south-african-power-plant-breakdowns-are-at-2022-high (accessed on 5 July 2023).
40. USAID. South Africa Energy Sector Overview. Available online: https://www.usaid.gov/powerafrica/south-africa#:~:text=According%20to%20the%20Ministry%20of,(MW)%20from%20all%20sources (accessed on 5 July 2023).
41. Macrotrends. South Africa Population 1950–2023. Available online: https://www.macrotrends.net/countries/ZAF/south-africa/population#:~:text=The%20current%20population%20of%20South,a%201%25%20increase%20from%202020 (accessed on 5 July 2023).
42. CEIC Data. South Africa National Government Debt. Available online: https://www.ceicdata.com/en/indicator/south-africa/national-government-debt#:~:text=in%20Apr%202023?-,South%20Africa%20National%20Government%20Debt%20reached%20263.9%20USD%20bn%20in,table%20below%20for%20more%20data (accessed on 5 July 2023).

43. Eberhard, A.; Shkaratan, M. Powering Africa: Meeting the financing and reform challenges. *Energy Policy* **2012**, *42*, 9–18. [CrossRef]
44. NERC. NESI Generation Statistics. Available online: https://nerc.gov.ng/resources/nesi-generation-statistics/ (accessed on 5 July 2023).
45. GlobalData. Egypt Power Market Size and Trend. Available online: https://www.globaldata.com/store/report/egypt-power-market-analysis/ (accessed on 5 July 2023).
46. Data. Egypt Population Trend. Available online: https://datacommons.org/place/country/EGY/?utm_medium=explore&mprop=count&popt=Person&hl=en (accessed on 5 July 2023).
47. DBSA. How Can We Reduce the Carbon Footprint in Africa. Available online: https://www.dbsa.org/article/how-can-we-reduce-carbon-footprint-africa#:~:text=South%20Africa%20has%20the%2014th,5%25%20rise%20in%20carbon%20emissions (accessed on 7 July 2023).
48. Bekun, F.V.; Emir, F.; Sarkodie, S.A. Another look at the relationship between energy consumption, carbon dioxide emissions, and economic growth in South Africa. *Sci. Total Environ.* **2019**, *655*, 759–765. [CrossRef] [PubMed]
49. Narayan, P.K.; Narayan, S. Carbon dioxide emissions and economic growth: Panel data evidence from developing countries. *Energy Policy* **2010**, *38*, 661–666. [CrossRef]
50. Our World in Data. South Africa: CO2 Country Profile. Available online: https://ourworldindata.org/co2/country/south-africa (accessed on 7 July 2023).
51. Njoh, A.J. Renewable energy as a determinant of inter-country differentials in CO_2 emissions in Africa. *Renew. Energy* **2021**, *172*, 1225–1232.
52. Walwyn, D.R.; Brent, A.C. Renewable energy gathers steam in South Africa. *Renew. Sustain. Energy Rev.* **2015**, *41*, 390–401. [CrossRef]
53. Belelie, M.D.; Burger, R.P.; Mkhatshwa, G.; Piketh, S.J. Assessing the impact of Eskom power plant emissions on ambient air quality over KwaZamokuhle. *Clean Air J.* **2019**, *29*, 29–37.
54. NREL. Life Cycle Assessment and Photovoltaic (PV) Recycling: Designing a More Sustainable Energy System. Available online: https://www.nrel.gov/state-local-tribal/blog/posts/life-cycle-assessment-and-photovoltaic-pv-recycling-designing-a-more-sustainable-energy-system.html (accessed on 7 July 2023).
55. Singh, S.; Bhatti, T.; Kothari, D. A review of wind-resource-assessment technology. *J. Energy Eng.* **2006**, *132*, 8–14. [CrossRef]
56. Musango, J.K.; Amigun, B.; Brent, A.C. Sustainable electricity generation technologies in South Africa: Initiatives, challenges and policy implications. *Energy Environ. Res.* **2011**, *1*, 124. [CrossRef]
57. Diab, R.D. *Wind atlas of South Africa*; Department of Mineral and Energy Affairs: Pretoria, South Africa, 1995.
58. Szewczuk, S.; Prinsloo, F. Wind Atlas for South Africa (WASA): Project overview and current status. In Proceedings of the CSIR 3rd Biennial Conference: 2010 Science Real and, Relevant, CSIR International Convention Center, Pretoria, South Africa, 30 August–1 September 2010.
59. Wind Atlas for South Africa (WASA) SANEDI. Available online: http://www.wasaproject.info (accessed on 7 July 2023).
60. EnergyFact. 27MW Sustainable Wind Power Generated. Available online: https://rb.gy/xvtdn (accessed on 7 July 2023).
61. Jain, S.; Jain, P. The rise of renewable energy implementation in South Africa. *Energy Procedia* **2017**, *143*, 721–726. [CrossRef]
62. Wind Atlas for South Africa Launched. Available online: https://www.esi-africa.com/industry-sectors/generation/wind-atlas-for-south-africa-launched/ (accessed on 7 July 2023).
63. Department of Mineral Resources and Energy. Independent Power Producers Procurement Programme (IPPPP): An Overview. Available online: https://www.ipp-projects.co.za/Publications/GetPublicationFile?fileid=2d03d621-6dc4-ec11-956e-2c59e59ac9cd&fileName=20220318_IPP%20Office%20Q3%20Overview%202021-22%20WEB%20VERSION.PDF (accessed on 11 July 2023).
64. Karamanski, S.; Erfort, G. Wind Energy Supply Profiling and Offshore Potential in South Africa. *Energies* **2023**, *16*, 3668. [CrossRef]
65. Gandhar, S.; Ohri, J.; Singh, M. A critical review of wind energy based power generation systems. *Asian J. Water Environ. Pollut.* **2020**, *17*, 29–36. [CrossRef]
66. Emetere, M.; Akinyemi, M. Climate Variability and Renewable Energy Planning. *IOP Conf. Ser. Earth Environ. Sci.* **2019**, *331*, 012036. [CrossRef]
67. Sebestyén, V. Renewable and Sustainable Energy Reviews: Environmental impact networks of renewable energy power plants. *Renew. Sustain. Energy Rev.* **2021**, *151*, 111626. [CrossRef]
68. Kunz, T.H.; Arnett, E.B.; Erickson, W.P.; Hoar, A.R.; Johnson, G.D.; Larkin, R.P.; Strickland, M.D.; Thresher, R.W.; Tuttle, M.D. Ecological impacts of wind energy development on bats: Questions, research needs, and hypotheses. *Front. Ecol. Environ.* **2007**, *5*, 315–324. [CrossRef]
69. Barré, K.; Le Viol, I.; Bas, Y.; Julliard, R.; Kerbiriou, C. Estimating habitat loss due to wind turbine avoidance by bats: Implications for European siting guidance. *Biol. Conserv.* **2018**, *226*, 205–214. [CrossRef]
70. Nazir, M.S.; Ali, N.; Bilal, M.; Iqbal, H.M. Potential environmental impacts of wind energy development: A global perspective. *Curr. Opin. Environ. Sci. Health* **2020**, *13*, 85–90. [CrossRef]
71. Kikuchi, R. Adverse impacts of wind power generation on collision behaviour of birds and anti-predator behaviour of squirrels. *J. Nat. Conserv.* **2008**, *16*, 44–55. [CrossRef]
72. Nonhebel, S. Renewable energy and food supply: Will there be enough land? *Renew. Sustain. Energy Rev.* **2005**, *9*, 191–201. [CrossRef]

73. Van de Ven, D.-J.; Capellan-Peréz, I.; Arto, I.; Cazcarro, I.; de Castro, C.; Patel, P.; Gonzalez-Eguino, M. The potential land requirements and related land use change emissions of solar energy. *Sci. Rep.* **2021**, *11*, 2907. [CrossRef] [PubMed]
74. Statista. Total Wind Energy Capacity in South Africa from 2013 to 2023. Available online: https://www.statista.com/statistics/1218114/south-africa-total-wind-energy-capacity/ (accessed on 24 June 2024).
75. Hernández-Callejo, L.; Gallardo-Saavedra, S.; Alonso-Gómez, V. A review of photovoltaic systems: Design, operation and maintenance. *Sol. Energy* **2019**, *188*, 426–440. [CrossRef]
76. Oliveira-Pinto, S.; Stokkermans, J. Assessment of the potential of different floating solar technologies–Overview and analysis of different case studies. *Energy Convers. Manag.* **2020**, *211*, 112747. [CrossRef]
77. Wang, P.; Yang, N.; Xie, L.; Xu, Y.; He, H.; Fu, Y.; Shen, X. Theoretical Analysis of InGaN Solar Energy Converters Based on Photon-Enhanced Thermionic Emission. *Energies* **2023**, *16*, 3483. [CrossRef]
78. Qiu, H.; Xu, H.; Ni, M.; Xiao, G. Photo-thermo-electric modeling of photon-enhanced thermionic emission with concentrated solar power. *Sol. Energy Mater. Sol. Cells* **2022**, *246*, 111922. [CrossRef]
79. Daniela-Abigail, H.-L.; Tariq, R.; El Mekaoui, A.; Bassam, A.; De Lille, M.V.; Ricalde, L.J.; Riech, I. Does recycling solar panels make this renewable resource sustainable? Evidence supported by environmental, economic, and social dimensions. *Sustain. Cities Soc.* **2022**, *77*, 103539. [CrossRef]
80. Hosseini, S.E. An outlook on the global development of renewable and sustainable energy at the time of COVID-19. *Energy Res. Soc. Sci.* **2020**, *68*, 101633. [CrossRef]
81. Ajayan, J.; Nirmal, D.; Mohankumar, P.; Saravanan, M.; Jagadesh, M.; Arivazhagan, L. A review of photovoltaic performance of organic/inorganic solar cells for future renewable and sustainable energy technologies. *Superlattices Microstruct.* **2020**, *143*, 106549. [CrossRef]
82. Hosenuzzaman, M.; Rahim, N.A.; Selvaraj, J.; Hasanuzzaman, M.; Malek, A.A.; Nahar, A. Global prospects; progress, policies, and environmental impact of solar photovoltaic power generation. *Renew. Sustain. Energy Rev.* **2015**, *41*, 284–297. [CrossRef]
83. Tawalbeh, M.; Al-Othman, A.; Kafiah, F.; Abdelsalam, E.; Almomani, F.; Alkasrawi, M. Environmental impacts of solar photovoltaic systems: A critical review of recent progress and future outlook. *Sci. Total Environ.* **2021**, *759*, 143528. [CrossRef] [PubMed]
84. Mustafa, R.J.; Gomaa, M.R.; Al-Dhaifallah, M.; Rezk, H. Environmental impacts on the performance of solar photovoltaic systems. *Sustainability* **2020**, *12*, 608. [CrossRef]
85. Xu, Z.; Li, Y.; Gao, G.; Xie, F.; Ju, R.; Yu, S.; Liu, K.; Li, J.; Wang, W.; Li, W.; et al. Scalable selective absorber with quasiperiodic nanostructure for low-grade solar energy harvesting. *APL Photonics* **2023**, *8*, 020801. [CrossRef]
86. Divya, A.; Adish, T.; Kaustubh, P.; Zade, P. Review on recycling of solar modules/panels. *Sol. Energy Mater. Sol. Cells* **2023**, *253*, 112151. [CrossRef]
87. Sharma, V.; Aziz, S.M.; Haque, M.H.; Kauschke, T. Effects of high solar photovoltaic penetration on distribution feeders and the economic impact. *Renew. Sustain. Energy Rev.* **2020**, *131*, 110021. [CrossRef]
88. Maka, A.O.; Alabid, J.M. Solar energy technology and its roles in sustainable development. *Clean Energy* **2022**, *6*, 476–483. [CrossRef]
89. Naicker, P.; Thopil, G.A. A framework for sustainable utility scale renewable energy selection in South Africa. *J. Clean. Prod.* **2019**, *224*, 637–650. [CrossRef]
90. Ayamolowo, O.J.; Manditereza, P.; Kusakana, K. South Africa power reforms: The Path to a dominant renewable energy-sourced grid. *Energy Rep.* **2022**, *8*, 1208–1215. [CrossRef]
91. Statista. Total Solar Energy Capacity in South Africa from 2013 to 2023. Available online: https://www.statista.com/statistics/1218115/south-africa-total-solar-energy-capacity/#:~:text=Total%20solar%20energy%20capacity%20in%20South%20Africa%202013-2023&text=During%20the%20period%20under%20review,an%20alternative%20source%20of%20electricity (accessed on 24 June 2024).
92. Umoh, V.; Davidson, I.; Adebiyi, A.; Ekpe, U. Methods and Tools for PV and EV Hosting Capacity Determination in Low Voltage Distribution Networks—A Review. *Energies* **2023**, *16*, 3609. [CrossRef]
93. Umoh, V.; Adebiyi, A.; Moloi, K. Hosting Capacity Assessment of South African Residential Low-Voltage Networks for Electric Vehicle Charging. *Eng* **2023**, *4*, 1965–1980. [CrossRef]
94. Moyo, A. New Portal Displays SA's Solar Installation Data. IT Web, 6 October 2023. Available online: https://www.itweb.co.za/article/new-portal-displays-sas-solar-installation-data/DZQ58MV8eDxvzXy2 (accessed on 29 February 2024).
95. 15 Biggest Solar Projects in South Africa. Solar Feeds. Available online: https://www.solarfeeds.com/mag/biggest-solar-projects-in-south-africa/ (accessed on 4 March 2024).
96. Weekly Data: South Africa's Unprecedented Rooftop Solar Boom. Energy Monitor. Available online: https://rb.gy/lv0yzp (accessed on 4 March 2024).
97. Rao, N.D.; Min, J.; Mastrucci, A. Energy requirements for decent living in India, Brazil and South Africa. *Nat. Energy* **2019**, *4*, 1025–1032. [CrossRef]
98. Tang, A.; Alsultany, F.H.; Borisov, V.; Mohebihafshejani, A.; Goli, A.; Mostafaeipour, A.; Riahi, R. Technical, environmental and ranking analysis of using solar heating: A case study in South Africa. *Sustain. Energy Technol. Assess.* **2022**, *52*, 102299. [CrossRef]
99. Adebiyi, A.A.; Lazarus, I.J.; Saha, A.K.; Ojo, E.E. Performance analysis of 8 kW Grid-tied solar photovoltaic power plant in Durban South Africa. *Int. J. Mech. Eng. Technol.* **2019**, *10*, 82–95. [CrossRef]

100. Redstone Key Facts. SR Energy. Available online: https://sr.energy/redstone/ (accessed on 4 March 2024).
101. Zhang, H.; Baeyens, J.; Degrève, J.; Cacères, G. Concentrated solar power plants: Review and design methodology. *Renew. Sustain. Energy Rev.* **2013**, *22*, 466–481. [CrossRef]
102. Shahabuddin, M.; Alim, M.A.; Alam, T.; Mofijur, M.; Ahmed, S.F.; Perkins, G. A critical review on the development and challenges of concentrated solar power technologies. *Sustain. Energy Technol. Assess.* **2021**, *47*, 101434. [CrossRef]
103. Pavlović, T.M.; Radonjić, I.S.; Milosavljević, D.D.; Pantić, L.S. A review of concentrating solar power plants in the world and their potential use in Serbia. *Renew. Sustain. Energy Rev.* **2012**, *16*, 3891–3902. [CrossRef]
104. Hafez, A.; Soliman, A.; El-Metwally, K.; Ismail, I. Solar parabolic dish Stirling engine system design, simulation, and thermal analysis. *Energy Convers. Manag.* **2016**, *126*, 60–75. [CrossRef]
105. Sonawane, P.D.; Raja, V.B. An overview of concentrated solar energy and its applications. *Int. J. Ambient Energy* **2018**, *39*, 898–903. [CrossRef]
106. Poullikkas, A.; Kourtis, G.; Hadjipaschalis, I. Parametric analysis for the installation of solar dish technologies in Mediterranean regions. *Renew. Sustain. Energy Rev.* **2010**, *14*, 2772–2783. [CrossRef]
107. Cheng, Z.-D.; Leng, Y.-K.; Men, J.-J.; He, Y.-L. Numerical study on a novel parabolic trough solar receiver-reactor and a new control strategy for continuous and efficient hydrogen production. *Appl. Energy* **2020**, *261*, 114–136. [CrossRef]
108. Wang, Q.; Pei, G.; Yang, H. Techno-economic assessment of performance-enhanced parabolic trough receiver in concentrated solar power plants. *Renew. Energy* **2021**, *167*, 629–643. [CrossRef]
109. Price, H.; Beltran, J.; Ramirez, R. Advances in parabolic trough solar power technology. *J. Sol. Energy Eng.* **2002**, *124*, 109–125. [CrossRef]
110. Department of Energy. *State of Renewable Energy in South Africa*; Department of Energy: Pretoria, South Africa, 2015.
111. Miller, J. Promoting Sustainable Development in South Africa: Environmental Regulation in Support of Renewable Energy. Master's Thesis, University of the Western Cape, Cape Town, South Africa, 2016.
112. Baker, L. Commercial-scale renewable energy in South Africa and its progress to date. In *Green Power for Africa: Overcoming the Main Constraints*; Pueyo, A., Bawakyillenuo, S., Eds.; IDS Bulletin: Brighton, UK, 2017; Volume 48.
113. Fluri, T.P. The potential of concentrating solar power in South Africa. *Energy Policy* **2009**, *37*, 5075–5080. [CrossRef]
114. Wang, Z. *Design of Solar Thermal Power Plants*; Academic Press: Cambridge, MA, USA, 2019; pp. 1–46.
115. Weinstein, L.A.; Loomis, J.; Bhatia, B.; Bierman, D.M.; Wang, E.N.; Chen, G. Concentrating solar power. *Chem. Rev.* **2015**, *115*, 12797–12838. [CrossRef] [PubMed]
116. Ummadisingu, A.; Soni, M. Concentrating solar power–technology, potential and policy in India. *Renew. Sustain. Energy Rev.* **2011**, *15*, 5169–5175. [CrossRef]
117. Achkari, O.; El Fadar, A. Latest developments on TES and CSP technologies–Energy and environmental issues, applications and research trends. *Appl. Therm. Eng.* **2020**, *167*, 114–136. [CrossRef]
118. Morningstar, W. *The Physics of Solar Concentration*; Stanford University: Stanford, CA, USA, 2017.
119. Shafiullah, M.; Ahmed, S.D.; Al-Sulaiman, F.A. Grid integration challenges and solution strategies for solar PV systems: A review. *IEEE Access* **2022**, *10*, 52233–52257. [CrossRef]
120. Khodayar, M.E.; Feizi, M.R.; Vafamehr, A. Solar photovoltaic generation: Benefits and operation challenges in distribution networks. *Electr. J.* **2019**, *32*, 50–57. [CrossRef]
121. Prăvălie, R.; Patriche, C.; Bandoc, G. Spatial assessment of solar energy potential at global scale. A geographical approach. *J. Clean. Prod.* **2019**, *209*, 692–721. [CrossRef]

energies

MDPI

Article

Investigating Published Research towards a Fossil-Energy-Free Agriculture Transformation

Athanasios T. Balafoutis [1], Magdalena Borzecka [2], Stelios Rozakis [3,*], Katerina Troullaki [3], Foteini Vandorou [1] and Malgorzata Wydra [2]

[1] Centre of Research and Technology Hellas (CERTH), Institute for Bioeconomy and Agritechnology (iBO), Ethnarchou Makariou 34, 16341 Ilioupoli, Greece; a.balafoutis@certh.gr (A.T.B.); f.vandorou@certh.gr (F.V.)
[2] Department of Bioeconomy and Systems Analysis, Institute of Soil Sciences and Plant Cultivation (IUNG-PIB), Czartoryskich 8, 24-100 Puławy, Poland; mborzecka@iung.pulawy.pl (M.B.); malgorzatawydra@iung.pulawy.pl (M.W.)
[3] Bioeconomy and Biosystems Economics Laboratory, Department of Chemical and Environmental Engineering, Technical University of Crete (TUC), Akrotiri Campus, 73100 Chania, Greece; atroullaki@tuc.gr
* Correspondence: srozakis@tuc.gr

Abstract: The defossilisation of the agricultural sector is driven by intense worldwide academic research on non-fossil, renewable and energy-efficient agriculture, and the acknowledgment of the need for sustainable farming practices. For this purpose, not only technical transformations but also socio-technical system changes towards sustainability need to take place in a co-evolutionary manner. This paper investigates structural and qualitative characteristics of the knowledge produced by research on fossil-energy-free agriculture. We provide evidence on the worldwide research directions, as well as investigate whether academic research and publicly funded research projects foster knowledge creation for the desired transformation. Bibliographic maps are constructed using a query-based methodology as social networks to investigate the efficiency of the EU-funded research to achieve the goals set for the 2050 EU Green Deal. The H2020-funded papers are further analysed with dictionary-based text analysis to quantify the relative emphasis of different types of knowledge in the text. This approach is eventually used to relate transformational capacity to project profiles in the European Union, to evaluate past funding schemes and to improve the shape of future research programs on renewable and sustainable agriculture.

Keywords: agricultural research; sustainable agriculture; fossil-energy-free; defossilisation; transformative knowledge; social network analysis; dictionary-based content analysis

1. Introduction

The world is currently grappling with an unprecedented ecological crisis, influenced greatly by our dependence on fossil fuels. The relentless burning of coal, oil, and natural gas has unleashed a cascade of environmental challenges, from climate change and rising sea levels to biodiversity loss and air pollution. Burning fossil fuels adds billions of tons of CO_2 to the atmosphere each year, making it the main cause of anthropogenic climate change. To combat this crisis, a fundamental transformation is imperative—defossilisation. This process entails shifting away from our reliance on fossil fuels and unsustainable resource exploitation. It demands a transition to cleaner, renewable energy sources, efficient energy use, responsible land use, conservation efforts, and a commitment to sustainable living. Simultaneously, the need arises to reimagine whole economic sectors, such as transportation and agriculture, to reduce carbon emissions.

Global agriculture mainly relies on fossil resources for covering most of its energy needs and supporting agricultural productivity, contributing a significant share to global warming through greenhouse gas (GHG) emissions. Simultaneously, it is vulnerable to the adverse effects of climate change, which can disrupt food production systems, threaten food security, and harm ecosystems.

A series of mitigation and adaptation solutions for the defossilisation of the agricultural sector are being developed, including advanced digital technologies and regenerative practices. Academic research on non-fossil and energy-efficient agriculture has been gaining momentum worldwide, driven by the growing recognition of the need for sustainable farming practices. The European Union (EU) has been particularly proactive in funding and promoting research in this field through various programs, including Horizon 2020 (H2020 in the rest of the paper).

However, in the agricultural sector, there is still a significant gap between such developments and the actual adoption and use of the available tools and practices by the EU farmers. The defossilisation of agriculture is not merely a technological challenge [1]. It is a complex process that entails a transition towards a more sustainable socio-technical system, for which social and technical transformations need to take place in a co-evolutionary manner. For defossilisation endeavours to contribute to overcoming currently unsustainable practices, relevant research and innovations must involve more than alternative raw materials and new technologies. Radical systemic innovations are needed including new institutions, infrastructure, user practices, and mental models [2].

Scientists in the field of sustainability transformations identify the systems' knowledge bases as strong leverage points for transition processes [3]. The knowledge needed for sustainability transitions is complex and transdisciplinary. It has been argued, though, that research and public policy primarily produce knowledge of a techno-economic nature, while lacking knowledge about systems, values, and genuine participation and learning of societies in transition processes. To induce transformative change in the face of wicked problems, a sustainability-dedicated knowledge base is necessary. To this end, three additional types of knowledge need to complement the prevailing techno-economic one: systems knowledge, normative knowledge, and transformative knowledge [4–6].

This paper intends to explore how and to what extent international- and European-level research outcomes attempt to transform global agriculture towards a defossilised era. For evaluating the contribution of research to the fossil-energy-free agriculture transition, the quality of the produced knowledge needs to be assessed as well. Based on this argument, the authors pose the following research questions: (RQ1) "How is research on fossil-energy-free agriculture progressing worldwide and in the EU, specifically?" and (RQ2) "Have academic research and publicly funded research projects fostered knowledge creation for the desired transformation?"

Guided by the above research questions, this work aims at investigating structural and qualitative characteristics of the knowledge produced by research on fossil-energy-free agriculture. To this end, we investigate Scopus indexed journal papers in the period from 2014 to 2020 using a query-based methodology, and bibliographic maps presenting the collected data are created using social network analysis (SNA) for illustration and interpretation purposes. This corpus of papers is then compared with a subset of papers which are funded by European H2020 projects. By focusing on the published outcome of projects funded under the H2020 framework, we investigate the efficiency of the EU-funded research to achieve the goals set for the 2050 EU Green Deal. The H2020-funded papers are further analysed with dictionary-based text analysis to quantify the relative emphasis of different types of knowledge in the text. Furthermore, we plotted the subset of the publications that have transformative character in the landscape of international research mapping in clusters derived from SNA to detect correlations of specific themes in the literature with more transformative potential.

The rest of the paper is organized as follows. Section 1 defines the subject matter, namely defossilisation technologies and strategies in agriculture, and innovative knowledge prerequisites for sustainability. Section 2 describes materials used, and the methodological steps for searching the literature and analysing the international and H2020-funded research corpus with SNA and dictionary-based text analysis, respectively. Next, the analysis results are presented and discussed in Section 3, while Section 4 presents the conclusions and further research directions.

1.1. Theoretical Context

1.1.1. Definition of Fossil-Energy-Free Agriculture and Fossil-Energy-Free Technologies and Strategies (FEFTSs)

Fossil-energy-free agriculture responds to a system where all energy needs for production of primary agricultural goods are derived from renewable sources. However, this system goes beyond just energy origin and attempts to rebuild the energy profile of intensive farms in Europe to reduce the energy consumption for the existing practices or implement new strategies that minimize the energy requirements to produce the same or even higher amounts of goods. The main agricultural subsectors, where energy is extensively used to maintain acceptable production levels, are arable farming, orchards, vineyards, open-field vegetables, greenhouses, and livestock facilities.

Today, such farms could be a reality, as industry and research entities have been developing novel technologies and strategies related to more sustainable energy production, its efficient use, and GHG emission reduction. Specifically for the EU, in the last 30 years, these technologies and strategies were applied in most economic sectors with very positive results, leaving agriculture out of scope because it was not considered a main energy consumer and GHG producer, and because the installation cost of such technologies was excessively high for small/medium farmers due to their low income and high prices of such innovative systems. The wide application of such technologies and strategies has increased their market share, allowing their optimisation for better efficiency and quality coupled with prices reduced to acceptable levels for most energy consumers. Therefore, agriculture has become one of the clients and technology has started to adapt to its needs.

In this paper, we introduce a new term, Fossil-Energy-Free Technologies and Strategies (FEFTSs) for farms, which refers to the tools required to address clean energy production and use in agriculture that can reduce both direct and indirect energy consumption. Renewable energy sources (RESs), biofuel production and use, agricultural machinery using biofuels, electrified implements, and others cover the direct energy reduction target by introducing locally produced energy for the farm's needs. Energy efficiency technologies, agricultural constructions management systems, smart farming technologies, conservation agriculture methods, and best energy management practices for rational use of energy, on the other hand, cover the indirect energy reduction goals by increasing the energy use efficiency through a reduction in agricultural inputs (i.e., fertilisers, pesticides, irrigation water, animal feed) and the introduction of circular use of resources. The benefits of FEFTSs are related to cleaner and more efficient energy production and use, resulting in economic, agronomic, and environmental benefits.

1.1.2. Innovation and Knowledge Dedicated to Sustainability

In classical innovation management theory, any innovation is desirable and inevitable as a sign of progress and growth [7]. The unrestrained pursuit of material and economic growth, however, has long been problematised as a driver of climate change and other ecological degradation. Hence, this interpretation of innovation has inherent tensions with our stated desire for a more sustainable world. Today, sustainability scientists argue for innovation systems dedicated to sustainability [6], where, primarily, innovations that contribute to sustainability transformations are desirable [2,3].

The same reasoning applies to the knowledge needed for sustainability transformations. Sustainability scholars argue for sustainability-dedicated knowledge systems [8], which are validated through the new paradigm of transdisciplinarity [9]. In this paradigm, science is in a dialectic process with society to co-produce 'socially robust knowledge' [10]. The types of knowledge that are required for sustainability transformations have been debated in the literature [4,6,9]. According to a common typology [3,4,8,11], sustainability science needs to foster the creation of the following types of knowledge:

- Systems knowledge (SK) refers to the understanding of complex biological, economic, and social systems, and their interactions [11]. It involves recognizing the relationships, feedback loops, and interdependencies that exist within and between these systems, and considers the nonlinear interactions and unintended consequences that can arise from various interventions and changes within a system.
- Normative knowledge (NK) deals with the ethical and cultural aspects of sustainability. It explores questions related to values, beliefs, and norms of different social groups, and helps in understanding the ethical and cultural implications of different policies, practices, attitudes, or interventions related to sustainability.
- Transformative knowledge (TK) refers to knowledge that has the power to leverage significant and positive change towards sustainability goals. Achieving sustainability goals often requires shifts in perspectives, values, behaviours, and institutions, and transformative knowledge plays a crucial role in facilitating such shifts. Key strategies associated with the generation of transformative knowledge are participation, empowerment, education, and communication [4].
- It has been argued, however, that researchers and policymakers primarily produce and prioritize knowledge of a techno-economic nature, which, while necessary to some extent, is only one part of the knowledge base essential for sustainability transformations [2,3,6,8,12]. In this context, techno-economic knowledge (TEK) [3] refers to understanding the interconnections between technology and economics and is decisive for innovation production and introduction into the market.

It is the combination of the four aforementioned types of knowledge that is arguably necessary for successful interventions in socio-technical systems [3], such as transition processes towards the defossilisation of agriculture. Whether the funding efforts in research have assisted the creation of sustainability-dedicated knowledge beyond purely techno-economic knowledge requires, however, further investigation [12].

2. Materials and Methods

2.1. Data Collection

2.1.1. The Global Literature Corpus

A literature review was implemented to investigate international- and European-level research outcomes targeted towards the defossilisation of the agricultural sector by introducing FEFTSs. Peer-reviewed papers related to FEFTSs were found following a Scopus database search through queries from 2014 to 2023. Only papers that were published from 2014 (the starting date of H2020 projects) and onwards were taken into consideration as we aimed to determine if there is a correlation between the international research and H2020-derived papers.

The search aimed at finding articles related to FEFTSs (as explained in Section 2.1) that can (or could be) used by agricultural practitioners. This procedure entailed the creation of a pool of available FEFTSs by using specific keywords based on a categorisation consisting of four main categories: Energy User/Consumer, Clean Energy Supply, Energy Efficiency Improvement, and Soil Carbon Sequestration, divided into specific sub-categories. The hierarchical categorisation and Level 1 and 2 subcategories are presented in Table 1.

With each addition of a new term, the results were better refined. The majority of the queries contained the word "agriculture" and then, based on the four main categories, corresponding keywords were used (i.e., the Renewable Energy Sources category contained the word "energy"). It should be highlighted that the queries searched for the aforementioned keywords in the title, keywords, and abstract of the paper only. The full list of queries used in this review is given in Table S1. Through this query-based search, initially, 6621 scientific publications were collected. This global literature corpus is onwards referred to as the "global corpus".

Table 1. FEFTSs categories and subcategories.

FEFTSs Category	Level 1 Sub-Category	Level 2 Sub-Category
Energy User/Consumer	Agricultural technology applications	heating and cooling of buildings process heat/cold lighting agricultural field practices vehicles tools energy sales to external consumers
Clean Energy Supply	Renewable Energy Sources	solar, wind, hydro geothermal bioenergy free energy
	Energy types	heating cooling electricity mechanical energy chemical energy
	Energy Technologies	photovoltaics solar thermal windmills hydropower heat pumps geothermal solid biomass conversion biogas/biomethane production liquid biofuels production
	Energy Storages	heat storage electricity storage cold storage intermediate bioenergy carriers
Energy Efficiency Improvement	Energy savings	efficient buildings efficient vehicles efficient tools precision agriculture precision livestock farming conservation agriculture
Carbon sequestration	Carbon sequestration	soil organic cover tillage (conservation agriculture + CTF) nutrient management crop diversification soil and water conservation techniques fire management grassland management

2.1.2. The European Literature Corpus

In order to perceive the impact of H2020-funded projects on agricultural defossilisation knowledge through scientific papers, first, we identified such projects in the CORDIS database. The methodology used for the identification of these projects is similar to the search queries methodology used for the scientific papers. By using the categorisation in Table 1, queries for the CORDIS database were created. The full list of queries used for the research projects is given in Table S2. Regarding the selection made amongst the available projects found, it was performed in relation to the total number of projects found from each query. Queries regarding solar energy and biomass yielded the majority of results, whereas aerothermal and hydro energy-related projects were scarce. For the purpose of our study and due to limited resources, we selected around 10% of the number of projects found per query, after reading their titles and short descriptions in the CORDIS database, in order to include the most relevant projects to all FEFTS types.

Additionally, we only kept research projects that were funded under the H2020 scheme and had published at least one peer-reviewed article. In total, 345 papers derived from those projects were collected. This corpus of papers derived from H2020 projects is onwards referred to as the "European corpus". The next sub-sections present how a combination of methods were employed to analyse the global and European corpora of papers on fossil-energy-free agriculture. Our methodological procedure is summarised in Figure 1.

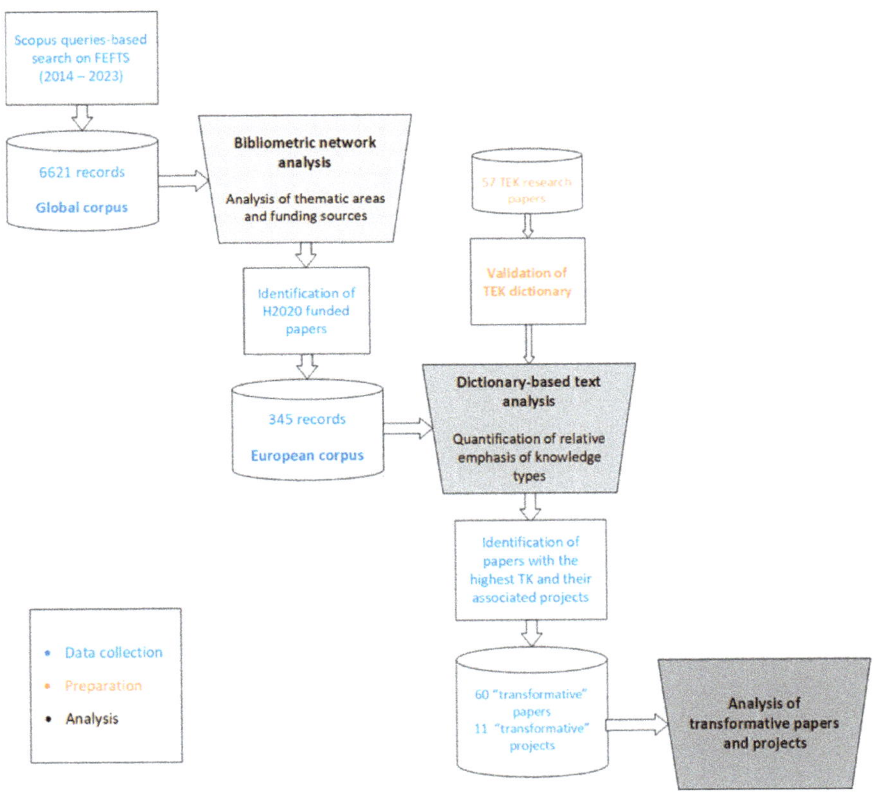

Figure 1. Methodological steps followed.

2.1.3. Social Networks for Bibliographic Analysis

Analysing the international literature corpus of FEFTSs may offer a bird's eye view on the evolution and dissemination of knowledge in this field. Bibliometric analysis techniques enable researchers to analyse such large volumes of research from the literature, mainly through the application of statistical indicators on bibliographic data. Bibliometric analysis provides valuable insights into the productivity, impact, structure, relationships, and evolution of research activities in various fields [13].

Bibliometric network analysis specifically focuses on analysing the relationships and patterns within bibliographic data (such as citations, co-authorships, co-citations, and keywords) to gain insights into the structure and dynamics of scholarly communication networks. It involves identifying national and international networks, as well as mapping the development of new transdisciplinary fields of science and technology through statistical indicators from the scientific literature on the productivity of individuals, groups, institutions, and countries. Bibliometric network analysis is widely used in academic research, policymaking, and information retrieval to identify influential papers, track research trends, evaluate the impact of research, and make strategic decisions related to research funding and collaboration. By visualizing and quantifying relationships between scholarly publications, researchers can gain valuable insights into the evolution and dissemination of knowledge within specific fields of study [14].

A bibliometric network is defined as the visual representation, by means of nodes and links, of complex meanings with multi-level influences that allows us to transform the quantitative bibliometric information into qualitative conclusions. There are different types of analysis that are used to determine the relatedness of the network's terms [15], including citation analysis, co-citation analysis, bibliographic coupling, co-occurrence analysis, and co-authorship analysis [16].

Specifically, the present investigation concerns the following:

- A keyword's co-occurrence network to identify thematic areas and the evolution in time of the international research on FEFTS. Analysing keywords assigned to publications can help identify emerging trends, topics, and the evolution of research areas over time. Nodes represent keywords, and their connections indicate co-occurrence in documents. The relatedness of items is determined based on the number of documents in which they occur together.
- A funding organisation's co-authorship network to see which organisations co-fund research on FEFTSs. Studying co-authorship patterns helps visualize collaborations among researchers. Nodes represent authors, and edges represent co-authorship relationships. The relatedness of items is determined based on their number of co-authored documents. Analysing this network can reveal important research collaborations and trends.

These networks are constructed with the help of the open-source software VOSviewer 1.6.20 (www.vosviewer.com, accessed on 4 March 2024) [14,15]. The global literature corpus of FEFTSs in csv format is inserted into the software, which creates the maps. VOSviewer 1.6.20 creates distance-based networks, meaning that the distance between two nodes approximately indicates relatedness. In VOSviewer 1.6.20, the weight of each node can be set to represent different variables, e.g., the node's number of occurrences, citations, or links. The constructed networks can be viewed with different visualisations to support different interpretations. We have used two types of visualisations:

- the network visualisation, whereby nodes are grouped in clusters of different colours based on their relatedness, in order to identify thematic areas of the literature;
- the density visualisation, whereby each point has a colour that indicates the density of items at that point, in order to emphasize network areas with greater productivity (e.g., of authored documents).

2.1.4. Dictionary-Based Text Analysis of Scientific Papers

The corpus of publications from H2020-funded projects is analysed with dictionary-based text analysis to investigate the relative emphasis of different types of knowledge in these publications. Dictionary-based text analysis is a natural language processing technique used to analyse and extract information from text using predefined dictionaries or lexicons [17]. These dictionaries contain lists of words or phrases, often categorized into different themes or sentiments. The goal of dictionary-based text analysis is to identify and classify words or phrases in a text document based on their presence in these dictionaries. The following steps are typically involved: (1) dictionary creation, (2) text pre-processing, (3) matching words, and (4) analysis and interpretation.

To quantify the relative emphasis of each type of knowledge in these publications, we use lexicons of indicator words available in the literature. Abson et al. [4] have identified vocabulary related to different aspects of sustainability based on an 'expert', top-down consensual process. They thus compiled lexicons for normative knowledge (NK), systems knowledge (SK), and transformative knowledge (TK). Following this publication, Bogner and Dahlke (2022) complemented the three lexicons with one for techno-economic knowledge (TEK), by analysing techno-economic papers and identifying word co-occurrence with the word "knowledge". They then used the four lexicons to assess the types of knowledge present in bioeconomy strategy papers.

In this paper, we use the four lexicons to assess the types of produced knowledge in fossil-energy-free agriculture papers. Specifically, the SK, TK, and NK dictionaries have been directly applied in our analysis without modifications. We decided, however, to validate the TEK dictionary because it had been compiled with a different methodology than the other three dictionaries and comprised significantly more keywords. We used the following approach:

(1) For this purpose, a sample of techno-economic papers was identified within the global corpus of papers on fossil-energy-free agriculture. This way, 57 papers that included the term "techno-economic" in their title were collected.
(2) These papers and the original TEK dictionary were entered in content analysis software (WordStat 9—https://provalisresearch.com/products/content-analysis-software/ accessed on 4 March 2024). For each keyword in the TEK dictionary, the software counted the number of papers in which it was present.
(3) We eliminated less significant keywords by testing two different thresholds; keywords that were present in at least the threshold number of papers were kept in the dictionary. Eventually, a threshold of 10 papers was applied. The authors eliminated a few extra keywords which were deemed as too generic, thus incapable of signifying any specific type of knowledge. In the end, 97 out of 174 keywords were kept in the modified TEK dictionary.

The four dictionaries that have been used in our analysis—including the modified TEK dictionary—are presented in Tables S3–S6. Next, we used the open-source content analysis software Yoshikoder v0.6.4 (https://www.yoshikoder.org/ accessed on 4 March 2024) to perform the dictionary-based text analysis. The four dictionaries, as well as the 345 H2020 papers, were entered in the software. To improve the accuracy of the text analysis, the following parts of the texts have been deleted: references, affiliations, and the journal's generic information.

2.1.5. Analysing Transformative Papers and Projects

Next, in order to check the correlation between the worldwide and the European research, we proceeded to characterize the EU-derived papers based on thematic areas of the global literature. To this end, we reviewed the titles, keywords, and abstracts of the most 'transformative papers' (those with the highest percentage of TK) and matched them to the previously presented SNA thematic clusters (or their sub-sections) to understand whether specific themes among the clusters relate better to the production of transformative knowledge. We also performed a concordance check with the Yoshikoder software to exclude any papers that were rated with high transformative knowledge but the keywords identified as 'transformative' do not have the intended meaning (for example, "machine learning" is not the intended form of learning to justify transformative knowledge production).

Based on the number of 'transformative' papers produced, the most 'transformative' projects also emerged. These projects were investigated for correlations that justify more transformative potential. The analysis was performed based on the available data regarding the project coordinator, partners, activity type of the coordinator, amount of the partners' budgets, funding scheme, and programme. Detailed characteristics of the analysed projects according to the above criteria is presented in Section 3.2.3.

3. Results and Discussion

3.1. *Analysing and Mapping International Research on Fossil-Energy-Free Agriculture*

3.1.1. Statistical Indicators

The total outcome of the queries used to obtain the global literature corpus included 6621 papers. In Figure 2, the distribution of the number of scientific papers on FEFTSs found for each Level 2 query is presented. It is evident that, regarding energy use in agriculture, biomass (including liquid biofuels and biogas/biomethane) and solar technologies are the dominating renewable energy sources, whereas heat pumps, precision agriculture, and e-storage provide evidence on the emphasis given to energy efficiency/saving technologies. It is interesting to point out that conservation agriculture is highly ranked, as it is a standalone innovative strategy that acts as an auxiliary pillar to assist defossilisation in both direct (reduction in fossil fuels consumed by field machinery) and indirect (reduction in agricultural inputs) energy consumption. This supports the fact that, in the past decade, there has been dedicated research directed to energy-related issues covering most important FEFTS types.

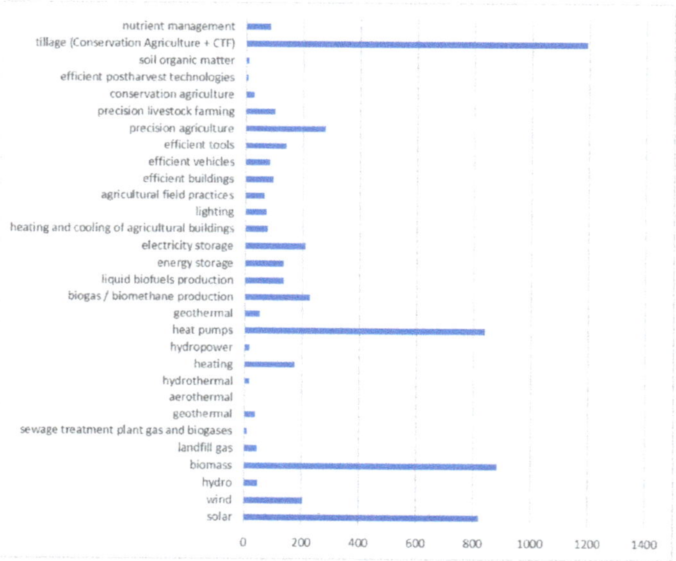

Figure 2. Distribution of the Level 2 queries results categorisation for the scientific papers.

3.1.2. Keywords Co-Occurrence Network

Through the network visualisation in VOSviewer 1.6.20, the keywords of the international literature are visualized in clusters, indicating thematic areas of research (Figure 3). An in-depth description of the observed clusters follows.

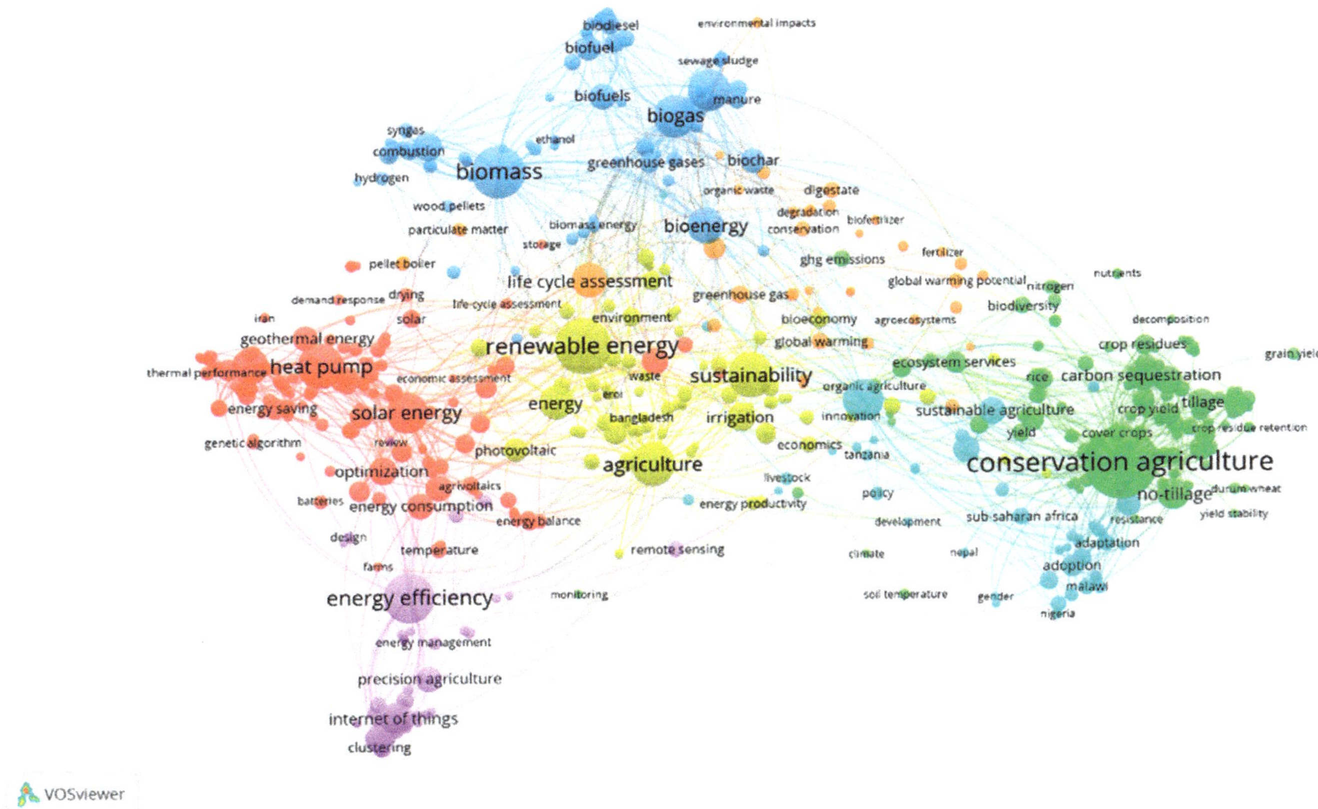

Figure 3. Keyword co-occurrence network of the global literature corpus of FEFTSs. The map has occurred with the following configurations on the software: a threshold of minimum 7 occurrences for a keyword to be included in the network; the value of 0.9 for the analysis resolution; the values of 1 and 0, respectively, for the attraction and repulsion parameters.

1. Blue Cluster: Biomass Conversion into Energy and circular by-products

This cluster of keywords refers to scientific articles related to the use of biomass of any kind for energy production, mainly heat and secondly power. However, since the clustering has been performed on recent papers of the last decade, the cluster also contains a lot of research activities that attempt to gain valuable by-products. More particularly, biomass remains the core of the cluster, but anaerobic digestion and biogas (acting as almost synonymous) play a significant role as large nodes of the network, also connected to bioenergy as a separate node. This is expected, since

- livestock facilities have identified their waste as a valuable feedstock for biogas production, leading researchers to work a lot with different waste qualities to optimize the biogas production and composition, and recently even with new technologies to upgrade biogas into biomethane (for direct use as fuel or injection into the natural gas grids), and
- agriculture has shifted significantly into using agricultural residues (e.g., straw) or other products (e.g., grass) as new feedstock for biogas production and a lot of research is conducted on the optimum conditions for existing digesters to host such feedstock.

The remaining digestate has also been considered as an important asset for farmers, as it can be further processed to produce bio-fertilisers and soil amendments (e.g., compost) and be returned into the land that originally produced the feedstock for direct (straw and grass) and indirect (feed for livestock) biogas production.

In smaller but significant nodes, the cluster contains other forms of biomass conversion into energy. Biofuels, mainly in their liquid form (e.g., biodiesel, bioethanol, bio-oil) have been playing an important role in the agricultural input of renewable energy, and research continues to identify ways to produce and use biofuels in a more sustainable way (e.g., biorefineries) through combining extra by-products (e.g., natural colorants). Gasification comes as a technology of huge potential to convert biomass of different kinds into gaseous fuels (syngas), and it has been continuously researched for decades, due to the fact that a single recipe with gasification specifications does not exist, leading to different solutions for other biomass types, including the idea of torrefaction to produce a middle product of unified characteristics to be later gasified. By-products, such as biochar, are also a part of the gasification process, increasing the research conducted in this sector.

2. Green Cluster: Conservation Agriculture and Its Impacts

Conservation agriculture is the key node of this network cluster. It is a well-known term that refers to an agricultural production system which favours minimum mechanical soil disturbance and, therefore, redirects agricultural practices from conventional mechanised farming. It follows three main pillars: (i) application of zero tillage; (ii) continuous soil coverage with either cover crops or residues; and (iii) crop rotation that allows soil to recover from intensive agriculture and remain healthy and fertile. These principles are given in the cluster image with many small nodes that refer to the principles themselves, which are very closely related to each other (e.g., no-tillage, conventional tillage, conservation tillage, zero tillage, etc.) and their impacts (e.g., nitrogen use efficiency, energy use efficiency, soil degradation, soil quality, crop yield, etc.). Research on conservation agriculture is continuous and, especially in Europe where the application of this production system is relatively low, there is much room for applied research to prove the benefits of its application.

An important keyword node, closely connected to conservation agriculture, is soil carbon sequestration, which is surrounded by smaller nodes concerning soil health, crop residues, tillage systems, and nutrients. This comes as a result of the huge potential of soils to act as carbon pools and naturally absorb a portion of the anthropogenic CO_2 to assist in the battle against climate change and adaptation and, at the same time, help soils to improve their characteristics and maintain their productivity for the years to come with less fertilising requirements (indirect energy input). Therefore, the fact that research is directed to this sector seems to be natural.

Finally, it is interesting to point out that ecosystem services (e.g., carbon farming) seem to have a significant role in recent research output. They offer a very good financial incentive for farmers, encouraging them to shift to conservation agriculture to simultaneously reduce direct (machinery fuels) and indirect (fertilisers reduction) energy consumption, and assist their soil to recover its structure and organic matter through the sustainability of underground flora and fauna.

3. Light Blue Cluster: Climate Change and Food Security

This cluster attempts to connect research for climate change and food security. Climate change is indeed directly connected to food security due to extreme climatic events (e.g., long drought periods or instant heavy rain, hail etc.) that are difficult to be confronted, especially in agricultural production systems that are not technologically advanced, like the ones in developing countries (e.g., sub-Saharan Africa). Therefore, research into climate change's impact on agriculture in these countries has significantly evolved, and sustainable agriculture and intensification are considered as important nodes of research to cover the gap between current production systems and future intense climatic conditions. Adaptation, resilience, and resistance close the loop of research actions to avoid food crises. It is interesting to stress the very close connection to conservation agriculture (green cluster), which proposes a sustainable production system that reduces human disturbance to the agricultural environment.

4. Orange Cluster: Life Cycle Assessment for Agricultural Systems

The process of life cycle assessment (LCA) has been applied in all production systems and tries to use data from all processes in the production line to identify the most important environmental impacts (including energy consumption) that accompany a product in its life cycle. Recently, LCA is complemented by life cycle costing (LCC) and social impact analysis, in order to have a holistic analysis of a product's impact. Agriculture has also been using this tool and research has grown significantly in this sector. This cluster refers to the application of LCA in agriculture, and the main outputs that interest researchers are the measurement of global warming potential and greenhouse gas emissions.

5. Purple Cluster: Energy Efficiency Through Smart Agricultural Technologies

This cluster connects energy efficiency in agricultural production systems with smart farming. Precision farming, Internet of Things (IoT), and wireless sensors are tools to receive a continuous data flow from a field during a production season in order to identify commonalities and differences on a spatial and temporal basis and assist in decision making regarding agricultural practices. Therefore, smart farming indeed has a huge potential to increase energy efficiency of a farm, as it helps reduce the number of agricultural practices (machinery fuel reduction) and more importantly the agricultural inputs (fertilisers, pesticides, water) applied in the field. Therefore, a lot of research is conducted in this direction, having energy efficiency in the core of the activity.

6. Red Cluster: Heat Pumps and Solar Technologies in Agriculture

Solar technologies have been applied in agricultural systems to a large extent, which can be clearly seen in this cluster. It is expected that optimisation of these systems is constantly performed, and research plays its role in this regard, to become more efficient and to reduce the energy consumption of farms and the environmental impact of agriculture. However, heat pumps and ground source heat pumps, which are the largest nodes of this cluster, are not considered so closely related to agriculture, and yet this result shows that probably most agricultural constructions (greenhouses, livestock buildings, silos, etc.) are gaining interest from engineering research to apply heat pumps for climate control (heating, ventilation, air conditioning). For both solar and heat pumps, the cluster shows an interest in research using simulation tools (e.g., TRNSYS) and computational fluid dynamics, but also in important issues to help or hinder adoption, which are energy, environmental, and economic evaluation. The solar technologies are in close vicinity to the yellow cluster, as they are a part of the renewable energy technologies.

7. Yellow Cluster: Renewable Energy for Sustainable Agriculture

This cluster occupies the central position in the network, connecting all the rest, because the scientific articles selected for this analysis focus mainly on clean energy production in agriculture. Therefore, it can be seen that there are numerous small nodes connected to the three main nodes of renewable energy, sustainability, and agriculture. These, in combination, explain the research attempts to change agricultural production to a more sustainable system in terms of energy and water use, and also to reduce the greenhouse gas emission share of agriculture, as this main node of the red cluster is exactly in the centre. Hence, all renewable energy applied in agriculture is always researched in terms of being environmentally friendly.

3.1.3. Funding Organisations Co-Authorship Network

Through the density visualisation in VOSviewer 1.6.20, the funding organisations of the international literature are visualized in a coloured network, whereby the red colour indicates areas with a high concentration of publications (Figure 4).

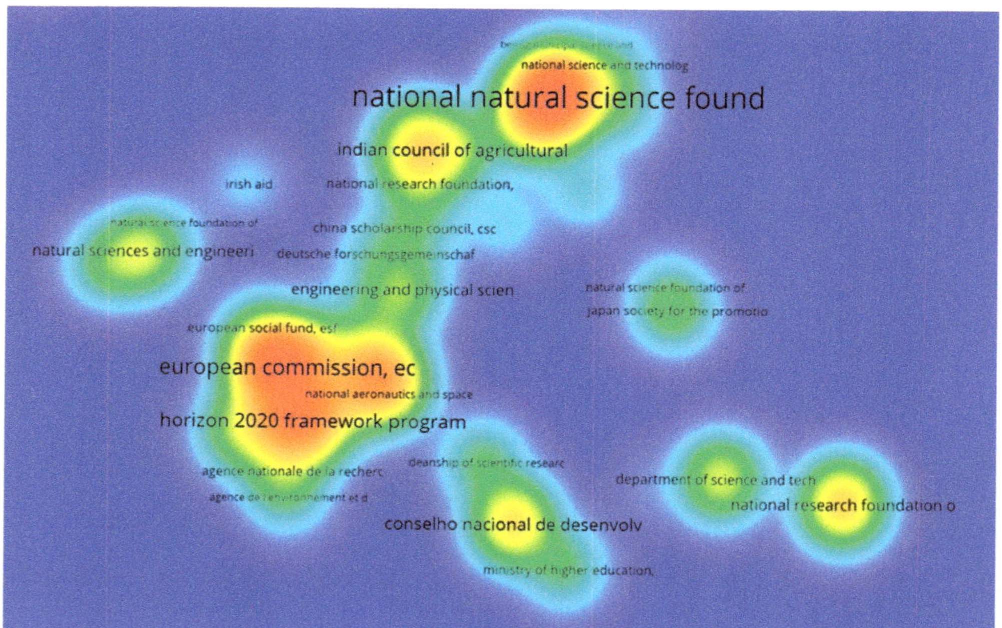

Figure 4. Funding organisations co-authorship network in the global literature corpus of FEFTSs. The density visualization has been applied, which provides a quick overview of the main areas in the network. Each point in the map has a color (in-between red and blue) that depends on the density of items at that point. The larger the number of items in the neighborhood of a point and the higher the weights of the neighboring items, the closer the color of the point is to red. Conversely, the smaller the number of items in the neighborhood of a point and the lower the weights of the neighboring items, the closer the color of the point is to blue. The map has occurred with the following configurations on the software: a threshold of minimum 2 documents and a threshold of minimum 100 citations for an organisation to be included in the network; the values of 1 and 0, respectively, assigned to the attraction and repulsion parameters.

Following the National Natural Science Foundation of China, the European Union (represented mainly by the nodes "European Commission", "Horizon 2020 framework program", and "7th framework program") concentrates a large number of funded publications on fossil-energy-free agriculture. Since the EU arises as a key actor in funding research on fossil-energy-free agriculture, in the next phase, we focus our analysis on the papers produced by H2020-funded projects to investigate the quality of the produced knowledge in these projects.

3.2. Focus on EU Results

3.2.1. Research Project Statistics

Our search in the CORDIS database according to the aforementioned methodology yielded 156 research projects, out of which 37 were funded under the H2020 framework and had published peer-reviewed articles. The results are presented in detail in Table 2.

Table 2. Analysis of the selected research projects.

Type of Research Project	Number
Total number of selected research projects acquired from CORDIS database	156
H2020 Projects	95
H2020 Projects that produced papers (peer-reviewed and conference papers)	40
H2020 Projects that produced peer-reviewed papers	37

https://ec.europa.eu/research/participants/docs/h2020-funding-guide/grants/applying-for-funding/find-a-call/what-you-need-to-know_en.htm accessed on 4 March 2024.

Different action types specify H2020 objectives, including Research and Innovation Actions (RIA), Innovation Actions (IA), Coordination and Support Actions (CSA), European Research Council (ERC), Marie Skłodowska-Curie Actions, SME Instruments, and Fast Track to Innovation (FTI). Regarding the action type of the 37 analysed projects, most are funded by RIA (14) and by IA (13), followed by seven CSA, one SME-2, one ERC-STG, and one ERA-NET—Cofund. This result is presented in Figure 5. The full list of research projects identified in CORDIS is available in Table S7.

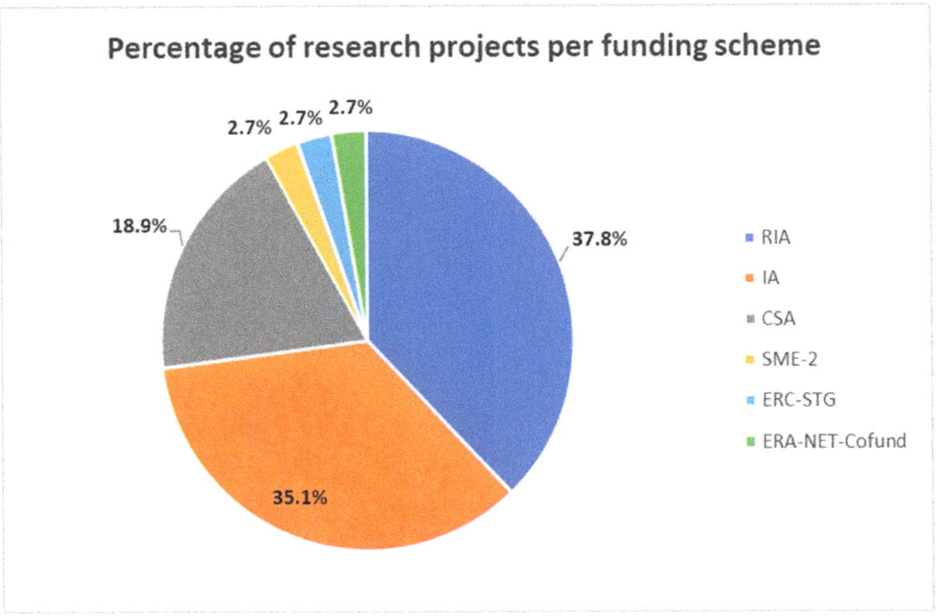

Figure 5. Pie chart of the percentage of research projects that produced peer-reviewed papers, per funding scheme.

By examining the scientific papers produced by the aforementioned projects, we acquired 345 peer-reviewed articles to be analysed further. The final numbers are presented in detail in Table 3.

Table 3. Analysis of the scientific papers derived from the chosen research projects.

Type of Paper Derived from Projects	Number
Papers derived with funding number and project name from Scopus	507
Irrelevant papers found	78
Conference papers	81
N/A papers	3
Peer-reviewed papers to be analysed	345

In their majority, these papers stem from RIA (220), followed by IA (85), CSA (25), ERA-NET—Cofund (13), ERC-STG (2), and SME-2 (1) projects. The results are presented in Figure 6. In Table S8, the full list of the research papers identified from the 37 research projects is presented. This result seems reasonable, as indeed RIA projects are dedicated to basic and applied research trying to bring novel ideas to prototyping, while IA projects focus more on demonstrating existing technologies, optimising them and bringing them closer to commercialization. On the other hand, CSA projects do not focus directly on research results but attempt to develop the conditions for specific scientific themes to become more known and integrated into real life, by developing mainly policy and data analysis papers. ERA-NET, ERC-STG, and SME-2 funding programmes are not represented with a large number of projects, which leads to a small number of published papers as well.

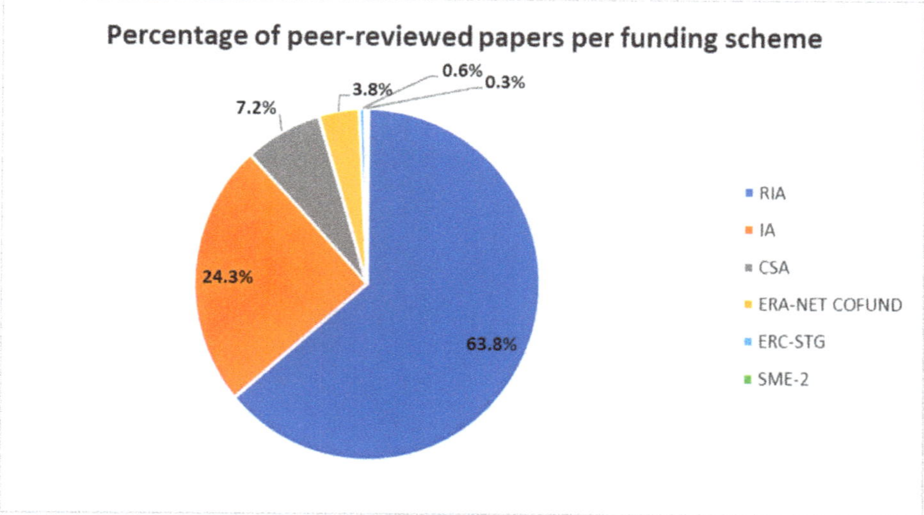

Figure 6. Pie chart of the percentage of peer-reviewed papers per funding scheme.

3.2.2. Lexicographic Focus to Explore the Quality of the Produced Knowledge

Next, the results of the dictionary-based text analysis of the European literature corpus are presented. For each paper, the content analysis software counted how many keywords from each dictionary are present. This resulted in a percentage allocated to each type of knowledge (normative, techno-economic, transformative, or systems knowledge) within each paper, as well as an average percentage among all papers. This analysis, presented in Figure 7, shows that research papers on fossil-energy-free agriculture are dominated by techno-economic knowledge, which averagely accounts for 58.5% of all dictionary words found in the papers and, therefore, outperforms the other categories altogether. Systems knowledge averagely accounts for 20.7%, normative knowledge for 12.5%, and transformative knowledge for 8.3% of the detected dictionary words.

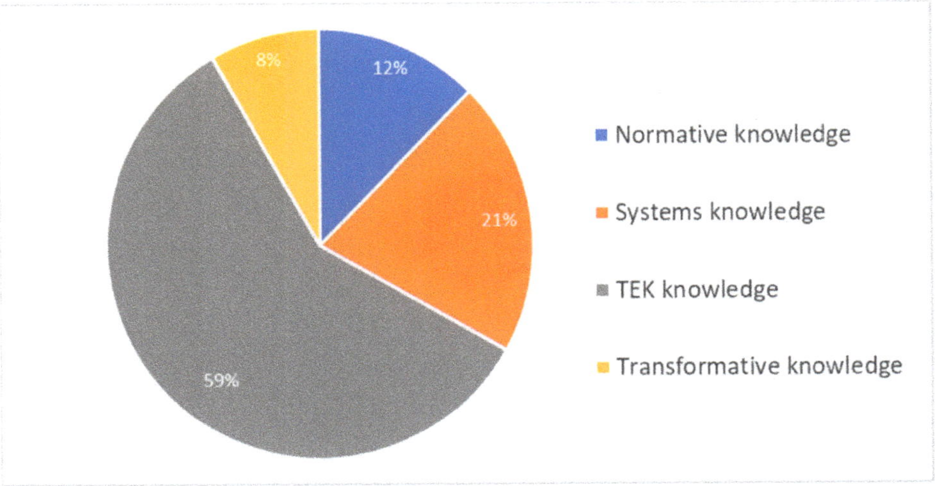

Figure 7. Relative presence of knowledge types in fossil-energy-free agriculture research papers.

A deeper examination of the content analysis results further reveals the dominance of TEK, even horizontally across the other knowledge types. Each knowledge type comprises four sub-categories, which represent different aspects of the main theme. For example, transformative knowledge consists of four elements: policy and decision making; participation; motivation; communication and education. Figure 8 shows in detail the relative presence of the sub-categories in the analysed papers. We may observe that, for each knowledge type, the sub-categories that are more relevant to the dominant techno-economic paradigm are the most prominent: "efficiency" in NK, "market" and "technology" in TEK, "policy and decision making" in TK, and "function and process" in SK. On the contrary, the least numerous sub-categories (less than 2% of total dictionary words) are "justice and ethics" in NK, "resilience" in SK, "motivation" in TEK, and "communication and education" in TK. Thus, the most transformative dimensions of each knowledge type are the most under-represented.

This outcome suggests that the EU-funded research products in fossil-energy-free agriculture have favoured the production of techno-economic knowledge and have hardly embraced a more transdisciplinary and transformative paradigm. While a sustainability-dedicated knowledge base requires all the four aforementioned knowledge types, our analysis shows that transformative and normative knowledge are rather poorly represented. This finding suggests that the produced knowledge, while improving technological status and economic performance, may not be transformative enough to trigger transitions to alter the regime.

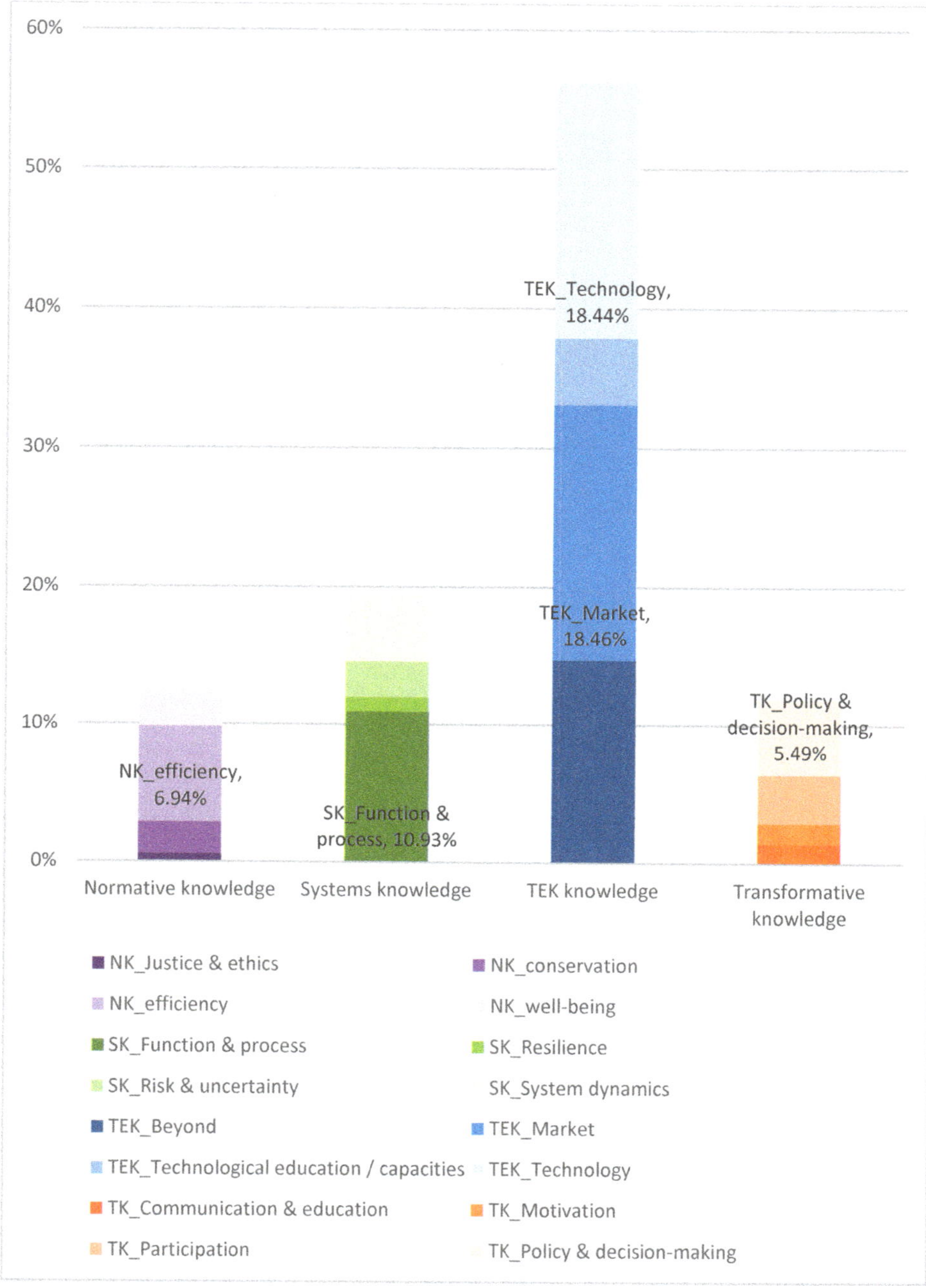

Figure 8. Knowledge type sub-categories in fossil-energy-free agriculture research papers.

3.2.3. Focus on Transformative Papers and Projects

Among the H2020-derived papers, which have been analysed with dictionary-based text analysis, we further examined those with the highest percentage of transformative knowledge. Specifically, 60 papers with the highest TK percentage were selected, ranging from 16.7% to 61.1% of transformative knowledge content. Through the concordance check performed with Yoshikoder, four of these papers have been excluded as irrelevant. The remaining 56 papers were analysed to identify potential correlations that explain their increased content of TK compared to the rest of the papers. The aforementioned papers are highlighted with yellow in Table S8.

An initial analysis shows that these papers are not concentrated in specific journals; rather, they are published in a variety of journals. The publishing year does not seem to be an important factor, as there is no tendency in more recent papers to contain higher (or lower) percentages of transformative knowledge. Unlike what the recent literature suggests [18], our analysis shows that research on sustainable agriculture has not yet shifted away from the techno-economic focus.

Instead, there seemed to be specific H2020 research projects that have produced a large number of 'transformative' papers. To further investigate this correlation, we associated each of these transformative papers with a corresponding H2020 research project. It should be mentioned that, from our analysis, we came across eight research projects that were not included in our original list of projects but were under the H2020 framework, as each of these projects has produced a paper jointly with another project that was already included in our analysis. The projects were removed from our analysis as they did not meet the criteria set in our methodology and the papers produced from them were already counted and analysed. These projects are GENIALG (two papers with the MacroFuels project), LANDSUPPORT (one paper with the SOILCARE project), LANDMARK (one paper with the SOILCARE project), CONSOLE (one paper with the LIFT project), INVADE (one paper with the SET-Nav project), TRANSrisk (one paper with the REEM project), and SolACE (one paper with the SOILCARE project). Furthermore, the NEFERTITI project co-produced one paper together with two other projects on our list (IOF2020 and SmartAgriHubs). For the purposes of our analysis, this paper was removed both from the NEFERTITI and IOF2020 projects, with the latter being chosen as it produced more papers than SmartAgriHubs. As a result, a list has been created with H2020 projects sorted by the high-TK papers they have generated, which is presented in Table 4.

Table 4. H2020 projects with high TK% papers (TK $\geq$ 25%).

Project Name [1]	Number of Papers	Project's High-TK/Project's Total # of Papers
LIFT	8	42.11%
SOILCARE	7	15.91%
IOF2020	5	42.11%
BioEcon	3	23.08%
SET-Nav	3	21.43%
WiseGRID	2	100.00%
MacroFuels	2	22.22%
REEEM	2	7.14%
inteGRIDy	1	6.25%
ISAAC	1	50%
Smart-AKIS	1	25%

[1] All of these projects are represented in this table with their acronyms but detailed information about them is given in Table S9.

The subjects of all projects with the largest number of TK papers and their characteristics are presented in Table S9. As this investigation is a fortiori policy-oriented, we meant to relate TK outcomes of projects to the action types that characterize H2020 projects, and the respective funding modes. In Table 5, it can be seen that the majority of the projects with a high TK are RIA projects, followed by IA and CSA. We observe that CSA projects have a higher percentage of transformative papers among their total produced papers (16%), while this percentage for RIA and IA projects is 3% and 5%, respectively.

Table 5. Number of research projects per action type for the different TK thresholds.

Action Type	TK ≥ 25%	TK ≥ 20%	TK ≥ 16.7%	% of Total Papers
RIA	5	5	6	2.7
CSA	3	3	4	16
IA	3	4	4	4.7

This can be explained by the fact the CSA funding primarily covers accompanying measures concerning policy dialogues and mutual learning exercises and studies, as well as complementary activities of strategic planning, networking, and coordination between programmes in different countries. Thus, research and innovation outcomes from various projects are compiled in the frame of CSA funding for dissemination, raising awareness, and communication to ultimately support policy making. On the other hand, RIA and IA projects aiming at producing new knowledge/technology and new or improved products, respectively, have resulted in a higher absolute number of TK papers, along with a much higher number of total produced papers compared with CSA projects.

Therefore, scientific papers with high TK can be correlated with RIA projects as these projects' main goals are to establish new knowledge and, at the same time, explore the feasibility of new/improved technology. Although the large part of research outcomes of RIA projects pertains to knowledge types such as systems knowledge, pointing out complex system interactions, and techno-economic knowledge, evaluating economic feasibility of technology as a prerequisite for innovation, strong concern for transformative knowledge can also be observed. In fact, RIA projects first undertake a thorough analysis to assess the status of the studied subject, and then new methodologies are produced, so as to improve the situation and introduce new technologies. For example, the LIFT project, evaluated as one of the most transformative projects in terms of its produced papers, investigated the potential benefits of the adoption of ecological farming in the EU through 30 case studies and then promoted the performance and sustainability of such systems across the EU.

Finally, we checked the correlation between the worldwide and the European research by characterizing the most 'transformative' EU-derived papers based on the previously identified clusters of the global literature. To this end, we reviewed the titles, keywords, and abstracts of these papers and matched them to the previously presented thematic clusters (or their sub-sections) to understand whether specific themes among the clusters relate better to the production of transformative knowledge. The papers were categorised according to the closest relation to the clusters, as shown in Table 6.

Table 6. Number of research papers related to the clusters defined from the international literature review.

Cluster Name	Number of Papers Mainly Related to the Cluster
Blue Cluster: Biomass conversion into energy and circular by-products	4
Green Cluster: Conservation agriculture and its impacts	8
Light Blue Cluster: Climate change and food security	17
Orange Cluster: Life cycle assessment for agricultural systems	1
Purple Cluster: Energy efficiency through smart agricultural technologies	6
Red Cluster: Heat pumps and solar technologies in agriculture	0
Yellow Cluster: Renewable energy for sustainable agriculture	18
Non-Related	3

It can be observed that the majority of TK papers devoted to the subject of fossil-energy-free agriculture refer to renewable energy systems (Yellow Cluster) and the cluster of biomass conversion into energy (Blue Cluster), which in reality is a sub-category of the former as they both refer to direct energy production systems in farms. These categories of FEFTSs cover the direct energy impact of agriculture, which was expected to occupy the interest of the TK papers under consideration. Climate change (Light Blue Cluster) follows, and conservation agriculture (Green Cluster) comes next, both of which are also closely interrelated, as they refer to strategies and practices of sustainable and ecological agricultural production. Smart agricultural technologies (Purple Cluster) close the analysis, showing that the three latter clusters cover all FEFTSs concerning a reduction in indirect energy consumption in agricultural practices. Life cycle assessment of agricultural systems (Orange Cluster) and heat pumps (Red Cluster) have negligible participation in the selected papers, showing that environmental impact is still not in the core of agricultural transformation and that heat pumps are very innovative and specific to buildings' FEFTSs, making them still non-relevant for the immediate FEFTS integration into European farms.

The exercise discussed above proves it is difficult to relate with accuracy these 56 most transformative papers of the H2020 projects under analysis to the seven clusters identified in the international literature. This may be a result of concurrently scattered efforts from various thematic areas to produce more transformative research, without, however, more organized endeavours. Papers of high TK are not always related to a specific topic, but they are wider in perspective, trying more to analyse the communicational, educational, motivational, participatory, and policy-making components of knowledge.

4. Conclusions

Research related to the defossilisation of agriculture shapes a global nexus interlinking major topics to prevailing technologies. In this work, a social network analysis of the global corpus revealed climate change, food security, conservation agriculture, and biomass conversion as major topics, as well as technology focusing on smart agriculture, heat pumps, and renewable energy for sustainable agriculture. Life cycle assessment is a crucial methodological tool to environmentally evaluate any technology or strategy that declared an impact on the defossilisation of agriculture, leading to forming a separate cluster.

Europe is one of the global funding pillars of research on defossilisation of agriculture along with China and the USA, with other actors remaining of minor importance. Based on this, this paper proceeded with further analysis of the European research outcome corpus to provide evidence of the effectiveness of such research efforts undertaken by the EU administration and taxpayers.

As a first step, the authors examined the nature of knowledge created out of the research activity and, more specifically, the outcome of EU-funded research since 2014. Published material in peer-reviewed journals were classified based on the knowledge type they provide as systemic, normative, transformative, or techno-economic, while having in mind the overall goal to achieve sustainability. Ideally, all these aspects should be present to build sustainability knowledge; nevertheless, only the presence of transformative knowledge ensures effectiveness and tangible results for the transition to sustainability. Our analysis showed that techno-economic knowledge prevails in research outputs, while transformative and normative knowledge are rather poorly represented. This does not necessarily mean that research is performed to produce more research, as significant research outcomes support instrumental knowledge that integrates into the incumbent system, improving technological status and economic performance, although it is not able to trigger transitions to alter the regime. It leads to incremental changes rather than fundamental transformation of the landscape.

We observe, though, that the transformative research outcome is significant in relative terms in CSA funding schemes and in absolute terms in certain RIA- and IA-funded projects. These latter have some common characteristics (Table S9): (i) they are coordinated by major academic actors in the European Research Area, (ii) they pronounced clear vision statements (LIFT: "how socio-economic and policy drivers impact on the development of ecological approaches to farming", SET-Nav: "support strategic decision making in

Europe's energy sector"), (iii) they explicitly adopted transdisciplinary approaches and team work (SoilCare), (iv) they are focused on social acceptance (ISAAC: "raising social awareness and identifying barriers to wider biogas and biomethane production by means of a participatory model with the aim to involve multiple actors"), (v) they are conducted through stakeholder analysis (IOF2020: "network of farmers, food industry stake-holders, technology providers and research institutions, ... inclusion and interaction of various actors involved in the transformation process...", SmartAgriHubs: "multi-stakeholder approach and covered a broad value-chain network across all EU member states"), and last but not least, (vi) they are explicitly oriented to strategy development (REEM: "focused not only on technology research, development and innovation, but also impact on society and environment based on specific case studies", BIO-ECON).

Limitations of the analysis concern the detection of knowledge types, as content analysis was performed by means of software due to the large volume of information to examine. While dictionary-based text analysis provides a quick and interpretable way to analyse textual data, it does have limitations. For instance, it might not consider the context of words in sentences, leading to potential inaccuracies in the analysis. Nonetheless, it serves as a valuable tool for extracting insights from large volumes of text data.

The obtained results do, however, constitute a basis for further discussions on the right approach to achieve the desired transformation through academic research and publicly funded research projects on defossilisation of agriculture. This analysis can be seen as a step forward to a better understanding of the correlation between research funding in Europe for agricultural defossilisation and the creation of transformative results in this sector. It could be beneficial to follow this research with a further analysis of other EU research funding programmes, such as Horizon Europe, to see what trends in generating transformative knowledge prevail, and what are the most prominent actors and regions in the network of European research on fossil-energy-free agriculture, thus identifying the right conditions for successful development of transformative research. Such analyses would contribute to making sure the European knowledge network generated by the EU research funding programmes creates a favourable environment for transformative knowledge production and diffusion to help EU agriculture defossilise at a faster pace.

Supplementary Materials: The following supporting information can be downloaded at: https://www.mdpi.com/article/10.3390/en17174409/s1, Table S1: Full list of queries for the 'global corpus' of literature; Table S2: Full list of queries for the 'European corpus' of literature; Table S3: Normative knowledge dictionary; Table S4: Systems knowledge dictionary; Table S5: Techno-economic knowledge dictionary; Table S6: Transformative knowledge dictionary; Table S7: Full list of H2020 projects used in the analysis; Table S8: Full list of papers produced by the analysed H2020 projects; Table S9: 'Transformative' projects analysis.

Author Contributions: Conceptualization, A.T.B., S.R. and M.B.; Methodology, A.T.B., K.T. and F.V.; Software, K.T., F.V. and M.W.; Validation, A.T.B. and M.B.; Formal Analysis, A.T.B., K.T. and F.V.; Investigation, A.T.B. and F.V.; Resources, A.T.B., F.V., M.B. and M.W.; Data Curation, A.T.B., F.V. and M.B.; Writing—Original Draft Preparation, A.T.B., K.T. and F.V.; Writing—Review and Editing, A.T.B., S.R., M.B. and M.W.; Visualization, F.V.; Supervision, A.T.B. and S.R.; Project Administration, A.T.B. and M.B.; Funding Acquisition, A.T.B. and M.B. All authors have read and agreed to the published version of the manuscript.

Funding: This research was funded by the European Union's Horizon 2020 research and innovation programme under grant agreement ID 101000496.

Data Availability Statement: The data supporting the findings of this study are available within the paper and its Supplementary Materials. Tables S1 and S2 contain the full lists of queries used to search the literature. Tables S3–S6 contain the updated sustainability knowledge dictionaries used in the study. The original dictionaries upon which our version is based are cited within the paper. Table S7 lists the H2020 projects included in the analysis and Table S8 lists the papers produced by these projects. Finally, Table S9 includes an analysis of the projects with the most 'transformative' papers. If extra data are needed to further clarify the methodological processes and the results, these may be made available by the corresponding author upon request.

Acknowledgments: This study has been developed as part of the Horizon 2020 AgroFossilFree project (www.agrofossilfree.eu) and we would like to thank the partner entities of the project for their contribution during its implementation.

Conflicts of Interest: The authors declare no conflicts of interest. The funders had no role in the design of the study; in the collection, analyses, or interpretation of data; in the writing of the manuscript; or in the decision to publish the results.

References

1. Eversberg, D.; Koch, P.; Lehmann, R.; Saltelli, A.; Ramcilovic-Suominen, S.; Kovacic, Z. The more things change, the more they stay the same: Promises of bioeconomy and the economy of promises. *Sustain. Sci.* **2023**, *18*, 557–568. [CrossRef]
2. Schlaile, M.P.; Urmetzer, S. Transitions to Sustainable Development. In *Decent Work and Economic Growth*; Filho, W.L., Azul, A.M., Brandli, L., Özuyar, P.G., Wall, T., Eds.; Encyclopedia of the UN Sustainable Development Goals; Springer International Publishing: Cham, Switzerland, 2019; pp. 1–16. [CrossRef]
3. Bogner, K.; Dahlke, J. Born to transform? German bioeconomy policy and research projects for transformations towards sustainability. *Ecol. Econ.* **2022**, *195*, 107366. [CrossRef]
4. Abson, D.J.; Von Wehrden, H.; Baumgärtner, S.; Fischer, J.; Hanspach, J.; Härdtle, W.; Heinrichs, H.; Klein, A.M.; Lang, D.J.; Martens, P.; et al. Ecosystem services as a boundary object for sustainability. *Ecol. Econ.* **2014**, *103*, 29–37. [CrossRef]
5. von Wehrden, H.; Luederitz, C.; Leventon, J.; Russell, S. Methodological Challenges in Sustainability Science: A Call for Method Plurality, Procedural Rigor and Longitudinal Research. *Chall. Sustain.* **2017**, *5*, 35–42. [CrossRef]
6. Urmetzer, S.; Lask, J.; Vargas-Carpintero, R.; Pyka, A. Learning to change: Transformative knowledge for building a sustainable bioeconomy. *Ecol. Econ.* **2020**, *167*, 106435. [CrossRef]
7. Robra, B.; Pazaitis, A.; Giotitsas, C.; Pansera, M. From creative destruction to convivial innovation—A post-growth perspective. *Technovation* **2023**, *125*, 102760. [CrossRef]
8. Urmetzer, S.; Schlaile, M.P.; Bogner, K.B.; Mueller, M.; Pyka, A. Exploring the Dedicated Knowledge Base of a Transformation towards a Sustainable Bioeconomy. *Sustainability* **2018**, *10*, 1694. [CrossRef]
9. Wiek, A.; Lang, D.J. Transformational Sustainability Research Methodology. In *Sustainability Science: An Introduction*; Heinrichs, H., Martens, P., Michelsen, G., Wiek, A., Eds.; Springer: Dordrecht, The Netherlands, 2016; pp. 31–41. [CrossRef]
10. Gibbons, M. Science's new social contract with society. *Nature* **1999**, *402*, 6761. [CrossRef] [PubMed]
11. ProClim. *Research on Sustainability and Global Change—Visions in Science Policy by Swiss Researchers*; ProClim-Forum for Climate and Global Change Swiss Academy of Sciences (SAS): Berne, Switzerland, 1997. Available online: https://scholar.google.com/scholar_lookup?title=Research%20on%20Sustainability%20and%20Global%20Change%20%20Visions%20in%20Science%20Policy%20by%20Swiss%20Researchers&publication_year=1997&author=ProClim (accessed on 6 November 2023).
12. Bogner, K. *Knowledge Networks in the German Bioeconomy: Network Structure of Publicly Funded R&D Networks*; Universität Hohenheim, Fakultät Wirtschafts-und Sozialwissenschaften: Stuttgart, Germany, 2019.
13. Zupic, I.; Čater, T. Bibliometric Methods in Management and Organization. *Organ. Res. Methods* **2015**, *18*, 429–472. [CrossRef]
14. van Eck, N.J.; Waltman, L. Visualizing Bibliometric Networks. In *Measuring Scholarly Impact: Methods and Practice*; Ding, Y., Rousseau, R., Wolfram, D., Eds.; Springer International Publishing: Cham, Switzerland, 2014; pp. 285–320. [CrossRef]
15. van Eck, N.J.; Waltman, L. Software survey: VOSviewer, a computer program for bibliometric mapping. *Scientometrics* **2010**, *84*, 523–538. [CrossRef] [PubMed]
16. Donthu, N.; Kumar, S.; Mukherjee, D.; Pandey, N.; Lim, W.M. How to conduct a bibliometric analysis: An overview and guidelines. *J. Bus. Res.* **2021**, *133*, 285–296. [CrossRef]
17. Reid, S.W.; McKenny, A.F.; Short, J.C. Synthesizing Best Practices for Conducting Dictionary-Based Computerized Text Analysis Research. In *Methods to Improve Our Field 14*; Hill, A.D., McKenny, A.F., O'Kane, P., Paroutis, S., Eds.; Research Methodology in Strategy and Management 14; Emerald Publishing Limited: Bingley, UK, 2023; pp. 43–78. [CrossRef]
18. Friedrich, J.; Bunker, I.; Uthes, S.; Zscheischler, J. The Potential of Bioeconomic Innovations to Contribute to a Social-Ecological Transformation: A Case Study in the Livestock System. *J. Agric. Environ. Ethics* **2021**, *34*, 24. [CrossRef]

energies

MDPI

Article

Analyze the Temporal and Spatial Distribution of Carbon Capture in Sustainable Development of Work

Fu-Hsuan Chen [1] and Hao-Ren Liu [2,*]

[1] Department of Political Science, National Taiwan University, No. 1, Sec. 4, Roosevelt Road, Taipei 10617, Taiwan; fhchen@mail.ntu.edu.tw

[2] Institute of Architectural Planning, National Taiwan University, No. 1, Sec. 4, Roosevelt Road, Taipei 10617, Taiwan

* Correspondence: d01544007@ntu.edu.tw; Tel.: +886-2-3366-5855; Fax: +886-2-2362-7651

Abstract: This study aims to analyze the temporal and spatial distribution of carbon capture technologies worldwide, examining the economic, social, and political developments reflected in related academic literature. By conducting a comprehensive analysis of over 40,000 related documents from 2004 to June 2024, as well as selecting 108 relevant articles from SSCI and SCI journals, the study explores the development of carbon capture technologies from different perspectives through keyword searches, trend analysis, and relevance ranking. The study finds that, in terms of temporal trends, significant progress has been made in carbon capture technologies since 2009, and their importance has surpassed that of carbon trading, becoming one of the core technologies in addressing climate change. Spatial trend analysis shows that North American and European countries are more inclined to prioritize "carbon capture" technologies, while Asian countries focus more on "carbon trading", reflecting regional differences in economic, policy, and technological development. Although carbon capture technologies hold immense potential for sustainable development, they also face numerous challenges, including balancing technological advancements with economic and policy frameworks. This balance is crucial to ensuring that carbon capture technologies can make a positive contribution to sustainable work, climate action, and environmental sustainability, further transforming the essence of sustainable efforts. To fully realize their benefits, it is essential to recognize and address these challenges.

Keywords: carbon capture; carbon trading; temporal trends; spatial distribution; sustainable development of work

1. Introduction

Around 1990, the academic community conducted extensive research activities on the impact of fossil fuel emissions on climate change and, as a result, the Clean Air Act Amendments were born, becoming a specific example [1]. The reaction in the American market to this initiative was significant and clear. In an era that highly values environmental issues, carbon capture technology, as a potential solution, is being widely discussed and researched [2]. Carbon capture, commonly referred to as Carbon Capture and Storage (CCS) or Carbon Capture, Utilization, and Storage (CCUS), is a critical technology aimed at reducing industrial and energy-related carbon dioxide (CO_2) emissions to mitigate climate change (Carbon Capture and Storage (CCS) is a key technology for reducing industrial and energy-related sources of carbon dioxide emissions to mitigate climate change. This process involves capturing carbon dioxide at the source, transporting it, and storing it in geological formations or using it in various industrial processes. Carbon Capture, Utilization, and Storage (CCUS) is a key technology that involves capturing carbon dioxide from industrial and energy-related sources, using it for various applications, and storing it safely to prevent its release into the atmosphere. This integrated approach not only includes reducing greenhouse gases but also explores the possibility of using carbon dioxide as a resource. Carbon Capture and Utilization (CCU) technology converts captured carbon dioxide into valuable products such as fuels, chemicals, and building materials, promoting climate action under Sustainable Development Goal (SDG 13).). This process includes capturing CO_2 emissions at the source, transporting them, and storing them in geological formations or using them in various industrial processes [3–5].

Carbon Capture and Utilization (CCU) technology provides an important framework for converting carbon dioxide emissions into valuable products, thereby promoting the Sustainable Development Goal of climate action (SDG 13). However, due to the high energy consumption characteristics of these technologies, their application may shift the burden to other sustainable development goals, such as human health and water scarcity issues. Combining CCU technology with negative carbon energy can mitigate these adverse effects to some extent. This underscores the importance of adopting integrated technology development strategies that align with multiple sustainable development goals [6]. Carbon Capture and Storage (CCS) is considered sustainable, with positive interactions with Sustainable Development Goal 13. However, it requires additional energy, which can have environmental impacts. Despite these challenges, CCS does not hinder the realization of other sustainable development goals, thus becoming a viable option for mitigating climate change [7]. Since around 2009, carbon capture has become an important technological option to help achieve global climate goals. A comprehensive review of relevant data and literature shows that understanding the economic, social, and political developments of carbon capture through carbon trading is crucial. This study, through global trends and spatial distribution approaches, using more than 40,000 literature records from 2004 to June 2024, demonstrates the temporal and spatial trend changes of carbon trading and carbon capture. These trends echo the statistical perspectives of academic research.

2. Materials and Methods

The discussion on carbon capture is multifaceted, focusing on technological advancements, economic feasibility, and policy frameworks. This study is conducted in two phases. First, assess the academic relevance through keywords, then conduct trend analysis based on the long-term characteristics of the carbon capture phenomenon revealed by past datasets in the literature system.

2.1. *Keyword Query*

Carbon capture, as a technology to reduce greenhouse gas emissions, is widely applied in European and American countries, whereas in Asia, the carbon trading mechanism receives more attention. Carbon trading is a market-based system based on carbon emission allowances, allowing companies to buy and sell carbon emission allowances according to their actual emissions, thereby effectively managing and controlling emissions. This system, by providing economic incentives, encourages companies to adopt carbon capture technologies to reduce emissions, thus achieving greenhouse gas reduction targets. In the carbon trading market, the application of carbon capture technologies helps companies achieve more stringent emission reduction standards and gain a competitive market advantage. Carbon trading also promotes cooperation between companies and drives the realization of global reduction targets. In this study, we conducted keyword searches for "carbon trading" and "carbon capture" using the databases Google Scholar, Web of Science, and Scopus. Through cross-referencing and relevance filtering, a total of 40,600 related documents were found. The documents reflect the extensive academic discussion on the important role of carbon trading and carbon capture technologies in mitigating climate change (see Table 1).

Table 1. A literature review of carbon trading and carbon capture.

Item	Research Content
Total Literature	40,600
Search Keywords	Carbon Trading, Carbon Capture
Database	Google Scholar, Web of Science, Scopus
Research Period	Multiple research periods (e.g., 2014–2022, 1992–2021)
Main Research Summary	The important role of carbon trading and carbon capture technologies in mitigating climate change
Mashari et al. [8]	Analyzed 506 articles on sustainable finance and carbon trading from 2014–2022, highlighting the need for further research connecting these two fields effectively.
Su and Yu [9]	Analyzed 4408 scholarly articles on carbon finance from 1992–2021, noting rapid publication growth and interest in carbon capture, economic growth, and carbon price forecasting.
Podder et al. [10]	Provided a comprehensive overview of carbon emissions from fossil fuel consumption and the importance of Carbon Capture and Storage (CCS), detailing mechanisms such as pre-combustion, post-combustion, and direct air capture.
Fujita [11]	Reviewed 626 articles on embodied carbon emissions and international pollution identification methods from 1994–2023, discussing the role of carbon trading in controlling emissions.
Kim [12]	Discussed accounting methods for early carbon sequestration investments and trades, emphasizing the need for standardized credit rating to ensure the integrity of carbon sequestration credit trading.
Otto and Gross [13]	Systematically reviewed 115 articles on communication practices regarding Carbon Capture and Storage (CCS), underlining the importance of effective stakeholder engagement.
Altman [3]	Discussed the extensive debate and research on incorporating carbon capture and storage technologies under the Clean Development Mechanism (CDM) of the Kyoto Protocol. (The Clean Development Mechanism (CDM) is an important component of the Kyoto Protocol aimed at promoting emission reductions while encouraging sustainable development in developing countries. It allows developed countries (Annex I countries) to fund projects in developing countries (non-Annex I countries) and receive Certified Emission Reductions (CERs) to meet their emission reduction commitments.)
Malan and Malan [14]	Analyzed challenges in comparing emission reductions and carbon sequestration, particularly issues related to the duration of storage for captured carbon dioxide.
Maitland [15]	Discussed carbon capture and storage technologies, emphasizing the need for tighter integration of materials and process design to optimize carbon dioxide capture and utilization processes.
Yu and Xu [16]	Scientometric analysis of Carbon Emission Trading (CET), highlighting the growth in publication numbers in the field and its major research areas, providing a clear picture of the current state and future trends. (Carbon Emission Trading (CET) is a market-oriented approach to reducing emissions by setting an emission cap and allowing entities to buy and sell allowances to reduce greenhouse gas emissions. This system is designed to incentivize businesses to reduce emissions cost-effectively while promoting sustainable development.)

In summary, these studies indicate that the body of literature on carbon trading and carbon capture is growing, reflecting their critical role in global climate change mitigation efforts. By analyzing trends in carbon capture and carbon trading from 2004 to 2024, this study conducted comprehensive statistical and spatial analysis of over 40,600 documents from January 2004 to June 2024 and selected 108 relevant SSCI and SCI papers to explore the development of carbon trading and carbon capture from different perspectives (as shown in Figure 1).

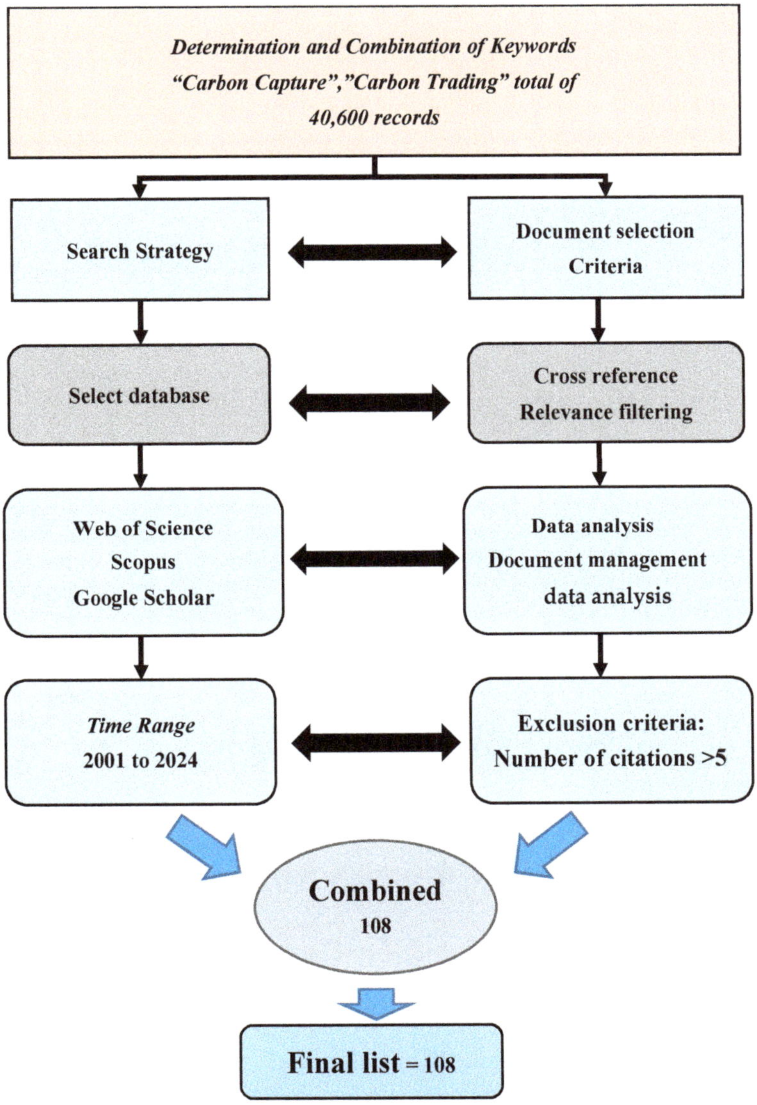

Figure 1. Flowchart of the literature selection procedure for carbon capture and carbon trading research.

2.2. Literature Screening and Analysis Steps

In the methodology section, our literature screening and analysis steps are as follows:

(1) Database Search:

- Using three major academic databases, Google Scholar, Web of Science, and Scopus, we conducted a comprehensive search with the keywords "carbon trading" and "carbon capture", initially obtaining 40,600 related documents.

(2) Screening Process

- Time frame: Limited to documents published from January 2004 to June 2024.
- Preliminary screening: Conducted based on titles, abstracts, and keywords to narrow down to approximately 40,000 documents to ensure the efficiency of the screening process.

(3) Cross-referencing and Relevance Filtering
- Utilized the cross-referencing features within the databases for mutual referencing and relevance filtering of the documents.

(4) Keyword Search
- For the preliminarily screened documents, performed more specific keyword searches, expanding the keyword range to CCS, CCU, "carbon trading market", and other related fields to enhance screening precision.

(5) Trend Analysis
- Used trend analysis tools to analyze the selected documents to determine research trends and development status of carbon capture and carbon trading.
- In the field of carbon capture technology research, the selection of appropriate keywords is paramount for capturing emerging trends and ensuring comprehensive coverage. For instance, in the realm of technological innovation, keywords such as "direct air capture", "metal-organic frameworks (MOFs)", "superbase-derived ionic liquids (SILs)", and "mixed matrix membranes" are instrumental in identifying the latest advancements. In terms of economic and policy aspects, exploring keywords like "economic feasibility", "policy framework", "carbon trading", and "quota allocation" aids in analyzing the economic and policy impacts. Regarding sustainability and environmental impact, terms like "sustainable solvents", "phase-change absorbents", and "nanoparticle-enhanced solvents" help locate relevant research projects. Additionally, regional and spatial distribution keywords such as "spatial distribution", "regional climate adaptability", and "site selection" should be used. For research on integrating carbon capture technology with renewable energy, terms like "renewable energy DAC", "wind power", and "U.S. 45Q tax credit" are essential. Through these strategies, researchers can effectively capture the latest trends in carbon capture technology, ensuring a comprehensive understanding of the field.

(6) Secondary Screening
- Peer-reviewed:
  Only peer-reviewed academic articles are selected, excluding any non-peer-reviewed documents such as conference abstracts, news reports, technical reports, and white papers.
- Publication source:
  Only articles published in journals indexed by SSCI, SCI and ESCI are considered. These journals possess high influence and reputation in their respective fields and can provide reliable research results.
- Relevance to the topic:
  Literature must directly relate to carbon capture technology. This includes, but is not limited to, the principles, methods, technological improvements, application cases, economic assessments, and environmental impact evaluations of carbon capture.
- Language:
  Only articles published in English-language journals are selected to ensure that researchers can fully comprehend and interpret the content.
- Rigorous research methods:
  Preference is given to articles employing strict and transparent research methods, such as those with clear experimental design, data analysis, and result interpretation.
- Citation frequency:
  The citation count of the literature is considered, with preference given to articles with high academic impact and citation rates. The minimum citation count is five. A secondary screening was conducted, focusing on documents with significant academic impact, high citation rates, and high journal impact factors, ultimately selecting 108 relevant SSCI, SCI and ESCI documents.

(7) Software Tools Utilization

- Literature management: Managed and organized literature using EndNote and Zotero.
- Data analysis: Performed statistical analysis, including trend analysis and spatial analysis, using Microsoft Excel and R.
- Data visualization: Generated charts and graphs using Tableau for a more intuitive presentation of the analysis results.

This systematic approach ensures the accuracy of literature screening and comprehensiveness of analysis, enabling us to comprehensively understand the research trends and development dynamics of carbon capture and carbon trading from 2004 to 2024.

3. Research Results: Temporal Trends

3.1. The Impact of Technology on Sustainable Development

For many years, due to the urgent need to reduce the adverse effects of carbon dioxide emissions on global climate, CCS has made significant progress. The Kyoto Protocol, signed in 1997 and effective in 2005, greatly increased interest in CCS technology and integrated it into the CDM to achieve carbon dioxide reduction targets [3]. The Kyoto Protocol has further highlighted the importance of CCS in global emission reductions since it came into effect in 2005 [3]. CCS has consistently been regarded as a potential yet underutilized solution that may play a crucial role in reducing the costs of achieving climate stability [17] and enhancing the flexibility of emission reduction strategies [4]. However, the widespread deployment of CCS still faces multiple challenges, including high capital costs, low capture efficiency, and the need for policy support and international cooperation [18,19].

As a supplementary method, CCU is gradually emerging, aiming to convert captured carbon dioxide into valuable products such as fuels, chemicals, and building materials, thus providing economic incentives to reduce carbon dioxide emissions [20,21]. Despite the fact that some CCS and CCU technologies have matured technically, their large-scale application is still constrained by economic, technical, and political barriers [18]. Scientific research has been continuously exploring innovative materials and methods, such as ionic liquids and advanced adsorbents, to enhance the efficiency and cost-effectiveness of carbon dioxide capture [19,22]. Therefore, controlling methane emissions can effectively reduce tropospheric ozone and radiative forcing, further supporting integrated reduction approaches [23]. In order to reduce pollutants and carbon dioxide emissions, advanced combustion technologies including gasoline compression ignition have been developed [24], although these technologies still face challenges [23]. Besides the challenge, the simultaneous presence of water and carbon dioxide also brings challenges, but the use of molecular sieves to capture hydrocarbons still shows potential in reducing industrial emissions [25]. CCU in recent years has made significant progress, introducing new materials and methods such as porous silica-based materials, ionic liquids, and membrane technology. These innovations have improved capture efficiency and reduced costs [26–28]. Research and development of new low-cost carbon capture technology, utilizing cold nitrogen refrigerants to minimize energy consumption and water resources usage, have also achieved breakthroughs [29]. Direct Air Capture (DAC) technology, such as the use of short-chain tetra (n-alkyl) ammonium hydroxide, has also become a potential method for capturing carbon dioxide directly from the atmosphere [30] (Direct Air Capture (DAC) is a cutting-edge technology aimed at mitigating climate change by removing carbon dioxide directly from the atmosphere. This method is gaining attention for its significant potential to address traditional carbon dioxide emission challenges and help achieve net zero emissions. Emission Trading Systems (ETS) are market-based mechanisms for reducing emissions by setting an emission cap and allowing the trading of emission allowances. These systems have been implemented globally with varying designs and impacts, aiming to encourage emission reductions and promote green technologies.). These technological developments are of great significance for transitioning to new sustainable energy sources while reducing greenhouse gas emissions [22,31].

3.2. Economic and Policy Considerations

This section primarily examines the temporal trends in carbon capture and carbon trading, alongside the significance of these technologies in global carbon reduction efforts. The interaction between carbon trading and carbon capture technologies is highly intricate, involving various domains such as economics, regulation, and technology. Mechanisms like Emission Trading Systems (ETS) offer a market-based approach to curbing greenhouse gas emissions by allowing for the trade of emission permits or credits. The formulation and execution of carbon trading mechanisms crucially influence the advancement and efficacy of Carbon Capture and Storage (CCS) technologies. These mechanisms set the marginal cost of carbon emissions, which is affected by the stringency of emission caps and the timing of their implementation. Consequently, fluctuations in carbon prices significantly impact the deployment and investment in CCS technologies. The marginal cost associated with carbon emissions is contingent on the rigidity of emission limits and the schedule of their establishment, which in turn directly influences the variability of carbon prices. This fluctuation has significant implications for the deployment and financial commitment towards CCS technology [32]. By linking different ETS systems, allowing emission permits from one system to be traded in another, overall compliance costs can be reduced, market liquidity enhanced, and spillover effects of emissions minimized, thereby promoting the expansion of CCS technology. However, the success of such linkages depends on the design of the ETS, political intentions, and existing economic and political relationships [33]. The Paris Agreement encourages countries to regularly review their climate policies and compliance progress, thereby promoting the development of carbon markets and indirectly supporting CCS technology [33]. Although such connections bring multiple benefits, the linkages between different ETS also introduce new challenges and uncertainties, such as price fluctuations, the need for transparent subsidy and cap-setting rules, which are crucial for maintaining market stability and promoting investment in CCS technology [34,35]. Additionally, for large-scale point source capture of carbon dioxide and its storage in geological formations, CCS technology is extremely costly, usually requiring national support or public–private partnerships to achieve [3]. This integration demonstrates the potential of CCS technology as part of a broader carbon trading scheme, thereby enhancing its economic viability. Moreover, innovative carbon trading schemes such as double carbon trading further balance emission reduction targets with social welfare, promote low carbon energy consumption, and support CCU [36]. The international CCS projects, which involve capturing carbon dioxide in one country and transporting it to another country for storage, also bring regulatory and liability challenges that must be addressed to promote global deployment [37]. To sum up, the interaction between carbon trading and CCS is extremely complex, which is crucial for achieving deep emission reductions and global emission targets. Through carefully designed and linked carbon markets, the necessary economic incentives and regulatory frameworks can be provided to support the comprehensive application of CCS technology, contributing to global climate change mitigation.

3.3. Trends in Carbon Trading and Carbon Capture

According to Figure 2, since 2004, carbon trading has received widespread attention, reaching its peak around 2008. However, its popularity has significantly declined thereafter. Conversely, carbon capture technology has seen significant growth since 2011, peaking in 2021, and continues to attract attention. Early carbon capture technology mainly relied on post-combustion capture using aqueous amine scrubbing, facing challenges of high regeneration energy and solvent degradation [28]. After 2011, innovation expanded to include pre-combustion and oxy-fuel combustion technologies, the latter utilizing pure oxygen to produce flue gases rich in carbon dioxide, thereby simplifying the carbon dioxide separation process [27].

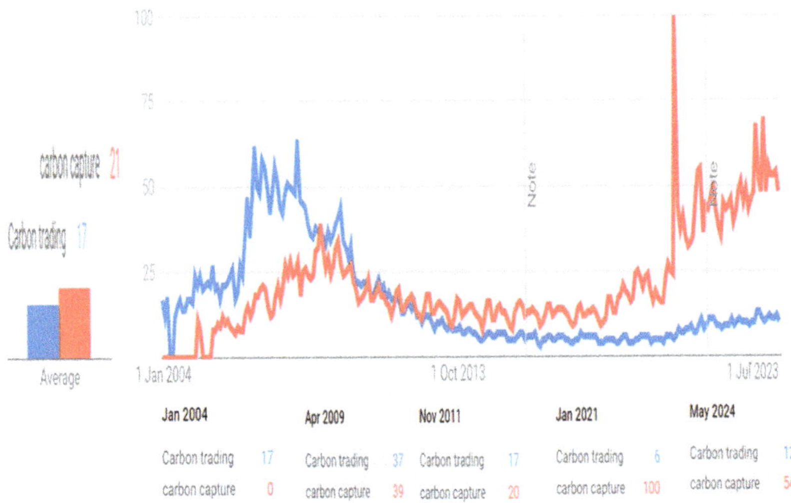

Figure 2. Trends in literature on carbon trading and capture (2004–June 2024).

In 2021, CCS made significant breakthroughs, addressing various challenges related to efficiency, cost, and scalability. One notable advancement was the development of mixed matrix membranes, greatly enhancing carbon dioxide permeability [38]. Direct air capture technology has also made progress in areas such as Metal–Organic Frameworks (MOFs) and superbasic-derived ionic liquids (SILs), demonstrating efficient carbon dioxide absorption performance and excellent cyclic stability [39] (Metal–Organic Frameworks (MOFs) are a class of crystalline materials with a porous structure formed by connecting metal ions or clusters with organic ligands. These materials are noted for their unique properties and wide range of applications. MOFs are known for their high surface area, tunable pore size, and multifunctionality, suitable for various industrial and scientific applications. Superalkaline Ionic Liquids (SILs) are ionic liquids that have received significant attention for their unique properties and diverse applications. These ionic liquids contain superalkaline cations, giving them high alkalinity and the ability to dissolve cellulose and other difficult-to-dissolve substances.). In addition, research focuses on more economical and sustainable solvents, such as phase-change absorbents and nanoparticle-enhanced solvents, to improve carbon dioxide absorption performance and reduce energy demand [22]. New developments also include a system called "reverse" SILP, which has improved carbon dioxide/N2 separation performance and carbon dioxide absorption capacity, overcoming the limitations of traditional SILP [40] (Supported Ionic Liquid Phase (SILP) materials represent a versatile and innovative approach in various fields such as catalysis, composites, and biocatalysis. These materials combine the unique properties of ionic liquids with the structural advantages of solid supports, enhancing stability, reusability, and efficiency.). Polymer Continuous Assembly (CAP) technology can produce ultra-thin membranes with high carbon dioxide permeability and excellent gas separation selectivity, further advancing membrane-based carbon dioxide capture methods [41]. Finally, the techno-economic evaluation of combining DAC with renewable energies such as wind power shows the potential to reduce costs and improve economic feasibility, especially with the support of the US 45Q tax credit incentive policies [42].

Overall, the average global search interest in "carbon trading" has fallen below "carbon capture". Carbon capture technology is gradually becoming one of the core strategies worldwide to address climate change. Various carbon reduction methods are being widely explored, particularly improvements in energy use technology and efficiency. For example, electrochemical, photochemical, and photo electrochemical processes can convert carbon dioxide into useful hydrocarbons [43,44]. Energy efficiency measures such as building lighting and power decarbonization are regarded as efficient emission reduction pathways [45]. The American industrial sector achieves targeted decarbonization through measures such as strategic energy management and smart manufacturing, which are crucial for achieving sustainable development [46]. Furthermore, combining renewable energy with storage technology provides a scalable energy solution while controlling atmospheric carbon dioxide levels [47]. According to the trend changes shown in Figure 2, we can observe that the research interest in "carbon trading" increased significantly from 2004 to 2011, peaking in 2008. This period of peak interest can be traced back to the formal implementation of the Kyoto Protocol in 2005, which prompted global attention to the carbon trading mechanism as an important market-based tool for achieving emission reduction targets. Many countries and companies began to explore and implement ETS, leading to a sharp increase in research interest. However, after 2011, the research enthusiasm for carbon trading dropped significantly, which may be related to the 2008 global economic crisis and the subsequent economic downturn. During economically difficult times, investment in and attention to environmental measures often decrease. Additionally, some carbon trading markets, such as the EU ETS, faced issues of excess allowances and plummeting carbon prices in their initial stages, further weakening market confidence and leading to a decline in research interest. At the same time, from 2011 onwards, research interest in CCS technology increased significantly. This shift in trend partly stems from the challenges and limitations faced by the carbon trading market, prompting a focus on alternative emission reduction technologies. Moreover, technological breakthroughs and cost reductions around 2011, such as the development of pre-combustion and oxy-fuel combustion technologies, greatly improved the efficiency and economics of the technology, further driving the rise of CCS research. In 2015, the signing of the Paris Agreement further reinforced the focus on CCS technology. The agreement called for strong measures by countries to achieve emission reduction targets, promoting CCS as a key technology for reaching these goals. Since then, research interest in CCS has continued to grow, reaching a new peak in 2021. This phenomenon can be attributed to several factors:

- Technological advancements: Innovative technologies such as hybrid matrix membranes, MOFs, and SILs have improved the efficiency, cost-effectiveness, and operability of CCS.
- Policy support: Policies like the US 45Q tax credit policy have provided economic incentives for CCS projects, boosting investor confidence.
- Strengthening of global climate policies: Under the framework of the Paris Agreement, countries have enhanced their climate targets, increasing the focus on CCS.

4. Research Results: Spatial Distribution

According to the trend index from 2004 to 2023, spatial distribution revealed regional preference differences in various climate-related themes. North American and European countries lean more towards "carbon capture and storage", while Asian countries show greater interest in "carbon trading" (see Figure 3). Figure 3 is a color-coded world map showing the distribution of global carbon capture and carbon trading activities from 2004 to June 2024. The intensity of the red shading indicates the level of carbon capture efforts in different regions: North America, Europe, and East Asia are shaded more deeply in red, indicating higher activity levels, while areas such as Africa, Central Asia, and parts of South America are shaded more lightly, indicating lower activity levels. The blue shading indicates the attention to carbon trading in different regions, with deeper blue shading in Asian regions, particularly in countries like China and India, indicating a higher interest in carbon trading activities. Regions like Greenland and central Sahara are colorless, indicating a lack of carbon capture or carbon trading activities. This map reveals that industrialized regions are more actively implementing carbon capture technologies, demonstrating their technological and economic capabilities; meanwhile, some Asian countries show significant activity in carbon trading. The significant disparities between developed and developing regions reflect the unequal response to climate change globally, underscoring the importance of promoting global collaboration and technology transfer.

Figure 3. Global carbon capture spatial distribution from 2004 to June 2024.

The different economic, policy, and social driving factors are the main reasons for the regional priority differences. In North America and Europe, CCS technology is seen as a key means to significantly reduce greenhouse gas emissions while still using fossil fuels. In the US, particularly in areas where coal is widely used, it is in a favorable position to apply DAC technology due to the mature policy environment, abundant funding opportunities, and low-carbon energy supply. However, even with a supportive regulatory environment, advancing CCS projects in the US still faces challenges related to underground resource ownership and state-level jurisdiction, which must be properly addressed to proceed smoothly [48].

In the European Union, established climate policies and a supportive political environment favor the deployment of CCS technology [48]. European countries are working to integrate CCS into a broader climate policy framework that includes not only the expansion of renewable energy but also cogeneration programs. In Germany, for example, the competition between CCS technology and renewable energy technology is a key consideration, while public opposition to carbon dioxide storage projects poses a significant obstacle [49]. The EU is particularly active in promoting maritime projects and is committed to clarifying the classification of carbon dioxide in international regulations [48]. At the same time, build a cost-effective global CCUS framework aimed at matching carbon sources with carbon sinks to achieve significant reductions in carbon dioxide emissions [50]. In addition, the EU focuses on subsidizing upstream segments in CCS technology innovation, benefiting suppliers and enhancing economic efficiency [51]. However, because the EU faces challenges in public planning and regulation, especially in the spatial reorganization of power generation and transmission systems, efficient installation of CCS-related infrastructure is necessary [52]. Some European countries, such as the Netherlands and the United Kingdom, have public attitudes towards industrial carbon capture and storage projects that are neutral to slightly positive, influenced by climate change, safety, and employment expectations [53]. The EU is also addressing regulatory and legal challenges supporting the development of the CCUS supply chain, ensuring its consistency with international law and creating a favorable environment for financial viability [54]. The results of the life cycle assessment of the potential benefits of integrating CCUS technology into the energy supply sector show a significant reduction in the global warming potential in countries such as Italy and Poland [55].

Asian countries such as China, Japan, and South Korea are actively promoting the development of carbon trading systems, which are seen as a market-oriented emission reduction strategy (see Figure 4). These countries have explored the potential of linking carbon emission trading systems with each other, demonstrating that a unified market can significantly enhance cost efficiency, promote economic benefits, and improve market liquidity and trading scale [56]. Asian emerging economies like China are developing carbon market mechanisms, creating domestic carbon credit demand and connecting with regional and global carbon markets [57]. Asian emerging economies such as China are developing carbon market mechanisms, creating domestic carbon credit demand and connecting with regional and global carbon markets [58]. However, developing countries in Asia still face challenges in public acceptance and stakeholder management in the deployment of CCS technology, which need to be strategically addressed to promote the adoption of CCS [59]. Despite facing challenges, China still possesses the potential to deploy DAC technology, thanks to its well-established policy and funding environment [60].

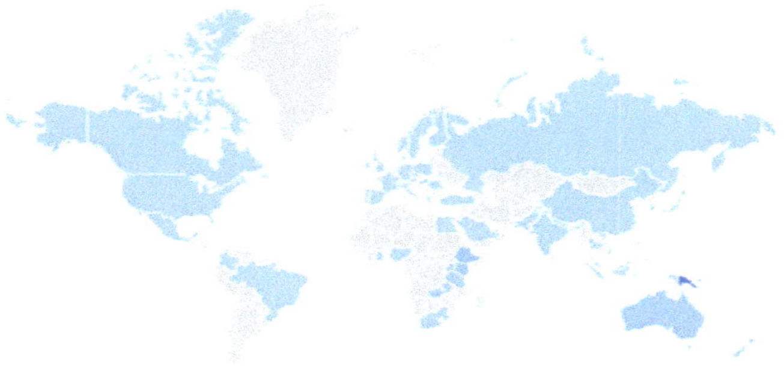

Figure 4. Global spatial distribution of carbon trading from 2004 to June 2024.

Figure 4 is a world map showing the trends in carbon trading distribution from 2004 to June 2024. The varying shades of blue represent the levels of carbon trading activities across different countries, with darker areas indicating more active carbon trading, such as in Europe and North America, while lighter or colorless areas such as most parts of Africa and some regions in Asia show fewer or non-existent carbon trading activities. The map reveals a slight increase in carbon trading activities among some emerging economies in Asia, indicating a growing participation in the carbon market in these regions. This distribution helps in understanding the global participation in the carbon market and identifying areas that may require policy intervention or support to enhance participation in carbon trading programs. As of 2011, many European, North American, and some Asian countries began to shift their focus to CCS, with a significant increase in attention towards achieving net-zero emissions targets. By 2022, this scope of attention had expanded to 55 countries, with Norway, Singapore, Denmark, Canada, and the UK being particularly prominent. Trend data indicate that the application of carbon capture technology has extended widely over time and space, with its influence gradually spreading from Europe and the Americas to Asia and Africa (see Figure 5). The United States has played a key role in the development and innovation of CCS technology, but it is not the only leader in this field. The United States has established a mature carbon dioxide injection policy framework, mainly applied to enhance oil recovery, and has made significant progress in CCS research [61,62]. However, the development of carbon dioxide capture and storage technology relies on global cooperation. As the largest greenhouse gas emitters, the

United States and China are collaborating in the field of CCS through initiatives such as the Clean Energy Research Center, leveraging the complementary strengths of both countries to accelerate the global deployment of CCS [63]. Norway and other countries have also participated in several important CCS projects, although some projects are facing cancellation or delays [64]. The European Union, Australia, and Canada are also actively contributing to the regulation and promotion of CCS, emphasizing the importance of international cooperation and a unified regulatory framework [65].

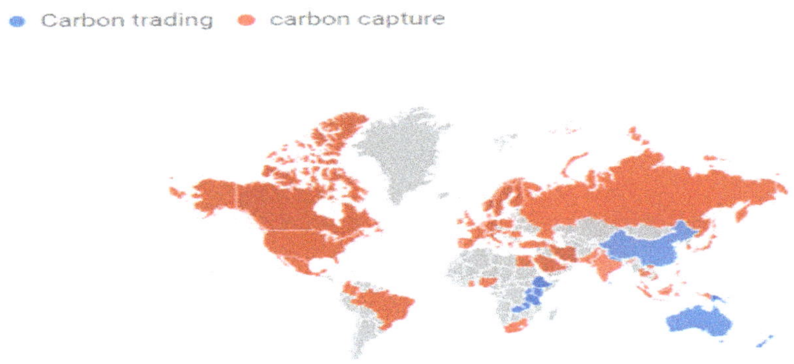

Figure 5. Global carbon trading and carbon capture spatial distribution 2004 to June 2024.

Effective CCS regional cooperation can be achieved through multi-party research integration and addressing various challenges. The integration of regional carbon markets into a unified national market helps strengthen resource integration and regulation. For example, China is committed to connecting its pilot regional markets to the national carbon market to establish a comprehensive spillover network, promoting information transfer and risk management [66]. A fair and efficient quota allocation plan can reduce economic costs and ensure the equitable distribution of emission reduction tasks, as demonstrated by China's multi-standard quota allocation plan [67]. In addition, regional climate adaptability is particularly important in the deployment of DAC technology, and site selection must be carefully made based on regional climate conditions to optimize performance and reduce costs [68]. The cooperation between major greenhouse gas emitters such as the United States and China on CCS projects fully demonstrates that through collaborative efforts, leveraging their respective comparative advantages, and ensuring mutual benefit, significant global impacts can be achieved [63]. Further, technological innovations, such as the application of imidazolium-based ionic liquids and bicarbonate mechanisms, can significantly enhance CO_2 capture efficiency [69,70]. The transition to a low-carbon economy also requires a skilled workforce capable of implementing and managing carbon capture technologies. The U.S. Department of Energy's Carbon Sequestration Research experience program demonstrates efforts to build a talent pipeline in CCUS, emphasizing sustainable development education that is diverse and inclusive. This program aims to prepare young leaders for the clean energy transition, highlighting the importance of workforce development in achieving sustainable work outcomes [71]. Figure 5 illustrates the geographical distribution of global carbon trading and carbon capture activities from 2004 to June 2024, with color coding to differentiate the level of participation by region. In the map, red areas indicate countries implementing carbon capture technologies, while blue areas show countries with established carbon trading systems. The main findings reveal that North America, Europe, and parts of Asia are predominantly marked in red, indicating a high level of participation in carbon capture activities in these regions. Meanwhile, countries marked in blue are relatively fewer and are concentrated mainly in Europe and North America, suggesting that carbon trading activities are somewhat limited but strategically significant in these areas. Overall, the adoption of carbon capture technologies is more widespread globally. This trend emphasizes the global focus on technological sequestration methods and the demand for related sustainable job roles, highlighting the importance of enhancing occupational safety training in carbon capture fields.

5. Conclusions

The aim of this research is to analyze the temporal and spatial distribution of global carbon capture and carbon trading technology, exploring the economic, social, and policy challenges in their development and implementation. The results indicate significant differences in the priorities of carbon capture and carbon trading across various regions, with differing temporal trends as well. Firstly, carbon trading research peaked in 2008 due to the Kyoto Protocol but declined after 2011 due to economic challenges and market limitations. This indicates that early policy promotion and market boom were followed by economic recession and market integration, affecting its prominence. Secondly, interest in carbon capture research has steadily increased since 2011, driven mainly by technological advances, policy support, and growing global attention to climate change. The establishment of technological innovations and policy frameworks has made carbon capture an important means of mitigating climate change. Spatially, North America and Europe have placed more emphasis on carbon capture technology, whereas Asia has focused on carbon trading. This phenomenon can be attributed to widespread coal use in North America and Europe, mature policy frameworks, and ample financial resources. In Asia, the emphasis on carbon trading is due to the need to coordinate emission reduction strategies and achieve carbon removal through technologies such as enhanced weathering and reforestation. A third research outcome shows the global carbon trading trend from 2004 to 2024 (Figures 4 and 5), further supporting the aforementioned spatial and temporal trends, particularly with the significant fluctuations brought about by changes in policy and market environments. However, differences between regions and uneven technological maturity have limited the promotion of carbon capture technology. Specifically, developed countries, because of their financial and technological strength, have made initial progress in carbon capture technology, gradually forming a substantial industry chain, whereas developing countries, due to lack of resources and technology, rely on international cooperation and assistance to promote related technology applications. Additionally, the operation of carbon trading markets reflects the role of policy support and market mechanisms, with regional market maturity having a positive impact on the promotion of carbon capture technology.

Therefore, this research reveals that integrating CCS technology with sustainable work practices involves multiple challenges in the economic, social, and policy dimensions. The economic challenges of CCS technology mainly stem from its high costs, encompassing phases such as CO_2 capture, separation, purification, and transportation. Without policy support and incentive measures, these costs are economically challenging to accept. Moreover, industries implementing CCS technology will face increased operating costs, potentially leading to a decline in their international competitiveness, especially when competitors do not bear similar costs. This scenario could shift employment and economic activities to regions with less stringent carbon policies. The absence of market mechanisms that penalize CO_2 emitters and incentivize the use of captured CO_2 products further complicates the economic feasibility of CCS technology. From a social perspective, the transition to CCS technology might lead to job losses in traditional industries. However, it also creates new job opportunities in emerging green sectors. Therefore, ensuring a just transition requires appropriate policy support to facilitate retraining and reemployment of workers. Additionally, transitioning to sustainable work modes (such as hybrid work models) must consider geographic inequalities and ensure that different regions and socioeconomic groups have equitable access to opportunities.

In conclusion, despite the comprehensive analysis of carbon capture and carbon trading trends presented in this research, there are limitations regarding data sources, regional complexity, and the time scope. Based on the findings and limitations of the research, for a deeper understanding of the differences in policy effectiveness across regions, future studies should conduct detailed regional analyses. Additionally, exploring the long-term impacts of policy changes on the adoption of CCS and carbon trading technologies has become an urgent need. Identifying methods to overcome barriers in CCS technology adoption, such as financing, technology, and market mechanisms, is crucial. Lastly, analyzing the impact of global economic trends on global carbon markets and carbon trading mechanisms will help formulate more comprehensive and effective strategies to promote the development and application of carbon reduction technologies.

Author Contributions: Writing—original draft, F.-H.C.; writing—review and editing, F.-H.C. and H.-R.L. All authors have read and agreed to the published version of the manuscript.

Funding: This research received no external funding.

Data Availability Statement: The data presented in this study are available on request from the corresponding author. The data are not publicly available due to privacy.

Conflicts of Interest: The author declares no conflicts of interest.

References

1. Vickers, V.L.; Hobbs, W.J.; Vemuri, S.; Todd, D.L. Fuel Resource Scheduling with Emission Constraints. *IEEE Trans. Power Syst.* **1994**, *9*, 1531–1538. [CrossRef]
2. Wu, C.; Xu, Y. Introducing Carbon Capture Science & Technology (CCST)! *Carbon Capture Sci. Technol.* **2021**, *1*, 100022. [CrossRef]
3. Altmann, P. Timmo Krüger, Das Hegemonieprojekt Der Ökologischen Modernisierung: Die Konflikte Um Carbon Capture and Storage (CCS) in Der Internationalen Klimapolitik. *Int. Sociol.* **2019**, *34*, 228–230. [CrossRef]
4. Chalmers, H. Fundamentals Point to Carbon Capture. *Nat. Clim. Chang.* **2019**, *9*, 348. [CrossRef]
5. Sun, S.; Lv, Z.; Qiao, Y.; Qin, C.; Xu, S.; Wu, C. Integrated CO_2 Capture and Utilization with CaO-Alone for High Purity Syngas Production. *Carbon Capture Sci. Technol.* **2021**, *1*, 100001. [CrossRef]
6. Ioannou, I.; Galán-Martín, Á.; Pérez-Ramírez, J.; Guillén-Gosálbez, G. Trade-Offs between Sustainable Development Goals in Carbon Capture and Utilisation. *Energy Environ. Sci.* **2023**, *16*, 113–124. [CrossRef]
7. Mikunda, T.; Brunner, L.; Skylogianni, E.; Monteiro, J.; Rycroft, L.; Kemper, J. Carbon Capture and Storage and the Sustainable Development Goals. *Int. J. Greenh. Gas. Control* **2021**, *108*, 103318. [CrossRef]
8. Mashari, D.P.S.; Zagloel, T.Y.M.; Soesilo, T.E.B.; Maftuchah, I. A Bibliometric and Literature Review: Alignment of Green Finance and Carbon Trading. *Sustainability* **2023**, *15*, 7877. [CrossRef]
9. Su, L.; Yu, W.; Zhou, Z. Global Trends of Carbon Finance: A Bibliometric Analysis. *Sustainability* **2023**, *15*, 6784. [CrossRef]
10. Podder, J.; Patra, B.R.; Pattnaik, F.; Nanda, S.; Dalai, A.K. A Review of Carbon Capture and Valorization Technologies. *Energies* **2023**, *16*, 2589. [CrossRef]
11. Wang, H.; Fujita, T. A Review of Research on Embodied Carbon in International Trade. *Sustainability* **2023**, *15*, 7879. [CrossRef]
12. King, D.M. Trade-Based Carbon Sequestration Accounting. *Environ. Manag.* **2004**, *33*, 559–571. [CrossRef]
13. Otto, D.; Gross, M. Stuck on Coal and Persuasion? A Critical Review of Carbon Capture and Storage Communication. *Energy Res. Soc. Sci.* **2021**, *82*, 102306. [CrossRef]
14. Marland, G.; Marland, E. Trading Permanent and Temporary Carbon Emissions Credits: An Editorial Comment. *Clim. Chang.* **2009**, *95*, 465–468. [CrossRef]
15. Maitland, G.C. Carbon Capture and Storage: Concluding Remarks. *Faraday Discuss.* **2016**, *192*, 581–599. [CrossRef]
16. Yu, D.; Xu, C. Mapping Research on Carbon Emissions Trading: A Co-Citation Analysis. *Renew. Sustain. Energy Rev.* **2017**, *74*, 1314–1322. [CrossRef]
17. Sun, S.; Wang, Y.; Zhao, X.; Zhang, C.; Wu, C. One Step Upcycling CO_2 from Flue Gas into CO Using Natural Stone in an Integrated CO_2 Capture and Utilisation System. *Carbon Capture Sci. Technol.* **2022**, *5*, 100078. [CrossRef]
18. Bui, M.; Bui, M.; Adjiman, C.S.; Adjiman, C.S.; Stolten, D.; Bardow, A.; Bardow, A.; Anthony, E.J.; Anthony, E.J.; Boston, A.; et al. Carbon Capture and Storage (CCS): The Way Forward. *Energy Environ. Sci.* **2018**, *11*, 1062–1176. [CrossRef]
19. Liu, E.; Lu, X.; Wang, D. A Systematic Review of Carbon Capture, Utilization and Storage: Status, Progress and Challenges. *Energies* **2023**, *16*, 2865. [CrossRef]
20. Desport, L.; Selosse, S. An Overview of CO_2 Capture and Utilization in Energy Models. *Resour. Conserv. Recycl.* **2022**, *180*, 106150. [CrossRef]
21. Gao, W.; Liang, S.; Wang, R.; Jiang, Q.; Zhang, Y.; Zheng, Q.; Xie, B.; Toe, C.Y.; Zhu, X.; Wang, J.; et al. Industrial Carbon Dioxide Capture and Utilization: State of the Art and Future Challenges. *Chem. Soc. Rev.* **2020**, *49*, 8584–8686. [CrossRef]
22. Ochedi, F.O.; Yu, J.; Yu, H.; Liu, Y.; Hussain, A. Carbon Dioxide Capture Using Liquid Absorption Methods: A Review. *Environ. Chem. Lett.* **2021**, *19*, 77–109. [CrossRef]
23. Dentener, F.; Stevenson, D.; Cofala, J.; Mechler, R.; Amann, M.; Bergamaschi, P.; Raes, F.; Derwent, R. The Impact of Air Pollutant and Methane Emission Controls on Tropospheric Ozone and Radiative Forcing: CTM Calculations for the Period 1990–2030. *Atmos. Chem. Phys.* **2005**, *5*, 1731–1755. [CrossRef]

24. Wu, C.; Huang, Q.; Xu, Z.; Sipra, A.T.; Gao, N.; de Souza Vandenberghe, L.P.; Vieira, S.; Soccol, C.R.; Zhao, R.; Deng, S. A Comprehensive Review of Carbon Capture Science and Technologies. *Carbon Capture Sci. Technol.* **2024**, *11*, 100178.
25. Iliyas, A.; Zahedi-Niaki, H.M.; Eić, M. One-Dimensional Molecular Sieves for Hydrocarbon Cold-Start Emission Control: Influence of Water and CO_2. *Appl. Catal. A Gen.* **2010**, *382*, 213–219. [CrossRef]
26. Amaraweera, S.M.; Gunathilake, C.A.; Gunawardene, O.H.P.; Dassanayake, R.S.; Cho, E.B.; Du, Y. Carbon Capture Using Porous Silica Materials. *Nanomaterials* **2023**, *13*, 2050. [CrossRef]
27. Araújo, O.D.Q.F.; De Medeiros, J.L. Carbon Capture and Storage Technologies: Present Scenario and Drivers of Innovation. *Curr. Opin. Chem. Eng.* **2017**, *17*, 22–34. [CrossRef]
28. Chong, F.K.; Foo, D.C.Y.; Eljack, F.T.; Atilhan, M.; Chemmangattuvalappil, N.G. Ionic Liquid Design for Enhanced Carbon Dioxide Capture by Computer-Aided Molecular Design Approach. *Clean. Techn Env. Policy* **2015**, *17*, 1301–1312. [CrossRef]
29. De, D.K.; Oduniyi, I.A.; Alex Sam, A. A Novel Cryogenic Technology for Low-Cost Carbon Capture from NGCC Power Plants for Climate Change Mitigation. *Therm. Sci. Eng. Prog.* **2022**, *36*, 101495. [CrossRef]
30. Mishra, M.K.; Smetana, V.; Hiti, E.A.; Wineinger, H.B.; Qu, F.; Mudring, A.-V.; Rogers, R.D. CO_2 Capture from Ambient Air via Crystallization with Tetraalkylammonium Hydroxides. *Dalton Trans.* **2022**, *51*, 17724–17732. [CrossRef]
31. Vaz, S.; Rodrigues De Souza, A.P.; Lobo Baeta, B.E. Technologies for Carbon Dioxide Capture: A Review Applied to Energy Sectors. *Clean. Eng. Technol.* **2022**, *8*, 100456. [CrossRef]
32. Scott, M.J.; Edmonds, J.A.; Mahasenan, N.; Roop, J.M.; Brunello, A.L.; Haites, E.F. International Emission Trading and the Cost of Greenhouse Gas Emissions Mitigation and Sequestration. *Clim. Chang.* **2004**, *64*, 257–287. [CrossRef]
33. Jevnaker, T.; Wettestad, J. Linked Carbon Markets: Silver Bullet, or Castle in the Air? *Clim. Law* **2016**, *6*, 142–151. [CrossRef]
34. Doda, B.; Taschini, L. Carbon Dating: When Is It Beneficial to Link ETSs? *J. Assoc. Environ. Resour. Econ.* **2017**, *4*, 701–730. [CrossRef]
35. Pan, C.; Shrestha, A.K.; Wang, G.; Innes, J.L.; Wang, K.X.; Li, N.; Li, J.; He, Y.; Sheng, C.; Niles, J.-O. A Linkage Framework for the China National Emission Trading System (CETS): Insight from Key Global Carbon Markets. *Sustainability* **2021**, *13*, 7459. [CrossRef]
36. Yan, Z.; Li, C.; Yao, Y.; Lai, W.; Tang, J.; Shao, C.; Zhang, Q. Bi-Level Carbon Trading Model on Demand Side for Integrated Electricity-Gas System. *IEEE Trans. Smart Grid* **2023**, *14*, 2681–2696. [CrossRef]
37. Gola, S.; Noussia, K. From CO_2 Sources to Sinks: Regulatory Challenges for Trans-Boundary Trade, Shipment and Storage. *Resour. Conserv. Recycl.* **2022**, *179*, 106039. [CrossRef]
38. Li, S.; Sun, Y.; Wang, Z.; Jin, C.; Yin, M.; An, Q. Rapid Fabrication of High-Permeability Mixed Matrix Membranes at Mild Condition for CO_2 Capture (Small 19/2023). *Small* **2023**, *19*, 2370133. [CrossRef]
39. Qiu, L.; Peng, L.; Moitra, D.; Liu, H.; Fu, Y.; Dong, Z.; Hu, W.; Lei, M.; Jiang, D.; Lin, H.; et al. Harnessing the Hybridization of a Metal-Organic Framework and Superbase-Derived Ionic Liquid for High-Performance Direct Air Capture of CO_2. *Small* **2023**, *19*, 2302708. [CrossRef]
40. Romanos, G.E.; Schulz, P.S.; Bahlmann, M.; Wasserscheid, P.; Sapalidis, A.; Katsaros, F.K.; Athanasekou, C.P.; Beltsios, K.; Kanellopoulos, N.K. CO_2 Capture by Novel Supported Ionic Liquid Phase Systems Consisting of Silica Nanoparticles Encapsulating Amine-Functionalized Ionic Liquids. *J. Phys. Chem. C* **2014**, *118*, 24437–24451. [CrossRef]
41. Fu, Q.; Kim, J.; Gurr, P.A.; Scofield, J.M.P.; Kentish, S.E.; Qiao, G.G. A Novel Cross-Linked Nano-Coating for Carbon Dioxide Capture. *Energy Environ. Sci.* **2016**, *9*, 434–440. [CrossRef]
42. Datta, A.; Krishnamoorti, R. Analysis of Direct Air Capture Integrated with Wind Energy and Enhanced Oil Recovery. *Environ. Sci. Technol.* **2023**, *57*, 2084–2092. [CrossRef]
43. Han, G.H.; Bang, J.; Park, G.; Choe, S.; Jang, Y.J.; Jang, H.W.; Kim, S.Y.; Ahn, S.H. Recent Advances in Electrochemical, Photochemical, and Photoelectrochemical Reduction of CO_2 to C^{2+} Products. *Small* **2023**, *19*, 2205765. [CrossRef]
44. Xie, Y. Preface: Special Topic on Photocatalytic or Electrocatalytic Reduction of CO_2. Sci. *China Chem.* **2020**, *63*, 1703–1704. [CrossRef]
45. Babacan, O.; De Causmaecker, S.; Gambhir, A.; Fajardy, M.; Rutherford, A.W.; Fantuzzi, A.; Nelson, J. Assessing the Feasibility of Carbon Dioxide Mitigation Options in Terms of Energy Usage. *Nat. Energy* **2020**, *5*, 720–728. [CrossRef]
46. Sundaramoorthy, S.; Kamath, D.; Nimbalkar, S.; Price, C.; Wenning, T.; Cresko, J. Energy Efficiency as a Foundational Technology Pillar for Industrial Decarbonization. *Sustainability* **2023**, *15*, 9487. [CrossRef]
47. Zhang, S.; Fan, Q.; Xia, R.; Meyer, T.J. CO_2 Reduction: From Homogeneous to Heterogeneous Electrocatalysis. *Acc. Chem. Res.* **2020**, *53*, 255–264. [CrossRef]
48. Gibbons, A.J.; Wilson, E.J. Deploying Carbon Capture and Storage in Europe and the United States: A Comparative Analysis. *J. Eur. Env. Plan. Law* **2007**, *4*, 343–352. [CrossRef]
49. Viebahn, P.; Vallentin, D.; Höller, S.; Fischedick, M. Integrated Assessment of CCS in the German Power Plant Sector with Special Emphasis on the Competition with Renewable Energy Technologies. *Mitig. Adapt. Strateg. Glob. Chang.* **2012**, *17*, 707–730. [CrossRef]
50. Wei, Y.-M.; Kang, J.-N.; Liu, L.-C.; Li, Q.; Wang, P.-T.; Hou, J.-J.; Liang, Q.-M.; Liao, H.; Huang, S.-F.; Yu, B. A Proposed Global Layout of Carbon Capture and Storage in Line with a 2 °C Climate Target. *Nat. Clim. Chang.* **2021**, *11*, 112–118. [CrossRef]
51. Aune, F.R.; Gaure, S.; Golombek, R.; Greaker, M.; Kittelsen, S.A.C.; Ma, L. Promoting CCS in Europe: A Case for Green Strategic Trade Policy? *Energy J.* **2022**, *43*, 71–102. [CrossRef]

52. Heitmann, N.; Bertram, C.; Narita, D. Embedding CCS Infrastructure into the European Electricity System: A Policy Coordination Problem. *Mitig. Adapt. Strateg. Glob. Chang.* **2012**, *17*, 669–686. [CrossRef]
53. Broecks, K.; Jack, C.; Ter Mors, E.; Boomsma, C.; Shackley, S. How Do People Perceive Carbon Capture and Storage for Industrial Processes? Examining Factors Underlying Public Opinion in the Netherlands and the United Kingdom. *Energy Res. Soc. Sci.* **2021**, *81*, 102236. [CrossRef]
54. Zhang, H. Regulations for Carbon Capture, Utilization and Storage: Comparative Analysis of Development in Europe, China and the Middle East. *Resour. Conserv. Recycl.* **2021**, *173*, 105722. [CrossRef]
55. Facchino, M.; Popielak, P.; Panowski, M.; Wawrzyńczak, D.; Majchrzak-Kucęba, I.; De Falco, M. The Environmental Impacts of Carbon Capture Utilization and Storage on the Electricity Sector: A Life Cycle Assessment Comparison between Italy and Poland. *Energies* **2022**, *15*, 6809. [CrossRef]
56. Ma, Z.; Cai, S.; Ye, W.; Gu, A. Linking Emissions Trading Schemes: Economic Valuation of a Joint China–Japan–Korea Carbon Market. *Sustainability* **2019**, *11*, 5303. [CrossRef]
57. Smits, M. The New (Fragmented) Geography of Carbon Market Mechanisms: Governance Challenges from Thailand and Vietnam. *Glob. Environ. Politics* **2017**, *17*, 69–90. [CrossRef]
58. Strefler, J.; Bauer, N.; Humpenöder, F.; Klein, D.; Popp, A.; Kriegler, E. Carbon Dioxide Removal Technologies Are Not Born Equal. *Environ. Res. Lett.* **2021**, *16*, 074021. [CrossRef]
59. Abdullah, M.R.T.L.; Siraj, S.; Ghazali, Z. An ISM Approach for Managing Critical Stakeholder Issues Regarding Carbon Capture and Storage (CCS) Deployment in Developing Asian Countries. *Sustainability* **2021**, *13*, 6618. [CrossRef]
60. Izikowitz, D.; Li, J.; Wang, E.; Zheng, B.; Zhang, Y.W. Assessing Capacity to Deploy Direct Air Capture Technology at the Country Level—An Expert and Information Entropy Comparative Analysis. *Environ. Res. Commun.* **2023**, *5*, 045003. [CrossRef]
61. Pollak, M.; Phillips, S.J.; Vajjhala, S. Carbon Capture and Storage Policy in the United States: A New Coalition Endeavors to Change Existing Policy. *Glob. Environ. Chang.* **2011**, *21*, 313–323. [CrossRef]
62. Qiu, H.-H.; Yang, J. An Assessment of Technological Innovation Capabilities of Carbon Capture and Storage Technology Based on Patent Analysis: A Comparative Study between China and the United States. *Sustainability* **2018**, *10*, 877. [CrossRef]
63. Yuan, J.-H.; Lyon, T.P. Promoting Global CCS RDD&D by Stronger U.S.–China Collaboration. *Renew. Sustain. Energy Rev.* **2012**, *16*, 6746–6769. [CrossRef]
64. Kintisch, E.U.S. Carbon Plan Relies on Uncertain Capture Technology. *Science* **2013**, *341*, 1438–1439. [CrossRef]
65. Wilson, E.J.; Morgan, M.G.; Apt, J.; Bonner, M.; Bunting, C.; Gode, J.; Haszeldine, R.S.; Jaeger, C.C.; Keith, D.W.; McCoy, S.T.; et al. Regulating the Geological Sequestration of CO_2. *Environ. Sci. Technol.* **2008**, *42*, 2718–2722. [CrossRef]
66. Liu, J.; Ding, X.; Song, X.; Dong, T.; Zhao, A.; Tan, M. Research on the Spillover Effect of China's Carbon Market from the Perspective of Regional Cooperation. *Energies* **2023**, *16*, 740. [CrossRef]
67. Tan, Q.; Zheng, J.; Ding, Y.; Zhang, Y. Provincial Carbon Emission Quota Allocation Study in China from the Perspective of Abatement Cost and Regional Cooperation. *Sustainability* **2020**, *12*, 8457. [CrossRef]
68. Sendi, M.; Bui, M.; Mac Dowell, N.; Fennell, P. Geospatial Analysis of Regional Climate Impacts to Accelerate Cost-Efficient Direct Air Capture Deployment. *One Earth* **2022**, *5*, 1153–1164. [CrossRef]
69. Huang, Y.; Cui, G.; Zhao, Y.; Wang, H.; Li, Z.; Dai, S.; Wang, J. Reply to the Correspondence on "Preorganization and Cooperation for Highly Efficient and Reversible Capture of Low-Concentration CO_2 by Ionic Liquids". *Angew. Chem. Int. Ed.* **2019**, *58*, 386–389. [CrossRef]
70. Zanatta, M.; Simon, N.M.; dos Santos, F.P.; Corvo, M.C.; Cabrita, E.J.; Dupont, J. Correspondence on "Preorganization and Cooperation for Highly Efficient and Reversible Capture of Low-Concentration CO_2 by Ionic Liquids". *Angew. Chem. Int. Ed.* **2019**, *58*, 382–385. [CrossRef]
71. Cao, X.E.; Tomski, P. The Sustainability Workforce Shift: Building a Talent Pipeline and Career Network. *Matter* **2023**, *6*, 2471–2475. [CrossRef]

energies

MDPI

Review

Wind Energy in Transition: Development, Socio-Economic Impacts, and Policy Challenges in Europe

Henryk Wojtaszek [1,*], Piotr F. Borowski [1,2], Mikołaj Handschke [1], Ireneusz Miciuła [3], Adam Stecyk [4], Anna Bielawa [5], Sławomir Ozdyk [6], Anna Kowalczyk [7] and Filip Czepło [8]

1 College of Economics and Management, SGMK Nicolaus Copernicus Superior School, 00-695 Warsaw, Poland; pborowski@sgmk.edu.pl (P.F.B.); mhandschke@sgmk.edu.pl (M.H.)
2 Department of Economics and Management, Khazar University, Baku 1096, Azerbaijan
3 Institute of Economics and Finance, University of Szczecin, 70-453 Szczecin, Poland; ireneusz.miciula@usz.edu.pl
4 Institute of Spatial Management and Socio-Economic Geography, University of Szczecin, 70-453 Szczecin, Poland; adam.stecyk@usz.edu.pl
5 Department of Enterprise Management, Institute of Management, University of Szczecin, 70-453 Szczecin, Poland; anna.bielawa@usz.edu.pl
6 Institute of Political Science and Security Studies, University of Szczecin, 70-453 Szczecin, Poland; slawomir.ozdyk@usz.edu.pl
7 Faculty of Entrepreneurship and Innovation, WSB Merito University in Warsaw, Łabiszyńska 25, 03-204 Warsaw, Poland; anna.kowalczyk@wsb.edu.pl
8 Polish Photovoltaic Instalation, 5 Politechniczna Str., 05-110 Jabłonna, Poland; filip.czeplo@gmail.com
* Correspondence: henrykwojtaszek@sgmk.edu.pl; Tel.: +48-782-950-360

Abstract: Wind energy has emerged as a strategic pillar in the global energy transition, offering both environmental and economic benefits. This comprehensive review explores the development of wind energy with a focus on the regulatory, socio-economic, and technological challenges that shape its deployment in Europe, particularly in Poland. The study highlights disparities between countries in terms of both total and per capita installed capacity, emphasizing the importance of equitable access to renewable energy. Denmark and Germany outperform larger economies like China and India in per capita terms, indicating the significance of effective policy frameworks and public engagement. The article presents detailed case studies of successful wind farm projects across the EU alongside economic evaluations including cost structures, return on investment, and local development impacts. Additionally, the role of innovation—such as floating offshore wind farms and AI-based energy management—is discussed in the context of improving efficiency and overcoming infrastructure and environmental barriers. The analysis is supported by quantitative comparisons, graphical representations, and policy reviews, culminating in practical recommendations for future growth. Wind energy's expansion depends on integrated strategies that combine policy reform, technological advancement, economic viability, and community participation.

Keywords: wind energy; renewable energy transition; energy policy; per capita wind capacity; floating offshore wind; community engagement; wind energy economics; Poland; European Union; energy infrastructure

1. Introduction

Wind energy is one of the key pillars of the energy transition and the pursuit of climate neutrality. Both globally and in Europe, countries are investing in the development of wind farms to enhance energy security, reduce greenhouse gas emissions, and support socio-economic growth [1–5].

Numerous literature reviews on wind energy have been conducted to date, focusing on various aspects of this sector. Particular attention has been paid to technological issues, including the development of wind turbines, energy storage solutions, and measures supporting the stability of power grids [6–8]. Research has also focused on analyzing energy potential and the administrative and legal barriers that hinder wind energy project development in Europe and globally [9–11]. Significant attention has also been devoted to social acceptance of wind farms, the impact of installations on local communities, and challenges related to public participation in renewable energy projects [12–14].

In particular, in the case of Poland, these topics have been extensively discussed in the scientific literature, covering both legal and administrative aspects [15–17], as well as detailed analyses of the current state and future development prospects of the wind energy market [18–22]. These publications provide a comprehensive overview of the challenges and potential inherent in the domestic wind energy sector.

However, this article stands out due to its comprehensive approach, integrating a detailed analysis of regulatory and administrative challenges of wind farm development with an assessment of their socio-economic impact at local and national levels [23–25]. The article also presents the latest technological innovations, such as floating wind farms, advanced energy management systems, and weather forecasting solutions, which have significant potential to shape the future of the industry [1,20,21]. Additionally, the article includes unique case studies of selected wind farms in Poland and European Union countries, illustrating practical solutions for effectively overcoming investment and social barriers (e.g., Potęgowo, Margonin, Horns Rev 3, or Gode Wind farms). Consequently, this work constitutes a valuable addition to the existing literature, providing not only a comprehensive overview of current issues but also practical recommendations for future development [1–21].

Modern energy policies focus on generating power from low- or zero-emission sources. To maintain its leadership in climate neutrality efforts and remain competitive in the global market, the European Union has introduced a range of initiatives and regulations aimed at accelerating the energy transition. Increasing the share of wind energy in each country's energy mix significantly strengthens energy security and positively impacts socio-economic development at both local and national levels. Recent policy documents clearly highlight the role of wind power as a crucial element of decarbonization and the energy transition. Programs such as the Net Zero Industry Act, Electricity Market Design, and the European Wind Power Action Plan aim to accelerate the growth of wind energy in Europe by simplifying administrative and legal procedures [6–12].

Wind energy already plays an essential role in Poland's energy mix and will remain one of its fundamental components. This year, onshore wind farms may become the second-largest source of electricity in Poland, surpassing lignite. The updated Polish Energy Policy aims to achieve 14.5 GW of installed wind turbine capacity by 2030. Meanwhile, the National Energy and Climate Plan projects approximately 15.8 GW from onshore wind farms and 5.9 GW from offshore wind power. The government recognizes the need to adjust these plans to the sector's dynamic growth, and implementing the right regulations could boost optimism within the industry. According to the Polish Wind Energy Association (PSEW), if the full potential is realized, wind power capacity could reach around 18 GW by 2030 [26–29] (Figure 1).

This forecast illustrates the dynamic growth of the wind energy sector in Poland, considering the objectives outlined in the updated Polish Energy Policy (PEP2040), as well as the long-term development potential up to mid-century. Starting from approximately

8.5 GW in 2024, a steady increase is projected, reaching around 15.8 GW by 2030. In the following decades, further substantial growth is anticipated, driven by technological advancements, expansion of transmission infrastructure, and innovative offshore wind solutions, such as floating wind farms, combined with advanced energy management systems.

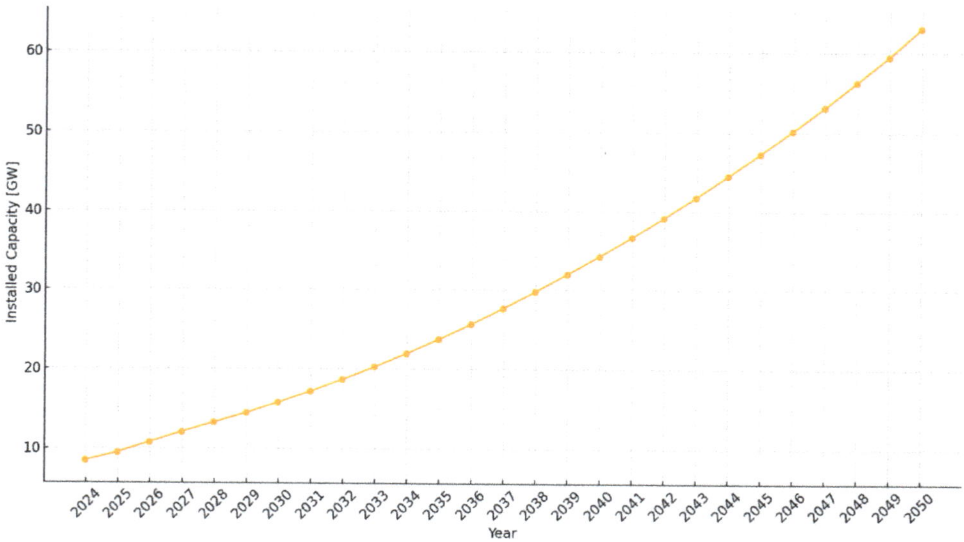

Figure 1. Forecast of Installed Wind Energy Capacity in Poland up to 2050 [GW]. Source: authors' own elaboration.

From 2030 to 2050, the growth rate may accelerate further due to the increasingly stringent climate objectives set by the European Union, stimulating member states to boost investments in renewable energy sources. Administrative process improvements and rising social acceptance will additionally facilitate this development. Consequently, the total installed wind capacity in Poland could exceed 60 GW by 2050, significantly contributing to national and European climate and energy targets. This long-term development strategy will also enhance energy security, create new employment opportunities, foster regional economic development, and support environmental sustainability through greenhouse gas emission reductions.

However, for onshore wind energy in Poland, location restrictions remain the biggest challenge. Increasing the minimum distance between wind turbines and residential buildings from 500 to 700 m last year has significantly reduced the availability of land for new investments, lowering potential installed capacity by 60–70%. Returning to a minimum distance of 500 m could open new opportunities for wind energy development while maintaining social acceptance and best practices. Such changes are not only expected by investors but also by local governments, which anticipate higher tax revenues from new wind farms [30–36].

Another obstacle to investment growth is the lengthy process of obtaining permits for onshore wind farms—one of the longest in Europe. Current administrative procedures are time-consuming, slowing down new projects and preventing full utilization of Poland's wind energy potential [37,38].

The main challenges for onshore wind energy development in Poland are location restrictions and lengthy administrative procedures. The introduction of a 700-m minimum distance requirement in 2023 significantly reduced potential installed capacity by approximately 60–70%. Reducing this distance to 500 m could accelerate industry growth, benefiting both investors and local governments through increased tax revenues [39,40].

Currently, the permitting process for wind farms in Poland is among the longest in Europe, taking up to 10 years. Simplifying these procedures and shortening project implementation timelines to 2–3 years would allow for more effective utilization of renewable energy sources (RES), in line with the EU's RED III directive [41–43].

Additional barriers include grid connection issues, as well as solutions such as cable pooling and direct power lines, which require further refinement. Collaboration between grid operators and the industry could enhance the efficiency of existing infrastructure [44].

Offshore wind energy is also developing, with new plans targeting 18 GW of installed capacity by 2040, marking significant progress compared to previous assumptions. The first wind farm in the Baltic Sea is set to start producing energy in 2026 [45–48].

2. Materials and Methods

This study employs a mixed-method approach combining qualitative and quantitative analyses to assess the development of wind energy, its challenges, and economic implications. The methodology is structured into three key sections: data collection, data analysis, and case study evaluation. The data for this research were gathered from multiple sources, including official reports and policy documents, such as those from the European Commission, Polish government energy policies, and regulatory frameworks, which were reviewed to assess policy and legislative impacts on wind energy development. Statistical data from the International Energy Agency (IEA), Eurostat, and the Polish Wind Energy Association (PSEW) were utilized to examine wind energy capacity, economic indicators, and employment statistics. Industry publications and scientific articles, including peer-reviewed journals and industry white papers, were consulted to incorporate the latest technological advancements and economic evaluations in wind energy. Additionally, selected case studies of wind farm projects in Poland and the European Union were analyzed to identify best practices and challenges in wind energy deployment [49–52].

The collected data were analyzed using both qualitative and quantitative techniques. Descriptive statistical analysis was employed to examine key economic and energy indicators using statistical methods to assess trends in wind energy development. Comparative analysis was conducted to evaluate regulatory policies, investment costs, and economic benefits of wind energy in Poland and other European countries. A SWOT (Strengths, Weaknesses, Opportunities, and Threats) analysis was performed to assess the current and future potential of wind energy development. Furthermore, economic viability was assessed using Levelized Cost of Electricity (LCOE) calculations, taking into account investment costs, operational expenses, and expected revenues [53].

A case study approach was used to examine specific wind energy projects, including Potęgowo Wind Farm in Poland (219 MW), where the analysis focused on its impact on local employment, tax revenues, and community acceptance; Margonin Wind Farm in Poland (120 MW), which was assessed for its financial and economic benefits for the local economy; Horns Rev 3 in Denmark (407 MW), where the socio-economic and environmental aspects of offshore wind energy were investigated; and Gode Wind in Germany (582 MW), where the study examined regulatory frameworks and technological innovations in offshore wind farms [54].

No human or animal intervention studies were conducted in this research. All data utilized were obtained from publicly available sources and official reports. Any proprietary datasets used have been appropriately cited, and the data processing methods adhere to ethical research guidelines. By integrating multiple data sources, analytical methods, and case study evaluations, this study provides a comprehensive overview of wind energy development, highlighting key challenges and opportunities for future expansion [55,56].

3. Dominant Countries in Wind Energy

In 2024, wind energy continues to play a pivotal role in the global energy transition, with leading countries achieving dominance through sustained investments in both onshore and offshore wind farms. China remains the undisputed leader in the global wind energy market, with an impressive installed capacity of 500 GW. This remarkable achievement underscores China's commitment to leveraging its vast geographical potential and investing in the rapid development of wind power infrastructure. The country's ability to scale both onshore and offshore wind farms at an unparalleled rate has positioned it as a global powerhouse in the renewable energy sector. Its aggressive policies and financial incentives have driven significant advancements, ensuring its leadership in both installed capacity and technological development [57,58].

The United States holds the second position globally, with a total installed wind power capacity of 160 GW. The U.S. wind energy sector is characterized by the extensive use of onshore wind farms, particularly in regions with high wind potential, such as Texas, the Midwest, and the Great Plains. These areas have capitalized on favorable wind conditions and robust state-level policies supporting renewable energy. Federal incentives, including production tax credits, have also played a key role in driving investments in the wind sector. Despite the focus on onshore wind, offshore wind projects are gaining traction, particularly along the eastern seaboard, where states such as New York and Massachusetts are pioneering large-scale offshore wind farms to diversify the country's renewable energy portfolio [54,59–66].

Germany, as Europe's leader in wind energy, has achieved an installed capacity of 73 GW, reflecting its strong commitment to renewable energy as part of its Energiewende (energy transition) strategy. Germany's efforts to expand both onshore and offshore wind capacity are driven by its ambitious climate goals, which aim to phase out coal and nuclear power while significantly reducing greenhouse gas emissions. Offshore wind farms in the North and Baltic Seas have become a cornerstone of Germany's energy policy, with state-of-the-art technology ensuring high efficiency and reliability. The German wind energy sector also benefits from robust grid infrastructure and innovative solutions for integrating intermittent renewable energy sources into the national grid [67–71].

India, with an installed capacity of 47 GW, continues to be a strong player in the wind energy market, especially in the context of Asia. The country has focused on scaling its wind energy capacity to meet its growing energy demand and reduce its dependence on fossil fuels. States such as Tamil Nadu, Gujarat, and Rajasthan are leading the way in onshore wind power development, supported by favorable policies and investments from both domestic and international stakeholders. India's renewable energy ambitions are aligned with its broader objectives of achieving energy security and fulfilling its commitments under the Paris Agreement [72–74].

Spain, with an installed capacity of 32 GW, has solidified its position as a leader in onshore wind energy. The country's commitment to renewable energy is evident in its well-established wind farms, which contribute significantly to its national energy mix. Spain has prioritized the development of wind energy to enhance energy independence and reduce reliance on imported fossil fuels. Meanwhile, the United Kingdom remains a global leader in offshore wind energy, with an installed capacity of 31 GW. The UK's innovative approach to floating offshore wind farms, combined with strong government support and a clear regulatory framework, has allowed it to maintain its leadership in this segment of the industry. Offshore projects such as Dogger Bank and Hornsea Wind Farm exemplify the UK's ambitious vision for renewable energy [75–77].

Brazil, with an installed capacity of 30 GW, stands out as the leader in Latin America's wind energy sector. The country has tapped into its vast wind resources, particularly in regions like the northeast, where consistent wind speeds create ideal conditions for wind farm development. Brazil's wind energy sector has benefited from competitive auctions and favorable financing mechanisms, which have attracted significant investments from private and public entities alike [78–80].

France, with 25 GW of installed capacity, is intensively developing its offshore wind sector, focusing on expanding its renewable energy capacity to meet ambitious climate targets. Offshore projects in the Atlantic and Mediterranean regions are expected to play a crucial role in France's energy transition strategy. The Netherlands is another key player in the European wind energy market, with cutting-edge offshore wind projects such as Hollandse Kust demonstrating its commitment to innovation and sustainability. Denmark, a pioneer in wind energy, continues to lead in terms of the highest share of wind energy in its national energy mix, with wind accounting for an impressive 56% of its electricity consumption. Denmark's early adoption of wind energy technologies and its focus on integrating renewable energy into the grid have made it a model for other countries to follow [72,74,81–83].

Europe, as a whole, plays a central role in the advancement of wind energy, leveraging advanced technologies, regulatory support, and collaborative efforts among member states. Countries like Germany, the United Kingdom, Spain, Denmark, and The Netherlands are at the forefront of innovation, setting global benchmarks for the deployment and integration of wind energy. European leadership in this sector is driven by a shared commitment to decarbonization and sustainable development, with wind energy serving as a cornerstone of these efforts (Figure 2) [84–86].

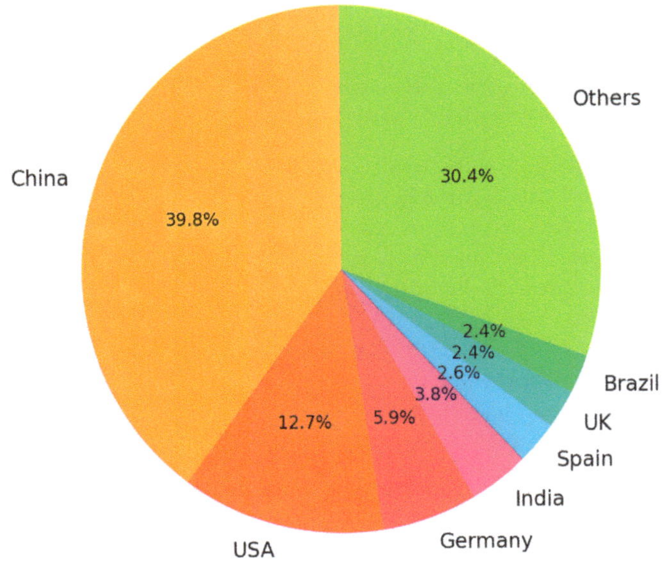

Figure 2. Installed Wind Power Capacity Worldwide (2024). Source: authors' own elaboration.

The global dominance of countries like China, the United States, and Germany in wind energy reflects their ability to capitalize on geographic potential, implement supportive policies, and foster technological innovation. Meanwhile, European nations, with their advanced infrastructure and commitment to renewable energy, continue to shape the future of the wind energy sector. The accompanying chart highlights the global distribution of installed wind power capacity in 2024, showcasing the countries that are driving growth and innovation in this critical area of renewable energy development [80,87–90].

The installed wind power capacity worldwide in 2024 is illustrated in the second chart, showcasing the countries with the largest capacities in this sector.

Although the absolute values of installed capacity confirm the dominance of countries such as China and the United States, it is equally important to analyze the installed wind power capacity per capita, which reflects actual access to renewable energy for citizens. This perspective reveals that smaller, highly developed nations, such as Denmark and Germany, achieve significantly higher per capita performance despite lower total capacity. A detailed analysis and comparison are presented below (Figure 3).

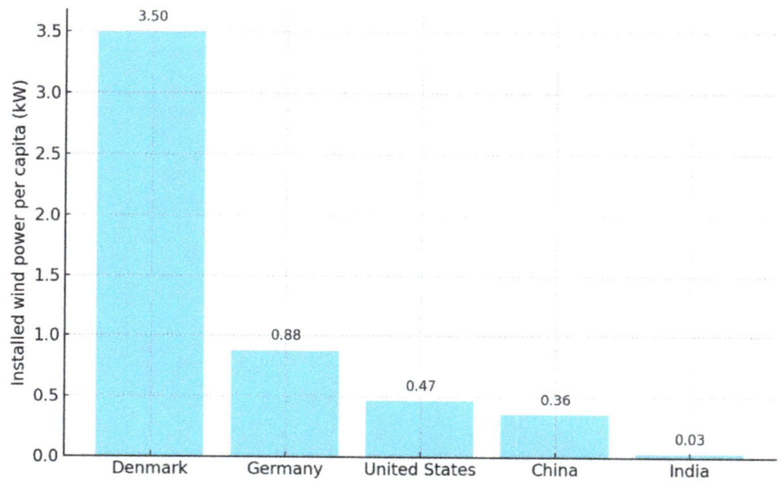

Figure 3. Comparison of wind energy availability per capita in Denmark, Germany, the United States, China, and India (2024). Source: authors' own elaboration.

In 2024, the global wind energy market was dominated by several key countries that stood out both in terms of total installed capacity and the pace of renewable infrastructure development. China led the way with an impressive 500 GW of installed capacity, significantly surpassing other nations. The United States ranked second with 160 GW, followed by Germany with 73 GW. Other major European contributors included Spain (32 GW), the United Kingdom (31 GW), France (25 GW), Denmark (21 GW), and the Netherlands (23.5 GW).

However, total installed capacity alone does not provide a full picture of investment efficiency. A more meaningful comparison requires analyzing installed capacity per capita, which accounts for population size and better reflects the actual availability of wind energy for citizens. In this regard, smaller, highly developed countries demonstrate significantly higher performance despite lower overall capacity.

Denmark, for example, with a population of approximately 6 million and 21 GW of installed wind power, achieves 3.50 kW per capita—the highest among the compared countries. Germany follows with 0.88 kW per capita, based on 73 GW of capacity and 83 million inhabitants. The United States reaches 0.47 kW per person with 160 GW of capacity and a population of around 340 million. China, despite its global leadership in total capacity, reaches only 0.36 kW per capita due to its large population of 1.4 billion.

Lastly, India, with 47 GW of installed capacity and the same population size as China, records the lowest per capita figure at just 0.034 kW.

The data clearly show that while China leads in total wind energy investments, on a per capita basis, it lags behind smaller, more developed countries. Nations like Denmark and Germany have integrated wind energy more effectively into their local systems, providing better access to clean energy for citizens. India, despite having a total installed capacity of 47 GW, has a per capita figure close to zero (0.034 kW), highlighting the imbalance between population size and the scale of renewable investments.

In conclusion, per capita analysis reveals substantial differences in wind energy accessibility between countries, allowing for a more balanced assessment of energy policies' effectiveness from a population-centered perspective.

Leading the ranking is China, which, with an impressive capacity of 500 GW, remains the undisputed global leader in wind energy. The United States follows in second place with 160 GW, reflecting the significant role of wind energy in its energy transition strategy. In third place is Germany, with a capacity of 73 GW, maintaining its position as the leader of the European wind energy market.

Further down the ranking, other European countries such as Spain (32 GW) and the United Kingdom (31 GW) demonstrate continued strong investments in both onshore and offshore wind farms. The list also includes Brazil (30 GW), which is characterized by a rapidly growing wind energy sector, and France (25 GW), which places a strong emphasis on developing offshore wind energy. The Netherlands (23.5 GW) and Denmark (21 GW) also feature prominently, with Denmark achieving the highest share of wind energy in its national electricity mix at 56%. These data highlight the pivotal role of wind energy in the global effort to transition towards more sustainable energy systems (Figure 4).

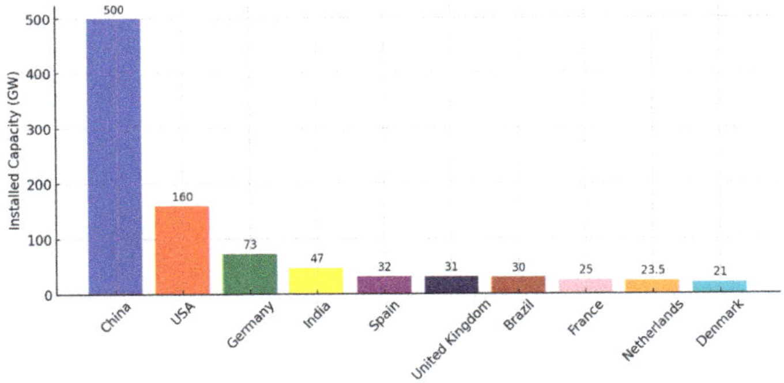

Figure 4. Installed Wind Power Capacity by Country (2024). Source: authors' own elaboration.

In 2024, the global wind energy sector was dominated by a single country, accounting for more than half of the total installed capacity, highlighting its massive investments in renewable energy. The second-largest contributor has a share more than three times smaller, emphasizing the scale of disparity between the leaders. The third position is held by a European country, which, although significantly behind the top economies, still plays a crucial role in the market. The remaining nations have considerably lower capacities, with their contributions to the global total not exceeding a few percent (Figure 4).

As a region, Europe maintains a strong position in the wind energy sector, although no single economy matches the global leaders. At the same time, emerging markets are showing increasing potential, steadily expanding their investments and share in wind energy production. The percentage analysis reveals the clear dominance of one country, a significant but much smaller presence of the second and third, and a relatively balanced distribution of smaller shares among the rest. These trends indicate a growing concentration of the sector in a few dominant hubs while new players gradually strengthen their position in the renewable energy market.

4. Challenges in the Development of Wind Energy

Wind energy faces several challenges, including location restrictions, complex administrative processes, social barriers, and the variability of weather conditions. In Poland, one of the main obstacles is the regulation regarding the minimum distance between turbines and residential buildings, which significantly limits available areas for investment. This regulation has greatly reduced the potential land for projects, thereby restricting the possible installed capacity. It is estimated that the current 700-m limit reduces the investment potential by approximately 60–70% compared to more flexible regulations, such as a minimum distance of 500 m.

Additionally, the lengthy process of obtaining permits for the construction of new wind farms remains a significant challenge. Administrative procedures in Poland are among the longest in Europe, causing delays in project implementation and increasing costs. The lack of digitization and the formal complexity of these processes further hinder the development of the sector.

Social aspects also pose barriers, such as the lack of acceptance from local communities due to concerns about noise, impact on the landscape, or potential decreases in property values near wind farms. To address these issues, it is crucial to conduct public consultations and educate communities about the benefits of wind energy.

Another challenge is the variability of weather conditions, which necessitates the development of energy storage systems and technologies that support grid stability. These include battery storage, hydrogen, and Power-to-X systems, which can mitigate the effects of intermittent energy generation.

Table 1 presents key economic aspects related to the development of wind energy, taking into account both investment and operational costs for onshore and offshore wind farms. The data also include estimated payback periods, potential revenues, and the impact on local economies, such as job creation and tax revenues. This analysis highlights the economic benefits of wind energy development and its importance for the sustainable development of regions where wind farms are established [88,91–93].

Table 1. Key economic aspects related to the development of wind energy [1–26,92].

Category	Details
Cost of building onshore wind farms	1.2–1.5 million EUR/MW
Cost of building offshore wind farms	2.5–3 million EUR/MW
Operational and maintenance costs onshore	30,000–50,000 EUR annually/MW
Operational and maintenance costs offshore	20–30% higher than onshore
Return on investment time (onshore)	8–12 years
Return on investment time (offshore)	12–15 years
Annual revenue of an onshore farm (50 MW)	6–8 million EUR annually
Jobs during the construction phase (per 1 MW)	10 jobs
Permanent jobs during the operational phase (per 1 MW)	0.3–0.5 jobs
Tax revenues per 1 MW of capacity	50,000–100,000 EUR annually

Source: authors' own elaboration.

An economic analysis of wind energy is presented in Table 1.

Table 1 outlines the financial specifics associated with the development and operation of wind farms, highlighting the differences between onshore and offshore installations as well as the economic impact, including job creation and tax revenues.

5. Social Acceptance of Wind Energy

Despite its numerous advantages, wind energy often faces concerns from local communities. The primary issues are related to landscape aesthetics, potential noise generated by turbines, and the impact on property values. An additional challenge is the lack of sufficient information about the benefits of such investments. To address these challenges, the European Wind Power Action Plan emphasizes improving transparency and community involvement in renewable energy projects. Proper information dissemination regarding economic and environmental benefits has been shown to significantly enhance local acceptance [75,83,88,93–95].

To increase social acceptance, implementing effective measures at various levels is crucial. Involving local communities in the planning process through early-stage public consultations can help dispel doubts and build trust. For example, in Denmark, residents have a say in selecting the locations of wind farms, which significantly increases their engagement and acceptance. The European Commission recommends early public consultations as part of the Wind Power Action Plan, highlighting that effective engagement methods, including public participation in site selection processes, are essential to achieving sustained local support [93,95–99].

Another essential step involves educational initiatives, such as information campaigns that highlight the environmental and economic benefits of wind farms. In Germany, open days organized at wind farms have gained popularity, allowing residents to learn about the technology and its impact on the surrounding environment.

Educational initiatives, such as open days at wind farms in Germany, play a critical role in social acceptance by enabling residents to directly observe the operation of wind turbines, thus significantly reducing public misconceptions about their environmental and technological impact [100–102].

A key element in supporting local communities is enabling them to share in the profits of wind farms through equity systems or compensation. In the United Kingdom, developers establish funds to support local initiatives, such as infrastructure development. Additionally, minimizing the impact on the landscape and environment—such as by using modern noise-reducing technologies and selecting locations distant from residential areas—can significantly influence acceptance. For instance, in The Netherlands, modern turbines with low noise levels are being utilized. The EU's Wind Power Action Plan further encourages economic benefit-sharing schemes, including financial compensation and local equity participation. Such measures have notably improved public acceptance in multiple European countries by creating direct economic benefits for affected communities [98,103].

Financial transparency also plays a critical role. Informing residents about planned investments and their impact on local economic development and infrastructure fosters trust and support for projects. According to the Wind Power Action Plan, transparency measures, such as public disclosure of financial data and project timelines via dedicated EU platforms, significantly increase local communities' trust in wind energy projects [102,104–106].

Practical successes in this area can be observed in Denmark, where local ownership of wind farms and profit-sharing with residents have yielded positive results. Similar solutions have been implemented in France, where educational and compensatory mechanisms were introduced. In Poland, good practices include municipalities where wind farms have funded the construction of roads and educational facilities, improving the image of wind energy among local communities. The EU's Net Zero Industry Act (NZIA) emphasizes legislative solutions, including the introduction of non-price criteria into renewable energy auctions, allowing projects to be evaluated based on environmental and community value. Such an approach further strengthens community acceptance by prioritizing projects beneficial to local economic development, employment, and environmental sustainability [75,107–109].

In summary, effective collaboration with residents, education, and financial transparency form the foundation for social acceptance of wind energy investments.

The Wind Power Action Plan, or the European Wind Power Action Plan, was developed in response to the challenges associated with accelerating the growth of the renewable energy sector. This document aims to streamline investment processes, increase transparency, and ensure the competitiveness of Europe's wind energy sector on the global market. The European Commission (EC) emphasizes the importance of an integrated approach that addresses technological, administrative, and regulatory aspects.

Table 2 outlines the key actions of the EC under the Wind Power Action Plan, which aim to support the sustainable development of wind energy, simplify procedures, and increase investments in innovative technologies [93,110–113].

Table 2. Key actions of the European Commission [25,114–116].

Action	Description
Proper auction design	New project qualification criteria: data security, environmental protection, delivery guarantees.
Contract indexing	Indexing of prices and auction tariffs by member states.
Investment financing	Funds for factories, infrastructure, and personnel related to wind energy.
Strengthening the role of the EIB	Risk mitigation tools, guarantees for the banking sector.
Information transparency	Publication of national auction plans on the EU platform.
Equal competition conditions	Monitoring of unfair trade practices, EC trade tools.
Acceleration of permits	Launching the EasyPermits tool for digitizing permit issuance processes.

Source: authors' own elaboration.

Wind energy constitutes a key pillar of the energy transition in Europe. Recognizing its importance, the European Commission introduced the Wind Power Action Plan—a comprehensive strategy designed to accelerate investments in wind farms. The plan focuses on streamlining administrative procedures, ensuring transparency of information, and implementing solutions to enhance the competitiveness of the European wind energy sector. A notable initiative under this plan is the EasyPermits 2023 tool, which digitizes permitting processes to reduce project implementation timelines. Additionally, contract indexing stabilizes the renewable energy investment market by providing price predictability.

Enhancing the competitiveness of Europe's wind energy sector is a crucial step toward achieving energy self-sufficiency and meeting climate goals. To achieve this, the Wind Power Action Plan outlines targeted actions addressing administrative, financial, and regulatory challenges, fostering a more efficient and sustainable energy transition.

The plan establishes proper auction design, introducing new qualification criteria for wind energy projects. These include requirements for data security, environmental protection, and delivery guarantees, ensuring high standards and reliable execution of projects. It also emphasizes indexing contracts, allowing EU member states to index prices and auction tariffs, which provides stability and predictability for investors.

Further, the investment financing action allocates funds for developing factories, infrastructure, and workforce training programs specific to the wind energy sector, driving innovation and economic growth. The Strengthening of the European Investment Bank (EIB) enhances this effort by introducing risk-reduction tools and guarantees for the banking sector, encouraging increased investments in wind energy projects.

The plan promotes transparency of information through the publication of national auction plans on an EU platform. This ensures stakeholders have access to clear and consistent information, fostering trust and broader participation in the sector. To ensure fair competition, a level playing field is maintained by monitoring unfair trade practices, with the European Commission providing tools to address these issues.

Finally, the acceleration of permits focuses on simplifying and expediting permitting processes through the launch of the EasyPermits tool. This digital solution significantly reduces delays in project implementation and facilitates faster deployment of wind energy infrastructure [84,117–119].

Collectively, these actions aim to enhance the efficiency, transparency, and competitiveness of the wind energy sector, contributing to Europe's broader goals of energy security, sustainability, and climate neutrality. By addressing critical barriers and fostering innovation, the Wind Power Action Plan ensures wind energy remains a cornerstone of Europe's energy transition (Table 3).

Table 3. Net-Zero Industry Act (NZIA).

Net-Zero Industry Act (NZIA)	Description
Goal	Increasing the production of net-zero emission technologies within the EU.
Key Actions	Simplifying regulations, improving the investment environment, and supporting green technologies.
Significance	Achieving climate neutrality and enhancing the EU's energy resilience.
Legislative Solutions	Allowing the inclusion of non-price criteria in renewable energy auctions (e.g., economic and environmental value).

Source: authors' own elaboration.

The introduction of the Net-Zero Industry Act (NZIA) by the European Union is a response to global challenges related to the decarbonization of industry. The key objective of this act is to increase the production of green technologies within the EU and to ensure a stable investment environment by simplifying regulations. The EU aims to accelerate the energy transformation while simultaneously supporting innovation and local supply chains, enhancing the EU's energy resilience by reducing dependence on imported fossil fuels and promoting self-sufficiency in energy through domestic green technologies.

Wind energy, a crucial element of the European Union's energy strategy, plays a significant role in these efforts. Achieving ambitious goals such as climate neutrality requires both international cooperation and the mobilization of resources at the national level. Strengthening the wind sector will not only contribute to achieving these climate goals but will also increase Europe's energy independence by reducing reliance on raw material imports. At the same time, investments in new technologies and infrastructure are expected to create numerous jobs, supporting economic development across the region.

The NZIA seeks to foster the development and production of net-zero emission technologies by implementing key actions such as improving the investment environment and allowing the inclusion of non-price criteria in renewable energy auctions, which means that decisions on winning bids can consider not just the cost but also the economic and environmental value they bring. This could prioritize projects that promise greater job creation locally, better environmental benefits like biodiversity conservation, or superior long-term sustainability.

These legislative solutions and actions underline the EU's commitment to a sustainable and environmentally friendly economy, positioning the NZIA as a robust framework designed to transition the EU toward a sustainable industrial base, supporting economic growth in a manner that aligns with environmental goals and global climate commitments.

6. Challenges and Needs of the Wind Energy Sector

The development of the wind energy sector faces several critical challenges that must be addressed to ensure its long-term sustainability and competitiveness. Key issues include supply chain bottlenecks, dependence on imported components, infrastructure shortages, and workforce deficits.

One of the most pressing concerns is the bottlenecks in supply chains, as Europe currently lacks sufficient production capacity for offshore wind foundations, cables, and steel towers. This limitation slows down the expansion of wind farms and increases reliance on non-European suppliers. Additionally, dependence on imports remains a significant challenge, with many key components such as cables, gearboxes, and steel towers being sourced from China. This reliance increases vulnerability to market fluctuations and geopolitical risks.

Another major obstacle is the shortage of infrastructure, including factories, ports, ships, and cranes necessary for the efficient development of both onshore and offshore wind energy projects. Without significant investments in industrial capacity and logistics, meeting Europe's ambitious wind energy targets will be difficult. Furthermore, the shortage of skilled workers highlights the need for greater investments in education and training programs to build a qualified workforce capable of supporting the growing wind sector.

To support the expansion of wind energy, it is essential to eliminate these key obstacles. Expanding infrastructure, increasing the number of local factories, and investing in workforce education are crucial steps toward meeting the rising demand for renewable energy. Strengthening Europe's domestic production capacity will enable the continent to position itself as a global leader in wind farm component manufacturing, enhancing its energy independence and economic resilience [120–123].

6.1. *Ambitious Energy Targets of the European Union*

The European Union has set bold targets to accelerate the deployment of wind energy as part of its broader strategy for energy transition and carbon neutrality. The primary objective is to increase the share of wind energy to 43% of Europe's electricity consumption by 2030, a significant leap from the current 17%. To achieve this, the EU plans to construct 35 GW of new wind farms annually by 2030.

Meeting these ambitious goals requires a substantial acceleration in renewable energy investments. Increasing the share of wind energy in Europe's power mix is essential for reducing CO_2 emissions and strengthening energy security. These efforts will not only contribute to environmental protection but also create thousands of new jobs and stimulate regional economic growth [124–127].

6.2. *Future Prospects*

Global projections indicate that installed wind power capacity could surge from 272 GW in 2023 to 3540 GW by 2030. Europe is expected to play a crucial role in achieving these targets, driven by regulatory support and technological innovation. The continent is planning a large-scale expansion of its transmission infrastructure and is actively supporting offshore wind projects, which are becoming an increasingly vital component of the region's electricity supply.

One of the most promising advancements in offshore wind technology is the development of floating wind turbines, which enable energy generation in deeper waters that were previously inaccessible to conventional turbines. Countries such as Germany, The Netherlands, and the United Kingdom are leading the way in offshore wind expansion while also implementing administrative reforms to streamline project approval processes. The goal is to reduce project implementation timelines from several years to just a few months, making wind energy deployment faster and more efficient.

On a global scale, the integration of renewable energy sources into existing power systems remains a key challenge. The successful implementation of advanced energy management systems and energy storage technologies will play a crucial role in stabilizing supply and ensuring the reliability of power grids.

Experts emphasize that the future growth of the wind energy sector depends on increased investment in research and development, particularly in next-generation turbine technologies that offer higher efficiency and lower operational costs. In developing countries, financial support mechanisms such as subsidies and low-interest loans will be essential for accelerating the adoption of wind energy.

By addressing these challenges and leveraging emerging opportunities, the wind energy sector can continue to expand, contributing to Europe's leadership in the global transition toward sustainable and resilient energy systems [128–131].

7. Results

7.1. Technological Analysis of Innovations in Wind Energy

7.1.1. New Technologies in Wind Energy

The development of technology in wind energy plays a key role in increasing turbine efficiency, improving energy supply stability, and integrating wind power into the electrical grid. Modern solutions not only enhance wind potential utilization but also minimize operational costs and reduce environmental impact. Innovations in this field focus on improving turbine design, advancing energy storage systems, implementing floating wind farms, and utilizing advanced predictive and grid management technologies.

Table 4 below presents key new technologies in wind energy, their applications, and their impact on the renewable energy sector.

The development of wind energy is closely linked to technological advancements that enhance turbine efficiency, optimize energy management, and reduce operational costs. Modern technological solutions are crucial for the further growth of this sector and its integration into the power grid.

One of the most significant trends in this field is the use of high-efficiency turbines with taller towers. Modern wind turbines feature larger rotors and longer blades, allowing for better utilization of wind potential, particularly in areas with lower wind speeds. The use of taller towers, reaching up to 200 m, enables access to more stable and stronger air currents. The introduction of new composite materials helps reduce the weight of turbines and their susceptibility to damage. Additionally, variable-speed technology optimizes energy production depending on atmospheric conditions.

A key challenge for wind energy is the variability of energy production due to changing weather conditions. Modern energy storage systems play a crucial role in stabilizing supply and integrating wind energy into the power grid. Various technologies are used for this purpose, including battery storage solutions such as lithium-ion, flow, and sodium/sulfur batteries, which allow excess energy to be stored and released during peak demand periods. Hydrogen systems are also gaining importance, enabling the conversion of surplus wind energy into hydrogen, which can be stored and used for power generation when needed. The development of artificial intelligence and predictive load management algorithms improves energy flow management and forecasts changes in production and consumption.

Table 4. New Technologies in Wind Energy.

Category	Description
High-efficiency turbines with taller towers	Modern wind turbines feature larger rotors and taller towers, optimizing wind energy potential. Key innovations include: - Taller towers (up to 200 m); - Longer rotor blades; - New composite materials reducing weight and damage susceptibility; - Variable speed technology optimizing energy production.
Energy storage systems and grid integration	Solutions for stabilizing wind energy generation fluctuations through energy storage. Key technologies include: - Battery storage (lithium-ion, flow, sodium/sulfur); - Hydrogen systems (Power-to-Gas); - AI and grid load prediction systems.
Floating wind farms as a new development direction	Floating wind farms enable wind energy utilization in deep-sea areas. Key aspects include: - Floating foundations instead of seabed-mounted structures; - Higher wind energy potential in deep waters; - Reduced spatial conflicts with fisheries and shipping, - Notable projects: Hywind Scotland, WindFloat Atlantic.
Advanced weather prediction and energy management systems	Advanced weather prediction and energy management technologies enhance wind farm efficiency include: - AI algorithms analyzing real-time meteorological data; - IoT-based monitoring systems for turbine diagnostics; - Satellite data and climate models for improved weather forecasts; - Smart grid management systems dynamically optimizing energy transmission.

Source: authors' own elaboration.

A new direction in wind energy development is floating wind farms, which allow for the utilization of strong and stable winds in deep-sea areas. Instead of traditional seabed-mounted foundations, turbines are installed on floating platforms, enabling deployment in locations with depths exceeding 50 m. In addition to increased efficiency, this solution minimizes spatial conflicts by reducing the impact on local ecosystems and other industries such as fisheries and maritime transport. The first successful implementations of this technology, such as Hywind Scotland and WindFloat Atlantic, demonstrate its effectiveness and open new investment opportunities (Figure 5).

Figure 5 provides a detailed illustration of a floating offshore wind turbine, highlighting its primary technological and structural components. At the top, a modern wind turbine is depicted, consisting of rotor blades designed to efficiently harness strong and consistent winds prevalent over open ocean waters. These blades drive a turbine, generating renewable electricity.

Below the turbine, a robust tower connects directly to a floating platform, engineered to maintain stability even under harsh maritime conditions such as strong winds, waves, and currents. The platform's design ensures minimal motion, which is crucial for both turbine operation and longevity.

Anchoring systems, clearly shown in the schematic, secure the platform to the seabed. These mooring lines ensure that the turbine maintains a consistent position, greatly reducing the risks associated with drifting. This anchoring mechanism typically consists of heavy-duty chains, cables, or synthetic ropes attached to anchors embedded in the ocean floor.

Moreover, the figure distinctly illustrates the undersea power cables, responsible for transferring the generated electricity from the floating platform to the shore. These subsea cables represent a critical infrastructure component, facilitating efficient energy delivery to the electrical grid on land.

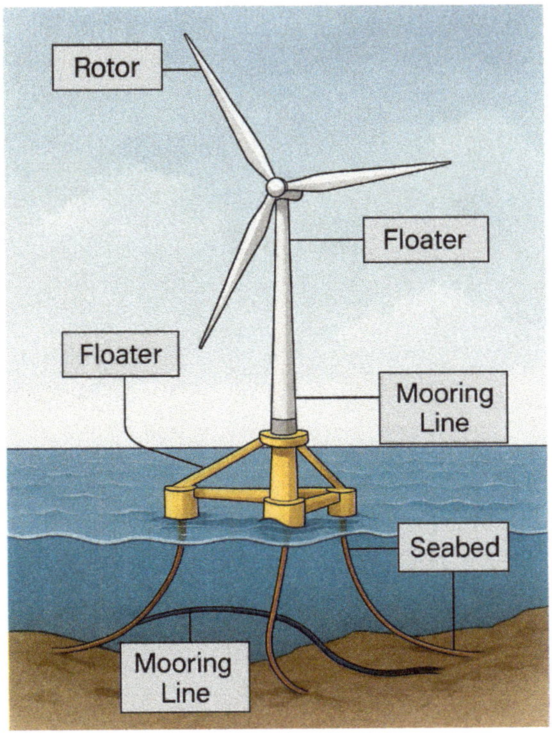

Figure 5. Enhanced Schematic of a Floating Offshore Wind Farm. Source: authors' own elaboration.

Floating wind turbine technology, exemplified by projects such as Hywind Scotland and WindFloat Atlantic, significantly expands possibilities for offshore wind deployment, particularly in deep-water locations (greater than 50 m), where fixed-bottom turbines are economically and technically unfeasible. Additionally, the floating turbine approach reduces environmental impacts, mitigates spatial conflicts with fishing and shipping industries, and opens substantial areas previously inaccessible for energy generation.

Overall, floating offshore wind farms represent a transformative step forward for renewable energy, enhancing the sustainability, flexibility, and economic viability of wind power generation in diverse maritime environments.

Another innovation in wind energy is the use of advanced weather prediction and energy management systems. Modern artificial intelligence algorithms analyze meteorological data in real-time, allowing turbines to adjust their operations according to current weather conditions. The development of IoT technology enables efficient monitoring of turbine technical conditions and predictive maintenance, minimizing the risk of failures and reducing operational costs. The use of satellite data and climate models improves the accuracy of weather forecasts, facilitating energy production planning and reducing energy losses. Smart grid management systems dynamically optimize energy transmission, adapting it to predicted changes in generation and demand.

Technological advancements in wind energy enable more efficient and stable use of this renewable energy source, contributing to the further decarbonization of the energy sector and increasing the share of renewable energy in the global energy mix.

7.1.2. Comparison of Wind Energy with Other Renewable Energy Sources

Wind energy is one of the most important renewable energy sources (RES), alongside solar and hydropower. Each of these technologies plays a crucial role in the energy transition, but their application depends on various factors such as geographical conditions, technological efficiency, investment costs, and environmental impact.

To better understand the position of wind energy within the renewable energy production structure, it is useful to compare it with other solutions based on key parameters such as energy production efficiency, supply stability, investment and operational costs, environmental impact, and integration with the power grid [132,133].

Table 5 below provides a comparison of the key aspects of wind, solar, and hydropower energy, allowing for an assessment of their advantages and limitations in the context of their application in Poland and Europe.

Table 5. Comparison of Key Aspects of Wind, Solar, and Hydropower Energy in the Context of Their Application in Poland and Europe.

Parameter	Wind Energy	Solar Energy	Hydropower
Investment Costs	Medium to high	Low to medium	High
Operational Costs	Medium	Low	Low to medium
Efficiency in Climate	High in windy areas	Dependent on sunlight	High, but requires a water reservoir
Development Potential in Poland	High on land and sea	High, especially in cities	Limited by water resources
Scalability	High, but requires large areas	High, with rooftop installation options	Limited by access to suitable rivers

Source: authors' own elaboration.

The comparative analysis of wind, solar, and hydropower energy highlights significant differences in costs, efficiency, and development potential for each technology. Solar energy stands out with the lowest investment and operational costs, making it particularly attractive for individual households. Wind energy, despite its moderate investment and operational costs, has the potential to generate substantial amounts of electricity, especially in areas with favorable wind conditions. Hydropower, while requiring high initial investments, offers stable and long-term energy production.

In terms of efficiency under different climatic conditions, wind energy performs best in regions with high average wind speeds, which in Poland primarily include coastal and central areas. Solar energy is most effective in locations with high sunlight exposure, but technological advancements now allow for its use even in less sunny regions. Hydropower provides a stable energy supply, but its efficiency is highly dependent on the availability of suitable rivers and water reservoirs.

The development potential of these technologies also varies based on geographic and technological factors. Wind energy, particularly offshore wind farms, has significant expansion potential and the capacity to deliver large amounts of electricity. Solar energy is highly scalable and can be utilized on both small scales, such as rooftop installations, and large-scale solar farms. Hydropower, though constrained by geographic limitations, still has development potential through the modernization of existing hydroelectric plants and the optimization of their efficiency [131,134].

7.1.3. Environmental Aspects and the Impact of Wind Energy on Ecosystems

Wind energy is one of the key sources of renewable energy, contributing to the reduction of greenhouse gas emissions and decreasing dependence on fossil fuels. Despite its many environmental benefits, the development of wind farms also presents certain environmental challenges that can impact both terrestrial and marine ecosystems.

One significant aspect is the effect of wind turbines on birds and other animal species, including the risk of collisions and habitat fragmentation. Migratory species are particularly vulnerable, as wind farms can obstruct their flight paths. In the case of offshore wind farms, seabed disturbances during construction and underwater noise generated during operation can disrupt marine ecosystems.

Another challenge is the noise produced by turbines, which can affect both humans and wildlife. However, technological advancements allow for the implementation of noise reduction solutions, such as sound-absorbing materials and optimized rotor blade design. Additionally, the issue of recycling wind turbine components, particularly rotor blades made from composite materials, remains a challenge due to the difficulty of processing these materials.

Table 6 below outlines the key environmental aspects of wind energy, its impact on ecosystems, and measures aimed at minimizing negative effects.

Table 6. Environmental Impact of Wind Farms and Mitigation Measures [89,127–130].

Category	Description	Mitigation Measures
Impact of Wind Farms on Birds and Other Animal Species	Wind turbines can pose a threat to birds and bats due to collisions with rotors and habitat fragmentation. Migratory species are particularly at risk. Wind farms can also influence animal behavior, deterring some species from nesting areas.	Conducting site assessments before construction to avoid migration areas. Using bird detection technology and automatic turbine shutdown systems. Painting turbine blades in contrasting colors to improve visibility for birds.
Effects of Offshore Wind Farms on Marine Ecosystems	Offshore wind farms disrupt the seabed during installation, potentially destroying habitats and altering sediment dynamics. Underwater noise generated by turbines may affect the behavior of marine mammals and fish.	Employing quieter installation methods for turbine foundations and creating artificial reefs around turbines to enhance biodiversity. Monitoring underwater noise levels and adjusting turbine operations during sensitive periods for marine wildlife.
Technologies for Noise Reduction and Other Negative Impacts	Noise from rotating turbine blades and internal mechanisms can impact both humans and terrestrial or marine animals. Various technologies help reduce noise emissions, such as the use of sound-absorbing materials and optimized blade design.	Using advanced noise-absorbing materials in turbine construction. Optimizing rotor blade shape based on owl wing structures to reduce noise emissions. Locating wind farms near natural sound barriers, such as forests.
Life Cycle Analysis of Wind Turbines and Component Recycling	Wind turbines have a lifespan of 20–30 years, after which the challenge of recycling components arises, particularly rotor blades made from fiber-reinforced composites. Recycling technologies are being developed, including repurposing materials for cement production and designing turbines with easier material recovery in mind.	Developing rotor blade recycling technology, e.g., repurposing them for cement production. Reusing mechanical and electrical components in new turbines or other industries. Designing turbines with recyclable materials to improve sustainability.

Wind energy plays a key role in the energy transition, contributing to the reduction of greenhouse gas emissions and decreasing dependence on fossil fuels. Despite its positive environmental impact, the development of wind farms presents certain ecological challenges that can affect both terrestrial and marine ecosystems.

One of the most significant environmental aspects is the impact of wind turbines on birds and other animal species. Their operation can lead to collisions with rotor blades and habitat fragmentation, posing a particular threat to migratory species for which wind farms may create obstacles along their flight paths. In offshore wind farms, seabed disturbances during turbine installation can result in habitat degradation for benthic organisms and alter sediment dynamics. Additionally, the noise generated by turbines during operation can disrupt the behavior of marine mammals and fish, interfering with their navigation and communication processes.

Noise pollution is also a concern on land, where it can affect both humans and wildlife. To address this issue, new technologies are being developed to reduce noise emissions, including the use of sound-absorbing materials and the optimization of rotor blade shapes. The placement of wind farms is also planned with consideration for natural sound barriers, such as forests, which can further reduce noise propagation.

Another important challenge is the recycling of wind turbine components, particularly rotor blades, which are made from fiber-reinforced composites that are difficult to process. In response, innovative recycling technologies are being developed, including repurposing blades for use in the cement industry and designing turbines with improved material recovery in mind. Additionally, mechanical and electrical components from decommissioned turbines can be reused in new installations or other industrial sectors.

Despite these challenges, the expansion of wind farms remains one of the most environmentally friendly directions in the energy sector. With modern solutions aimed at minimizing their environmental impact, it is possible to achieve a balance between producing clean energy and preserving ecosystems.

7.2. Analysis of Successful Wind Farm Projects in Poland and Selected European Union Countries

7.2.1. Examples of Successful Wind Farm Projects in Poland and Their Impact on Local Communities

The development of wind energy in Poland has brought numerous benefits to local communities, including economic growth, infrastructure improvements, and increased environmental awareness. Several large-scale wind farm projects have successfully integrated renewable energy production with regional development, fostering social acceptance through consultations and reinvestment in local economies. Below is an overview of selected wind farms in Poland, highlighting their capacity, benefits, and community engagement efforts (Table 7).

Table 7. Successful Wind Farm Projects in Poland and Their Impact on Local Communities.

Wind Farm Name	Location	Capacity	Benefits	Community Engagement
Potęgowo Wind Farm	Pomeranian Voivodeship, Poland	219 MW (largest wind farm in Poland)	New jobs, infrastructure development, revenue for the municipal budget	Local consultations and reinvestment of profits into regional development
Margonin Wind Farm	Greater Poland Voivodeship, Poland	120 MW, 60 wind turbines	Significant tax revenues enabled road modernization and public facility construction	Informational campaigns by local authorities to increase public acceptance
Jasna Wind Farm	Pomeranian Voivodeship, Poland	132 MW, developed following sustainable principles	New jobs, increased environmental awareness, active community consultations	Meetings with residents and environmental education initiatives

Source: authors' own elaboration.

Many successful wind farm projects have been developed in Poland, significantly contributing to the growth of local communities. One of the largest and most impressive projects is the Potęgowo wind farm in the Pomeranian Voivodeship, with a capacity of

219 MW, making it the largest wind farm in the country. This investment has led to the creation of new jobs, the development of local infrastructure, and substantial revenues for the municipal budget. Additionally, thanks to local consultations and the reinvestment of profits into regional development, the project has gained broad public acceptance. Another example is the Margonin wind farm in the Greater Poland Voivodeship, consisting of 60 turbines with a total capacity of 120 MW. The significant tax revenues generated by the farm have enabled road modernization and the construction of public facilities, significantly improving the quality of life for local residents. The community was actively engaged through informational campaigns aimed at increasing public acceptance of the project. Meanwhile, the Jasna wind farm, also located in the Pomeranian Voivodeship, is a 132 MW investment developed in accordance with the principles of sustainable development. In addition to benefits such as job creation and increased environmental awareness, this wind farm was established with active community involvement through informational meetings and educational initiatives. All these examples demonstrate that the development of wind energy in Poland not only contributes to the country's energy transition but also brings tangible economic and social benefits to the residents of the regions where these wind farms are located (Figure 5).

The map (Figure 6) presents the geographic locations of the largest and most significant wind energy projects in Poland: the Potęgowo, Margonin, and Jasna wind farms. These farms represent crucial centers for renewable energy development, substantially contributing to the increase of renewable energy in Poland's national energy mix. Their strategic positions in the northern and central-western regions of the country allow effective utilization of local wind conditions and existing energy infrastructure. Implementation of these projects is essential for achieving both national and European climate targets. Additionally, these developments provide considerable benefits to local communities, including job creation, infrastructure improvements, and increased municipal tax revenues.

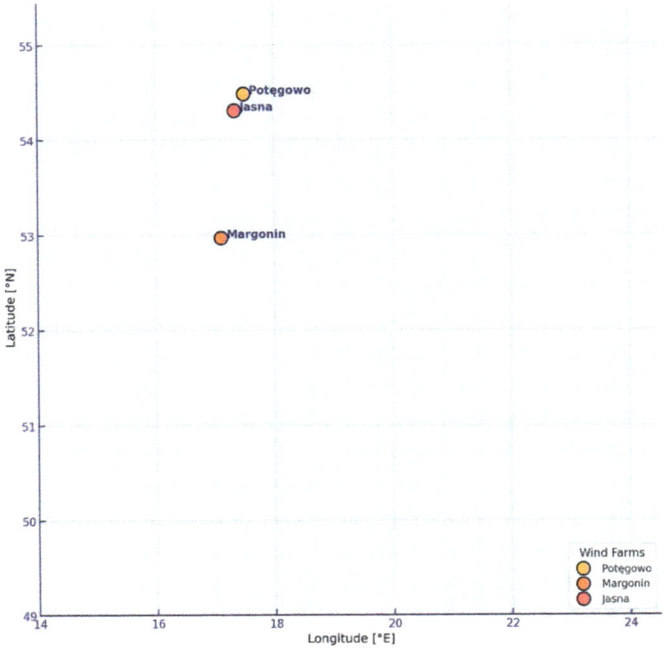

Figure 6. Geographic Location of Key Wind Energy Projects in Poland. Source: authors' own elaboration.

7.2.2. Selected Examples of Successful Wind Farm Projects in the European Union [135–137]

The European Union has been a global leader in wind energy development, with numerous large-scale projects that contribute to the transition toward renewable energy. These wind farms not only provide clean electricity but also bring economic and social benefits to local communities. Investments in offshore and onshore wind farms have helped reduce carbon emissions, create jobs, and improve regional infrastructure. Many projects have also fostered public acceptance through community engagement initiatives, education programs, and participatory investment models. The table below presents selected examples of successful wind farms in the EU, highlighting their capacity, benefits, and community involvement (Table 8).

Table 8. Successful Wind Farm Projects in the European Union.

Wind Farm Name	Location	Capacity	Benefits	Community Engagement
Horns Rev 3	Denmark	407 MW (offshore in the North Sea)	Increased share of renewables in Denmark, CO_2 emission reduction	Government support for local fishing communities through economic compensation programs
Gode Wind	Germany	582 MW (offshore off the German coast)	Collaboration with the scientific sector to minimize environmental impact	Regional governments invested in environmental education and awareness programs
Whitelee	Scotland, United Kingdom	539 MW (largest onshore wind farm in Europe)	Local budget revenues supporting education and infrastructure	Open tourist trails, renewable energy education center
Middelgrunden	Denmark	40 MW (first citizen-financed wind farm)	Participatory model—Copenhagen residents could invest in turbines and earn profits	Strong public support due to the cooperative ownership model

Horns Rev 3 is a Danish wind farm with a capacity of 407 MW, located in the North Sea. Its construction significantly increased Denmark's share of renewable energy and contributed to reducing CO_2 emissions. To mitigate the impact of the investment on local communities, the Danish government introduced compensation programs supporting fishermen and related economic sectors. Another example is the German wind farm Gode Wind, with a capacity of 582 MW, situated off the German coast. This project was developed in collaboration with the scientific sector, allowing for the implementation of solutions that minimize the farm's impact on the marine environment. As part of community engagement efforts, German regional governments allocated funds for environmental education and information campaigns, raising public awareness of the benefits of renewable energy. Whitelee, located in Scotland, was the largest onshore wind farm in Europe before Brexit, with a total capacity of 539 MW. Its operation contributed to increased local budget revenues, which were used to finance education and infrastructure. In support of local communities, open tourist trails and an educational center promoting renewable energy were established. An innovative approach was also adopted by the Danish wind farm Middelgrunden, with a capacity of 40 MW, which was the world's first wind farm partially financed by citizens. Through a participatory model, Copenhagen residents had the opportunity to invest in turbines and earn profits from energy production, leading to widespread public support and greater acceptance of wind energy. Each of these projects demonstrates that the development of wind farms in the European Union not only supports the energy transition but also brings economic and social benefits to local communities.

8. Discussion

This review article has examined the development of wind energy specifically within the context of Poland and the European Union, identifying key regulatory, social, and technological barriers as well as economic opportunities. The analysis revealed that wind energy plays a pivotal role in Europe's energy transition strategy but faces significant region-specific challenges.

In Poland, restrictive zoning regulations, especially the "10H" rule requiring substantial distances between turbines and residential buildings, significantly limit investment opportunities. Adjusting regulations back to a 500-m minimum distance could substantially increase the available area for wind farm installations, boosting the country's renewable capacity by an estimated 60–70%. Furthermore, excessively lengthy administrative procedures—among the longest in Europe—require urgent digitization and simplification to expedite the implementation of wind projects in line with EU renewable energy directives.

Social acceptance emerged as another critical factor influencing wind energy deployment across Europe. Practical case studies, such as the Margonin and Potęgowo wind farms in Poland, highlighted effective approaches for increasing local acceptance, including community profit-sharing schemes, transparent communication, and targeted educational initiatives. The European Union's Wind Power Action Plan reinforces these strategies by advocating early-stage community involvement and financial transparency as essential for securing sustained local support.

Technologically, the review identified promising advancements such as floating offshore wind farms and integrated energy storage solutions, essential for managing the intermittent nature of wind-generated electricity. Case analyses from Denmark (Horns Rev 3) and Germany (Gode Wind) demonstrate that proactive management of environmental impacts and structured cooperation with affected sectors (e.g., fisheries) significantly enhance project feasibility and public acceptance.

Economically, wind energy investments were found to generate measurable local benefits, including increased employment, infrastructure improvements, and higher municipal revenues, making them attractive for regional development strategies. However, overcoming initial high investment costs requires stable financing mechanisms, such as Power Purchase Agreements (PPAs), and continued EU financial support.

To fully leverage wind energy's potential, policymakers in Poland and across the EU must prioritize regulatory reform, accelerated permitting, and robust community engagement models. Technological innovation should remain a central focus, particularly in enhancing turbine efficiency, expanding offshore capacity, and integrating renewable energy into existing power grids.

9. Conclusions

This review article has thoroughly analyzed the current state and future prospects of wind energy development, particularly in Poland and across the European Union. The findings underscore the pivotal role of wind energy in the global transition towards a low-carbon and sustainable energy system. While remarkable growth has been observed in total installed capacity, this metric alone does not adequately capture the effectiveness, efficiency, or equity of energy distribution.

The study highlights that per capita analysis offers a more nuanced understanding of wind energy deployment. Countries like Denmark and Germany, despite their smaller size, have succeeded in achieving significantly higher wind energy accessibility per citizen than larger economies such as China and India. This observation emphasizes the importance of integrating population-based metrics into national and regional energy assessments to ensure equitable access to clean energy.

Poland remains a promising but challenging environment for wind energy investments. Regulatory barriers, including restrictive zoning laws and prolonged permitting procedures, have historically constrained growth. However, the recent policy amendments—such as restoring the 500-m distance rule and initiating legislative reforms—indicate a positive shift towards unlocking the country's wind energy potential. Offshore wind development, with a projected capacity reaching 18 GW by 2040, also signals a significant step forward in aligning with EU climate goals.

Social acceptance has emerged as a central factor in the successful implementation of wind energy projects. Case studies from Poland, Germany, and Denmark demonstrate that community engagement, transparency, and economic benefit-sharing are vital to securing long-term support. Innovative approaches—such as energy cooperatives and local reinvestment initiatives—have proven effective in building trust and encouraging citizen participation.

Technological advancements such as floating offshore wind farms, improved turbine efficiency, and smart grid integration offer new opportunities to overcome environmental and technical challenges. These innovations, supported by targeted EU funding and regulatory frameworks like the Wind Power Action Plan and the Net Zero Industry Act (NZIA), are essential to accelerating the energy transition. Investments in research and development, workforce education, and domestic manufacturing capacity will further strengthen Europe's competitiveness in the renewable energy sector.

Economic analysis confirms that wind energy contributes significantly to local development, generating jobs, increasing municipal revenues, and enhancing energy security. However, initial capital intensity and supply chain vulnerabilities—particularly dependence on imported components—must be addressed through diversified financing mechanisms, industrial policy coordination, and robust infrastructure planning.

In summary, wind energy is no longer just an alternative but a strategic pillar of future energy systems. Realizing its full potential requires coordinated actions across policy, technology, economics, and society. The path forward lies in removing systemic barriers, leveraging innovation, and placing citizens at the heart of the energy transition. Only through this integrated approach can wind energy continue to grow sustainably and equitably, helping to meet the urgent climate objectives of the coming decades.

Author Contributions: Conceptualization, H.W., P.F.B. and I.M.; methodology, H.W., P.F.B., A.S., A.K. and I.M.; software, M.H., A.K., A.S. and F.C.; validation, A.B., S.O. and F.C.; formal analysis, H.W., M.H., A.B., S.O., I.M. and P.F.B.; investigation, H.W., I.M. and A.B.; resources, A.S., S.O., A.B. and M.H.; data curation, A.B., S.O., A.S. and A.K.; writing—original draft preparation, H.W., P.F.B., M.H. and I.M.; writing—review and editing, H.W., F.C., A.S. and I.M.; visualization, H.W., A.S. and A.B.; supervision, M.H. and P.F.B.; project administration, H.W., A.B. and I.M. All authors have read and agreed to the published version of the manuscript.

Funding: Co-financed by the Minister of Science under the "Regional Excellence Initiative".

Data Availability Statement: Data sharing is not applicable to this article.

Conflicts of Interest: Author Filip Czepło was employed by the Polish Photovoltaic Instalation. The remaining authors declare that the research was conducted in the absence of any commercial or financial relationships that could be construed as a potential conflict of interest.

References

1. Hassan, Q.; Viktor, P.; Al-Musawi, T.J.; Ali, B.M.; Algburi, S.; Alzoubi, H.M.; Jaszczur, M. The renewable energy role in the global energy transformations. *Renew. Energy Focus* **2024**, *48*, 100545. [CrossRef]

2. Adelekan, O.A.; Ilugbusi, B.S.; Adisa, O.; Obi, O.C.; Awonuga, K.F.; Asuzu, O.F.; Ndubuisi, N.L. Energy transition policies: A global review of shifts towards renewable sources. *Eng. Sci. Technol. J.* **2024**, *5*, 272–287. [CrossRef]
3. Keramidas, K.; Fosse, F.; Diaz Vazquez, A.; Dowling, P.; Garaffa, R.; Després, J.; Wojtowicz, K. *Global Energy and Climate Outlook 2021: Advancing Towards Climate Neutrality*; Publications Office of the European Union: Luxembourg, 2021.
4. De La Peña, L.; Guo, R.; Cao, X.; Ni, X.; Zhang, W. Accelerating the energy transition to achieve carbon neutrality. *Resour. Conserv. Recycl.* **2022**, *177*, 105957. [CrossRef]
5. Bogdanov, D.; Ram, M.; Aghahosseini, A.; Gulagi, A.; Oyewo, A.S.; Child, M.; Breyer, C. Low-cost renewable electricity as the key driver of the global energy transition towards sustainability. *Energy* **2021**, *227*, 120467. [CrossRef]
6. Tagliapietra, S.; Zachmann, G.; Edenhofer, O.; Glachant, J.-M.; Linares, P.; Loeschel, A. The European union energy transition: Key priorities for the next five years. *Energy Policy* **2019**, *132*, 950–954. [CrossRef]
7. Pérez, M.D.L.E.M.; Scholten, D.; Stegen, K.S. The multi-speed energy transition in Europe: Opportunities and challenges for EU energy security. *Energy Strategy Rev.* **2019**, *26*, 100415. [CrossRef]
8. Krug, M.; Di Nucci, M.R. Citizens at the heart of the energy transition in Europe? *Renew. Energy Law Policy Rev.* **2020**, *9*, 9–27. [CrossRef]
9. Potrč, S.; Čuček, L.; Martin, M.; Kravanja, Z. Sustainable renewable energy supply networks optimization—The gradual transition to a renewable energy system within the European Union by 2050. *Renew. Sustain. Energy Rev.* **2021**, *146*, 111186. [CrossRef]
10. Allen, M.L.; Allen, M.M.; Cumming, D.; Johan, S. Comparative capitalisms and energy transitions: Renewable energy in the European Union. *Br. J. Manag.* **2021**, *32*, 611–629. [CrossRef]
11. Ćetković, S.; Buzogány, A. Varieties of capitalism and clean energy transitions in the European Union: When renewable energy hits different economic logics. *Clim. Policy* **2016**, *16*, 642–657. [CrossRef]
12. Bointner, R.; Pezzutto, S.; Grilli, G.; Sparber, W. Financing innovations for the renewable energy transition in Europe. *Energies* **2016**, *9*, 990. [CrossRef]
13. Solorio, I.; Jörgens, H. Contested energy transition? Europeanization and authority turns in EU renewable energy policy. *J. Eur. Integr.* **2020**, *42*, 77–93. [CrossRef]
14. Mey, F.; Diesendorf, M. Who owns an energy transition? Strategic action fields and community wind energy in Denmark. *Energy Res. Soc. Sci.* **2018**, *35*, 108–117. [CrossRef]
15. Gallo, A.B.; Simões-Moreira, J.R.; Costa, H.K.M.; Santos, M.M.; Dos Santos, E.M. Energy Storage in the Energy Transition Context: A Technology Review. *Renew. Sustain. Energy Rev.* **2016**, *65*, 800–822. [CrossRef]
16. Cantarero, M.M.V. Of Renewable Energy, Energy Democracy, and Sustainable Development: A Roadmap to Accelerate the Energy Transition in Developing Countries. *Energy Res. Soc. Sci.* **2020**, *70*, 101716. [CrossRef]
17. Igliński, B.; Iglińska, A.; Koziński, G.; Skrzatek, M.; Buczkowski, R. Wind energy in Poland–history, current state, surveys, renewable energy sources Act, SWOT analysis. *Renew. Sustain. Energy Rev.* **2016**, *64*, 19–33. [CrossRef]
18. Talarek, K.; Knitter-Piątkowska, A.; Garbowski, T. Wind parks in Poland—New challenges and perspectives. *Energies* **2022**, *15*, 7004. [CrossRef]
19. Woźniak, M. Landscape Protection vs. Onshore Wind Energy Investments in Poland—A Legal Perspective. *Int. J. Semiot. Law-Rev. Int. Sémiotique Jurid.* **2024**, *38*, 1199–1218. [CrossRef]
20. Bórawski, P.; Bełdycka-Bórawska, A.; Jankowski, K.J.; Dubis, B.; Dunn, J.W. Development of wind energy market in the European Union. *Renew. Energy* **2020**, *161*, 691–700. [CrossRef]
21. Aydin, O.; Igliński, B.; Krukowski, K.; Siemiński, M. Analyzing wind energy potential using efficient global optimization: A case study for the City Gdańsk in Poland. *Energies* **2022**, *15*, 3159. [CrossRef]
22. Brauers, H.; Oei, P.-Y. The political economy of coal in Poland: Drivers and barriers for a shift away from fossil fuels. *Energy Policy* **2020**, *144*, 111621. [CrossRef]
23. Dyka, I.; Harasymiuk, J. Legal barriers to the development of onshore wind power plants and the design of wind turbine tower pile foundation. *Arch. Civ. Eng.* **2024**, *70*, 53–71. [CrossRef]
24. Dawid, L. German support systems for onshore wind farms in the context of Polish acts limiting wind energy development. *J. Water Land Dev.* **2017**, *34*, 109. [CrossRef]
25. International Energy Agency (IEA). Renewables 2023: Analysis and Forecasts to 2028. Available online: https://www.iea.org/reports/renewables-2023 (accessed on 2 February 2025).
26. Danny Harvey, L.D. A Bottom-Up Assessment of Recent (2016–2020) Energy Use by the Global Iron and Steel Industry Constrained to Match a Top-Down (International Energy Agency) Assessment. *Energy* **2024**, *293*, 130675. [CrossRef]
27. Hattori, T.; Nam, H.; Chapman, A. Multilateral Energy Technology Cooperation: Improving Collaboration Effectiveness Through Evidence from International Energy Agency Technology Collaboration Programmes. *Energy Strategy Rev.* **2022**, *43*, 100920. [CrossRef]
28. Yu, Y.; Li, C.; Fu, Y.; Yang, W. A Group Decision-Making Method to Measure National Energy Architecture Performance: A Case Study of the International Energy Agency. *Appl. Energy* **2023**, *330*, 120285. [CrossRef]

29. Marzouk, O.A. Summary of the 2023 Report of TCEP (Tracking Clean Energy Progress) by the International Energy Agency (IEA), and Proposed Process for Computing a Single Aggregate Rating. *E3S Web Conf.* **2025**, *601*, 00048. [CrossRef]
30. Caiafa, C.; Hattori, T.; Nam, H.; de Coninck, H. International Technology Innovation to Accelerate Energy Transitions: The Case of the International Energy Agency Technology Collaboration Programmes. *Environ. Innov. Soc. Transit.* **2023**, *48*, 100766. [CrossRef]
31. Carey, B.; Yasukawa, K. Geothermal Collaboration Through the International Energy Agency Geothermal Technology Collaboration Programme. In Proceedings of the World Geothermal Congress, Beijing, China, 15–17 September 2023; pp. 1–17.
32. Mueller, T. When Policy Entrepreneurs Drift Between Levels: The Creation of the International Renewable Energy Agency. *Glob. Policy* **2023**, *14*, 588–599. [CrossRef]
33. White, S.M.; Marszal-Pomianowska, A.J.; Jin, G.Y.; Madsen, H.; Candanedo, J.; Harmelink, M.; Gori, V. A Data Sharing Guideline for Buildings and HVAC Systems: Energy in Buildings and Communities Technology Collaboration Programme. 2023. European Commission—Energy Policy. Available online: https://commission.europa.eu/topics/energy_en (accessed on 1 February 2025).
34. Energy Policy of Poland Until 2040 (PEP2040). Available online: https://www.gov.pl/web/climate/energy-policy-of-poland-until-2040-epp2040 (accessed on 11 February 2025).
35. Eurostat—Renewable Energy Statistics. Available online: https://ec.europa.eu/eurostat/statistics-explained/index.php?title=Renewable_energy_statistics (accessed on 11 February 2025).
36. Polish Wind Energy Association (PSEW)—Market Reports. Available online: https://psew.pl/en/ (accessed on 11 February 2025).
37. CMS Expert Guide to Renewable Energy—Poland. Available online: https://cms.law/en/int/expert-guides/cms-expert-guide-to-renewable-energy/poland (accessed on 3 February 2025).
38. Energy Transition—Poland's Renewable Energy Development. Available online: https://energytransition.org/2024/01/will-the-new-polish-government-launch-a-renewable-energy-revolution/ (accessed on 11 February 2025).
39. Poland's Wind Energy Policy and Legislative Changes. Available online: https://www.politico.eu/article/poland-parliament-law-wind-power-civic-coalition-law-and-justice/ (accessed on 4 February 2025).
40. China Built out Record Amount of Wind and Solar Power in 2024. Available online: https://apnews.com/article/b33750 3abfacfd9b7829fd7bbcd507e9 (accessed on 11 February 2025).
41. China's Solar, Wind Power Installed Capacity Soars in 2024. Available online: https://www.reuters.com/business/energy/chinas-solar-wind-power-installed-capacity-soars-2024-2025-01-21/ (accessed on 6 February 2025).
42. China to Roll Back Clean Power Subsidies after Boom. Available online: https://www.reuters.com/business/energy/china-roll-back-clean-power-subsidies-after-boom-2025-02-09/ (accessed on 7 February 2025).
43. Wind Power in the United States. Available online: https://en.wikipedia.org/wiki/Wind_power_in_the_United_States (accessed on 8 February 2025).
44. U.S. Renewable Energy Factsheet. Center for Sustainable Systems, University of Michigan. Available online: https://css.umich.edu/publications/factsheets/energy/us-renewable-energy-factsheet (accessed on 9 February 2025).
45. Wind Power Capacity Forecast 2022–2030. Incorrys. Available online: https://incorrys.com/power-generation/wind-power-generation/wind-power-capacity-forecast-2022-2030/ (accessed on 10 February 2025).
46. Biden-Harris Administration Approves Eleventh Offshore Wind Project in U.S. History. U.S. Department of the Interior. Available online: https://www.doi.gov/pressreleases/biden-harris-administration-approves-eleventh-offshore-wind-project-us-history (accessed on 9 February 2025).
47. Vineyard Wind, America's First Large-Scale Offshore Wind Farm, Delivers Full Power from 5 Turbines to the New England Grid. Mass.gov. Available online: https://www.mass.gov/news/vineyard-wind-americas-first-large-scale-offshore-wind-farm-delivers-full-power-from-5-turbines-to-the-new-england-grid (accessed on 9 February 2025).
48. New York's Offshore Wind Projects. New York State Energy Research and Development Authority (NYSERDA). Available online: https://www.nyserda.ny.gov/All-Programs/Offshore-Wind/Focus-Areas/NY-Offshore-Wind-Projects (accessed on 6 February 2025).
49. Offshore Wind. Avangrid. Available online: https://www.avangrid.com/aboutus/renewables/ourbusiness/offshorewind (accessed on 9 February 2025).
50. Offshore Wind Projects. Northeast Ocean Data Portal. Available online: https://www.northeastoceandata.org/offshore-wind-projects/ (accessed on 11 February 2025).
51. Massachusetts Activities. Bureau of Ocean Energy Management. Available online: https://www.boem.gov/renewable-energy/state-activities/massachusetts-activities (accessed on 11 February 2025).
52. Offshore Wind. Massachusetts Clean Energy Center (MassCEC). Available online: https://www.masscec.com/our-focus/offshore-wind (accessed on 8 February 2025).
53. Germany's Energiewende—The Easy Guide. Available online: https://www.cleanenergywire.org/easyguide (accessed on 11 February 2025).
54. Germany Added Nearly 20 GW of Renewable Power Capacity in 2024. Available online: https://www.enerdata.net/publications/daily-energy-news/germany-added-nearly-20-gw-renewable-power-capacity-2024.html (accessed on 5 February 2025).

55. Germany Adds More Than Double Offshore Wind Capacity in 2024 Compared to Previous Year. Available online: https://www.offshorewind.biz/2025/01/09/germany-adds-more-than-double-offshore-wind-capacity-in-2024-compared-to-previous-year/ (accessed on 4 February 2025).
56. Offshore Wind Power Yield from German North Sea up 8% in 2024. Available online: https://www.cleanenergywire.org/news/offshore-wind-power-yield-german-north-sea-8-2024 (accessed on 11 February 2025).
57. Germany's Top Clean Energy Source Set to Slow Growth Pace in 2024. Available online: https://www.reuters.com/business/energy/germanys-top-clean-energy-source-set-slow-growth-pace-2024-maguire-2024-08-22/ (accessed on 11 February 2025).
58. India's Renewable Energy Capacity Hits New Milestone. Press Information Bureau, Government of India. Available online: https://pib.gov.in/PressReleaseIframePage.aspx?PRID=2073038 (accessed on 11 February 2025).
59. An Overview of Wind Energy Development and Policy Initiatives In India. Available online: https://www.ncbi.nlm.nih.gov/pmc/articles/PMC8760121/ (accessed on 11 February 2025).
60. Gujarat and Tamil Nadu Have Offshore Wind Energy Potential of 70 GW. Mercom India. Available online: https://mercomindia.com/gujarat-tamil-nadu-offshore-wind-energy-potential-70-gw/ (accessed on 11 February 2025).
61. India to Add Record Renewables This Year as Green Push Gains Momentum. Reuters. Available online: https://www.reuters.com/business/energy/india-gets-386-bln-financial-commitment-expand-renewable-capacity-2024-09-16/ (accessed on 11 February 2025).
62. Wind Energy in Spain. Asociación Empresarial Eólica. Available online: https://aeeolica.org/en/about-wind-energy/wind-energy-in-spain/ (accessed on 11 February 2025).
63. Renewable Energies Generated 56% of Spain's Electricity Mix in 2024. Red Eléctrica de España. Available online: https://www.ree.es/en/press-office/press-release/news/press-release/2025/01/renewable-energies-generated-56-per-cent-spains-electricity-mix-2024 (accessed on 11 February 2025).
64. Wind Energy Lobby Says Spain Needs More Wind Power, Faster. Reuters. Available online: https://www.reuters.com/business/energy/wind-energy-lobby-says-spain-needs-more-wind-power-faster-2024-10-03/ (accessed on 11 February 2025).
65. Dogger Bank Wind Farm: The World's Largest Offshore Wind Farm. Dogger Bank Wind Farm. Available online: https://doggerbank.com/ (accessed on 11 February 2025).
66. Hornsea Wind Farm. Wikipedia. Available online: https://en.wikipedia.org/wiki/Hornsea_Wind_Farm (accessed on 11 February 2025).
67. Wind Britain's Top Electricity Source in 2024. Reuters. Available online: https://www.reuters.com/world/uk/wind-britains-top-electricity-source-2024-2025-01-07/ (accessed on 11 February 2025).
68. The Details of Renewables' Record-Breaking Year. Energy Magazine. Available online: https://energydigital.com/articles/how-renewable-energy-broke-records-in-2024 (accessed on 11 February 2025).
69. Getting UK Offshore Wind Back into Fighting Shape. Ember. Available online: https://ember-energy.org/latest-insights/uk-offshore-wind/ (accessed on 11 February 2025).
70. The North Sea | Offshore Wind Energy. National Grid Group. Available online: https://www.nationalgrid.com/stories/energy-explained/north-sea-future-clean-energy-powerhouse (accessed on 11 February 2025).
71. Wind Energy in Spain—Statistics & Facts. Statista. Available online: https://www.statista.com/topics/9046/wind-energy-in-spain/ (accessed on 11 February 2025).
72. Wind Power Outlook for Brazil and US. Available online: https://reglobal.org/wind-power-outlook-for-brazil-and-us/ (accessed on 11 February 2025).
73. Wind Energy Generation in Northeast Brazil Sets New Record. Available online: https://braziliannr.com/2024/07/30/wind-energy-generation-in-northeast-brazil-sets-new-record/ (accessed on 11 February 2025).
74. Promoting Renewable Energy Through Auctions: The Case of Brazil. Available online: https://www.worldbank.org/en/topic/energy/publication/livewire (accessed on 11 February 2025).
75. Financing Wind Power Energy Industry in Brazil. Available online: https://www.abeeolica.org.br/wp-content/uploads/2017/07/FINANCING-WIND-POWER-ENERGY-INDUSTRY-IN-BRAZIL-v4-1-2.pdf (accessed on 11 February 2025).
76. Renewable Energy Auctions: Brazil. Available online: https://www.netzeropathfinders.com/best-practices/renewable-energy-auctions-brazil (accessed on 11 February 2025).
77. Brazil's Power Gen Expands at Record Rate in 2024. Available online: https://www.argusmedia.com/en/news-and-insights/latest-market-news/2646593-brazil-s-power-gen-expands-at-record-rate-in-2024 (accessed on 11 February 2025).
78. Auction-Based Policy of Brazil's Wind Power Industry. Available online: https://www.mdpi.com/1996-1073/17/24/6450 (accessed on 11 February 2025).
79. Brazil Launches Expansion of Offshore Wind Energy. Available online: https://www.renewable-energy-industry.com/news/world/article-6591-brazil-launches-expansion-of-offshore-wind-energy (accessed on 11 February 2025).
80. Fostering Wind Power Through Auctions: The Brazilian Experience. Available online: https://ppp.worldbank.org/public-private-partnership/library/fostering-wind-power-through-auctions-brazilian-experience (accessed on 11 February 2025).

81. Annual Wind–Energy Report 2022. Available online: https://abeeolica.org.br/wp-content/uploads/2023/08/WIND-ENERGY-REPORT-2022-1.pdf (accessed on 11 February 2025).
82. France Awards Two Floating Offshore Wind Farms in Mediterranean. Power Engineering International. Available online: https://www.powerengineeringint.com/renewables/wind/france-awards-two-floating-offshore-wind-farms-in-mediterranean/ (accessed on 11 February 2025).
83. France Launches Tender Procedure for 2.3–2.9 GW of Offshore Wind. Enerdata. Available online: https://www.enerdata.net/publications/daily-energy-news/france-launches-tender-procedure-23-29-gw-offshore-wind.html (accessed on 11 February 2025).
84. France Announces Winners of World's First Commercial-Scale Floating Offshore Wind Auction. WindEurope. Available online: https://windeurope.org/newsroom/news/france-announces-winners-of-worlds-first-commercial-scale-floating-offshore-wind-auction/ (accessed on 11 February 2025).
85. The Netherlands: Hollandse Kust Zuid Offshore Wind Farm. Vattenfall. Available online: https://ignitisrenewables.com/career/?gad_source=1&gad_campaignid=22418915919&gbraid=0AAAAAoQ6yHTB7hUKINI3PCnVjRMoWT0vg&gclid=Cj0KCQjw5ubABhDIARIsAHMighbBSTgF8QjUfCklDZWMULfFS9NO2WvJ-BTioWVEwnOV2BqhGh7c5q0aAp9mEALw_wcB (accessed on 11 February 2025).
86. Offshore Wind Energy in the Netherlands. Government of the Netherlands. Available online: https://www.government.nl/topics/renewable-energy (accessed on 11 February 2025).
87. Denmark | IEA Wind TCP. International Energy Agency Wind Technology Collaboration Programme. Available online: https://iea-wind.org/about-iea-wind-tcp/members/denmark/ (accessed on 11 February 2025).
88. Denmark—Countries & Regions. International Energy Agency. Available online: https://www.iea.org/countries/denmark (accessed on 11 February 2025).
89. Wind Power in Denmark—Wikipedia. Wikipedia. Available online: https://en.wikipedia.org/wiki/Wind_power_in_Denmark (accessed on 11 February 2025).
90. Another Record-Breaking Year for Solar and Wind Power in Denmark. State of Green. Available online: https://stateofgreen.com/en/news/another-record-breaking-year-for-solar-and-wind-power-in-denmark/ (accessed on 11 February 2025).
91. Denmark is Showing the World What is Possible: How the North Sea Became the 'Green Power Plant of Europe'. The Guardian. Available online: https://www.theguardian.com/delivering-the-energy-transition/2025/feb/04/denmark-showing-world-possible-north-sea-green-power-plant-europe (accessed on 11 February 2025).
92. Global Wind Report 2024. Global Wind Energy Council. Available online: https://www.gwec.net/reports/globalwindreport/2024 (accessed on 11 February 2025).
93. Wind Power by Country. World Population Review. Available online: https://worldpopulationreview.com/country-rankings/wind-power-by-country (accessed on 11 February 2025).
94. Germany's Wind Power Capacity. Our World in Data. Available online: https://ourworldindata.org/grapher/cumulative-installed-wind-energy-capacity-gigawatts (accessed on 11 February 2025).
95. Kowalik, J. Barriers for Renewable Energy Sources Development in Poland on the Example of Wind Energy. *Environ. Sci. Eng. Technol.* **2015**, *9*, 52–63.
96. Ellis, G.; Ferraro, G. *The Social Acceptance of Wind Energy*; European Commission-JRC Science for Policy Report: Brussels, Belgium, 2016.
97. Hübner, G.; Leschinger, V.; Müller, F.J.; Pohl, J. Broadening the Social Acceptance of Wind Energy—An Integrated Acceptance Model. *Energy Policy* **2023**, *173*, 113360. [CrossRef]
98. Fournis, Y.; Fortin, M.J. From Social 'Acceptance' to Social 'Acceptability' of Wind Energy Projects: Towards a Territorial Perspective. *J. Environ. Plan. Manag.* **2017**, *60*, 1–21. [CrossRef]
99. Khorsand, I.; Kormos, C.; MacDonald, E.G.; Crawford, C. Wind Energy in the City: An Interurban Comparison of Social Acceptance of Wind Energy Projects. *Energy Res. Soc. Sci.* **2015**, *8*, 66–77. [CrossRef]
100. Yiridoe, E.K. Social Acceptance of Wind Energy Development and Planning in Rural Communities of Australia: A Consumer Analysis. *Energy Policy* **2014**, *74*, 262–270.
101. Westerlund, M. Social Acceptance of Wind Energy in Urban Landscapes. *Technol. Innov. Manag. Rev.* **2020**, *10*, 49–62. [CrossRef]
102. Caporale, D.; Sangiorgio, V.; Amodio, A.; De Lucia, C. Multi-Criteria and Focus Group Analysis for Social Acceptance of Wind Energy. *Energy Policy* **2020**, *140*, 111387. [CrossRef]
103. Klok, C.W.; Kirkels, A.F.; Alkemade, F. Impacts, Procedural Processes, and Local Context: Rethinking the Social Acceptance of Wind Energy Projects in the Netherlands. *Energy Res. Soc. Sci.* **2023**, *99*, 103044. [CrossRef]
104. Michalak, A.; Wolniak, R. The innovativeness of the country and the renewables and non-renewables in the energy mix on the example of European Union. *J. Open Innov. Technol. Mark. Complex.* **2023**, *9*, 100061. [CrossRef]
105. Van Hoey, G.; Bastardie, F.; Birchenough, S.; De Backer, A.; Gill, A.; De Koning, S.; Hodgson, S.; Chai, S.M.; Steenbergen, J.; Termeer, E.; et al. *Overview of the Effects of Offshore Wind Farms on Fisheries and Aquaculture*; Publications Office of the European Union: Luxembourg, 2021.

106. Miciuła, I.; Stępień, P. Analysis of the Global Market of Energy Resources. In *Eurasian Economic Perspectives*; Eurasian Studies in Business and Economics; Bilgin, M., Danis, H., Demir, E., Can, U., Eds.; Springer: Cham, Switzerland, 2019; Volume 10.
107. Igliński, B.; Iglińska, A.; Cichosz, M.; Kujawski, W.; Buczkowski, R. Renewable energy production in the Łódzkie Voivodeship. The PEST analysis of the RES in the voivodeship and in Poland. *Renew. Sustain. Energy Rev.* **2016**, *58*, 737–750. [CrossRef]
108. Tagliapietra, S.; Veugelers, R.; Zettelmeyer, J. Rebooting the European Union's Net Zero Industry Act. Bruegel Policy Brief, 2023. Available online: https://www.bruegel.org/policy-brief/rebooting-european-unions-net-zero-industry-act (accessed on 11 February 2025).
109. Jones, C. The Net-Zero Industry Act and the Reform of the Green Deal State Aid Rules: A Convincing Reaction to the Inflation Reduction Act? European University Institute, 2023. Available online: https://hdl.handle.net/1814/75591 (accessed on 11 February 2025).
110. Wolf, A. Framework Conditions for Net-Zero Industry Clusters in Europe. *Intereconomics* **2024**, *59*, 267–275. [CrossRef]
111. Cucculelli, M.; Giampaoli, N.; Renghini, M. Identification of STEP and NZIA Technologies through Text Mining: An Empirical Analysis of Patent Data. Money and Finance Research Group (Mo.Fi.R.)—Univ. Politecnica Marche—Dept. Economic and Social Sciences. 2024. Available online: https://ideas.repec.org/p/anc/wmofir/188.html (accessed on 11 February 2025).
112. Dabiri, J.O.; Greer, J.R.; Koseff, J.R.; Moin, P.; Peng, J. A New Approach to Wind Energy: Opportunities and Challenges. In Proceedings of the Physics of Sustainable Energy III (PSE III): Using Energy Efficiently and Producing it Renewably, Berkeley, CA, USA, 8–9 March 2014; Volume 1652, pp. 51–57.
113. Roga, S.; Bardhan, S.; Kumar, Y.; Dubey, S.K. Recent Technology and Challenges of Wind Energy Generation: A Review. *Sustain. Energy Technol. Assess.* **2022**, *52*, 102239. [CrossRef]
114. Clifton, A.; Barber, S.; Bray, A.; Enevoldsen, P.; Fields, J.; Sempreviva, A.M.; Ding, Y. Grand Challenges in the Digitalisation of Wind Energy. *Wind. Energy Sci.* **2023**, *8*, 947–974. [CrossRef]
115. Oei, P.-Y.; Hainsch, K.; Löffler, K.; von Hirschhausen, C.R.; Holz, F.; Kemfert, C. A new climate for Europe: 2030 climate targets must be more ambitious. *DIW Wkly. Rep.* **2019**, *9*, 365–372.
116. Hainsch, K.; Göke, L.; Kemfert, C.; Oei, P.-Y.; von Hirschhausen, C.R. European green deal: Using ambitious climate targets and renewable energy to climb out of the economic crisis. *DIW Wkly. Rep.* **2020**, *10*, 303–310.
117. Scarlat, N.; Dallemand, J.-F.; Monforti-Ferrario, F.; Banja, M.; Motola, V. Renewable energy policy framework and bioenergy contribution in the European Union–An overview from National Renewable Energy Action Plans and Progress Reports. *Renew. Sustain. Energy Rev.* **2015**, *51*, 969–985. [CrossRef]
118. Archer, C.L.; Jacobson, M.Z. Evaluation of Global Wind Power. *J. Geophys. Res. Atmos.* **2005**, *110*, D12110. [CrossRef]
119. Hanifi, S.; Liu, X.; Lin, Z.; Lotfian, S. A Critical Review of Wind Power Forecasting Methods—Past, Present and Future. *Energies* **2020**, *13*, 3764. [CrossRef]
120. Pryor, S.C.; Barthelmie, R.J. A Global Assessment of Extreme Wind Speeds for Wind Energy Applications. *Nat. Energy* **2021**, *6*, 268–276. [CrossRef]
121. Lu, X.; McElroy, M.B. Global Potential for Wind-Generated Electricity. In *Wind Energy Engineering*; Academic Press: Cambridge, MA, USA, 2023; pp. 47–61.
122. Olabi, A.G.; Wilberforce, T.; Elsaid, K.; Salameh, T.; Sayed, E.T.; Husain, K.S.; Abdelkareem, M.A. Selection Guidelines for Wind Energy Technologies. *Energies* **2021**, *14*, 3244. [CrossRef]
123. Voldsgaard, A. Analysing Renewable Energy Finance as an Evolving Complex System: Lessons from Offshore Wind. Ph.D. Thesis, University College London, London, UK, 2023.
124. Kaldellis, J.K.; Boulogiorgou, D. Renewable Energy: Wind Energy. In *Living with Climate Change*; Letcher, T.M., Ed.; Elsevier: Amsterdam, The Netherlands, 2024; pp. 513–557.
125. Del Amo Garcia, A. Analysis of a Renewable Hybrid Plant Using Solar and Wind Energy Conversion Systems. Bachelor's Thesis, Universitat Politècnica de Catalunya, Barcelona, Spain, 2024.
126. Jurasz, J.; Dąbek, P.B.; Kaźmierczak, B.; Kies, A.; Wdowikowski, M. Large Scale Complementary Solar and Wind Energy Sources Coupled with Pumped-Storage Hydroelectricity for Lower Silesia (Poland). *Energy* **2018**, *161*, 183–192. [CrossRef]
127. Hassan, Q.; Nassar, A.K.; Al-Jiboory, A.K.; Viktor, P.; Telba, A.A.; Awwad, E.M.; Barakat, M. Mapping Europe Renewable Energy Landscape: Insights into Solar, Wind, Hydro, and Green Hydrogen Production. *Technol. Soc.* **2024**, *77*, 102535. [CrossRef]
128. Iskandarova, M.; Dembek, A.; Fraaije, M.; Matthews, W.; Stasik, A.; Wittmayer, J.M.; Sovacool, B.K. Who Finances Renewable Energy in Europe? Examining Temporality, Authority and Contestation in Solar and Wind Subsidies in Poland, the Netherlands and the United Kingdom. *Energy Strategy Rev.* **2021**, *38*, 100730. [CrossRef]
129. Chowdhury, N.E.; Shakib, M.A.; Xu, F.; Salehin, S.; Islam, M.R.; Bhuiyan, A.A. Adverse environmental impacts of wind farm installations and alternative research pathways to their mitigation. *Clean. Eng. Technol.* **2022**, *7*, 100415. [CrossRef]
130. Msigwa, G.; Ighalo, J.O.; Yap, P.S. Considerations on environmental, economic, and energy impacts of wind energy generation: Projections towards sustainability initiatives. *Sci. Total Environ.* **2022**, *849*, 157755. [CrossRef]

131. Sayed, E.T.; Wilberforce, T.; Elsaid, K.; Rabaia, M.K.H.; Abdelkareem, M.A.; Chae, K.J.; Olabi, A.G. A critical review on environmental impacts of renewable energy systems and mitigation strategies: Wind, hydro, biomass and geothermal. *Sci. Total Environ.* **2021**, *766*, 144505. [CrossRef]
132. Mello, G.; Dias, M.F.; Robaina, M. Evaluation of the environmental impacts related to the wind farms end-of-life. *Energy Rep.* **2022**, *8*, 35–40. [CrossRef]
133. Jobert, A.; Laborgne, P.; Mimler, S. Local acceptance of wind energy: Factors of success identified in French and German case studies. *Energy Policy* **2007**, *35*, 2751–2760. [CrossRef]
134. Lacal-Arántegui, R. Globalization in the wind energy industry: Contribution and economic impact of European companies. *Renew. Energy* **2019**, *134*, 612–628. [CrossRef]
135. Bartoszewicz-Burczy, H. Barriers for Large Integration of PV and Onshore Wind Energy in the Distribution Network on the Selected European Union Electricity Markets. *Stud. Ecol. Bioethicae* **2022**, *20*, 67–77. [CrossRef]
136. Astolfi, D. Wind Turbine Drivetrain Condition Monitoring through SCADA-Collected Temperature Data: Discussion of Selected Recent Papers. *Energies* **2023**, *16*, 3614. [CrossRef]
137. Faraggiana, E.; Ghigo, A.; Sirigu, M.; Petracca, E.; Giorgi, G.; Mattiazzo, G.; Bracco, G. Optimal floating offshore wind farms for Mediterranean islands. *Renew. Energy* **2024**, *221*, 119785. [CrossRef]

energies

MDPI

Review

A Multi-Level Perspective on Transition to Renewable Energy in the Indonesian Transport Sector

Ferry Fathoni [1,2], Jon C. Lovett [1,*] and Muhammad Mufti Rifansha [3]

1 School of Geography, Faculty of Environment, University of Leeds, Leeds LS2 9JT, UK; gyff@leeds.ac.uk
2 PT Pertamina (Persero), Jakarta 10110, Indonesia
3 Department of Mechanical Engineering, Institut Teknologi Bandung, Bandung 40132, Indonesia; 13123170@mahasiswa.itb.ac.id
* Correspondence: j.lovett@leeds.ac.uk

Abstract

A transition from fossil fuels to renewable energy is underway to achieve net-zero emissions. The institutional arrangements in Indonesia's energy transportation sector are crucial for various stakeholders involved in the energy transition. This study combines historical institutionalism with a multi-level perspective to analyze how policy formulation, critical junctures, and path dependence shape institutional changes toward sustainable mobility. The evolution of institutional arrangements can be categorized into three phases: the establishment of fuel-oil-based infrastructure and dependency (1970–2003); the diversification of cleaner fuels through compressed natural gas and biofuels (2004–2014); and the development of affordable and clean energy, focusing on biofuels and electrification (2015 to present). In parallel, a quantitative total cost of ownership analysis of vehicles using different fuel types demonstrates how institutional reforms, fiscal incentives, and regulatory support reshape the economic feasibility of low-carbon technologies. Landscape pressures—such as global decarbonization, fuel import dependence, and energy security challenges—interact with niche innovations, including biofuels, electric vehicles, and hybrid systems, to drive systemic transformation. The findings indicate that institutional changes, supported by quantitative economic evidence and technology diffusion, play a pivotal role in realigning Indonesia's transport energy regime toward a more resilient, inclusive, and sustainable transition.

Keywords: transportation energy; net-zero transition; sustainable development; path dependence; institutional change

1. Introduction

The goal of reaching net-zero emissions has been the main spotlight of climate policy nowadays [1]. The net-zero concept emphasizes maintaining global emissions within certain limits; thus, excess emissions must be captured and removed through either natural or technological reservoirs [2]. To achieve the ambitious target of net-zero emissions, international cooperation is very important. Common problems such as global warming, climate change, and achieving net-zero emissions will require comprehensive solutions [3]. Nevertheless, national governments are responsible for deciding on how to commit to global agreements on emissions reduction and how to integrate these global standards into their policies and laws. They have the authority to shape institutions and actors within their borders through the laws and regulations they enforce [4].

The development of institutions and the formulation of policies are shaped by a series of events that occur over time. Many studies on the history of institutional development emphasize the significance of "path dependence" in understanding the background for forming institutions and organizational actors. Path dependence refers to formulated and implemented decisions dependent on previous decisions or experiences. Institutions and actors are formed by "lock-in" experiences that lead to positive feedback or increasing returns for adopting pre-selected technologies, which affect political and social decision-making behavior [5–8]. Political and social behavior play a significant role in policymaking for indigenous cultures, as they provide the necessary insight into the specific challenges these cultures encounter [9–11].

This study aims to contribute to an understanding of the HI processes that govern energy transition in the Indonesian transport sector. The MLP framework has been used to examine changes in the socio-technical landscape, triggered by multi-level and cross-border actors, both global and local, as well as niche innovations involved in formulating regime policies related to the sustainable development of affordable and clean energy. This study emphasizes key events, referred to as exogenous shocks, which serve as critical junctures that alter institutions and create path-dependent processes. Additionally, it explores endogenous factors that contribute to gradual changes and the establishment of institutional stability.

This study examines the impact of the global agenda, technology advancement, the orientation of transportation infrastructure development, and national environmental policies on the transition to net-zero emissions (NZE). The energy sector for transportation in Indonesia presents an intriguing case study due to its supply chain's intersection with global politics and trade [12], the adoption of eco-friendly technologies [13], and the issue of the threat to the preservation of the environment [14]. We view TCO 2025 as a key indicator of institutional arrangements rather than a primary driver of change. Its significance lies in the impact of underlying policy and governance choices related to pricing, taxes, incentives, infrastructure, and risks [15,16]. This research uses Indonesia as a case study to address the following research question: How do institutional legacies influence energy transitions? How do niche innovations disrupt or align with socio-technical regimes? What multi-level interactions drive or hinder institutional change toward sustainability in Indonesia's transportation sector? Indonesia is the largest archipelagic country in Southeast Asia and presents a significant case study due to its challenges in the transportation sector. The country's high reliance on fossil fuels, particularly imported fuels, makes transportation a major contributor to national energy consumption and carbon emissions. However, Indonesia also has vast potential for renewable energy, alongside a large transportation market, positioning it strategically to drive the transition to sustainable energy in the region.

2. Literature Review

2.1. Historical Institutionalism Approach

Historical institutionalism (HI) analyzes a series of historical events arranged chronologically and elucidates the impacts of the instances when important institutional changes occur and create new pathways for historical development as critical junctures. Institutional development usually occurs gradually in the period between critical junctures and is influenced by path dependence and the "lock-in" effect [6]. In HI, the "lock-in" effect refers to a situation where previous institutional choices limit current and future policy options, even when those earlier choices are no longer the best or most efficient. This lock-in occurs due to path-dependence, where initial decisions or developments create self-reinforcing mechanisms that make it increasingly challenging to implement change over time [17].

HI has been used extensively for analyses related to institutions across national–global borders by combining several political, economic, social, and technological contexts. The HI approach considers the factors of ideas, individual behavior, and structural influences that interact within an institution based on a series of events over a certain period of time [18]. HI does not only test institutions or processes at any single time because institutions, as a complex set, change over time. HI theory offers a framework with three essential characteristics, namely, the substantive agenda, arguments based on time sequences, and comprehensive emphasis [10]. First, HI focuses its questions on substantive matters that attract broad public and research interest. Second, HI explores transformations and processes of various scales based on time. Third, HI combines the impact of institutions and processes over a period rather than just at one time of testing, emphasizing macro-context analysis [10].

Public policymaking as a change in the institutionalization process varies in each case. In the context of HI, these variations are influenced by path-dependence, which asserts that the timing and sequence of events are crucial; numerous social outcomes are possible; and even minor, unpredictable events can lead to significant consequences. Additionally, reversing a course of action can often be very difficult, if not impossible, regardless of the consequences. Consequently, critical junctures play a significant role in political development and shape the broader framework of social life [19].

Path dependence influences political phenomena where the inheritance of past experiences "locks in" and limits the choices available to the formation of institutions and actors [20]. This institutional policy inheritance was originally a solution to past institutional problems. However, because it provides many benefits, the dominant actor tries to maintain institutional stability by maintaining the policy [19]. HI approaches institutional stability over a long period of time, creating a dichotomy related to institutional change. On the one hand, institutional change is slow, gradual, and takes place in a limited scope [20] but remains transformative [21,22]. Under other conditions, a radical exogenous shock at the "critical juncture" can change equilibrium and create new arrangements [23].

The HI approach examines the process of institutional change, which is often characterized by opposing dichotomies [24]. The previous institutional legacy consisted of long-standing policies aimed at addressing a specific issue, driven by the benefits obtained by the dominant actor [8,19,25]. This longevity of the policy led to institutional stability, resulting in endogenous institutional change. Endogenous institutional change arises from internal institutional processes that are slow, gradual, and transformative within the existing institutional framework [21,22].

On the other hand, dynamic institutional change can be triggered by external sources affecting the institution, known as exogenous shocks [24]. An exogenous radical shock [8] that occurs at a critical point [26] can disrupt the balance and lead to new institutional arrangements. HI elucidates the mechanisms of institutional inventions through displacement, replacing old rules with new rules, and layering, which occurs when new rules are placed above or next to existing regulations [27]. Repeated interactions can affect equilibrium, and if they continue, they will affect institutional stability [28]. Historical institutionalism also analyzes the comparison of cross-national public policies as a result of national political institutions [18]. Historical institutionalism is a complement to sociological institutionalism or socio-technical analysis in energy transition studies [29] toward the sustainability of national energy security.

In the context of Southeast Asia, transportation policies are entrenched in long-standing institutional arrangements. These include the dominance of state-owned enterprises (SOEs), centralized governance structures, and subsidy regimes [30–32]. For instance, Indonesia's historical reliance on fossil fuel subsidies has entrenched petroleum-based transportation, which has impeded the diffusion of more sustainable alternative fuels [30]. The historical institutionalism approach has allowed scholars to examine how institutional embeddedness interacts with emerging niche innovations, particularly in the fields of biofuels and electric vehicles (EVs) in Indonesia [4,30]. This interaction significantly shapes the pace and direction of socio-technical transitions [33].

2.2. Multi-Level Perspective Framework

Recent research has utilized the multi-level perspective (MLP) framework to investigate socio-technical transitions in the energy transition toward sustainability [34,35]. MLP analyzes evolutions in socio-technical systems by examining interaction at three levels: niche, regime, and landscape. The first level involves radical technological advancements in micro-level niches, which act as precursors to a sustainable transition. These micro-niches are where novel innovations and inventions occur.

Radical novel innovations at the niche level often encounter resistance from established socio-technical systems at the second level, which can hinder energy transition. The meso-level encompasses established regulations, institutions, and technologies that shape "socio-technical regimes". The literature highlights that socio-technical regimes experience various "lock-in" mechanisms that impede change due to the entrenchment of existing structures, interests, and stability [34,36]. The energy transition in emerging economies suffers from financial and technological lock-ins due to reliance on fossil fuel infrastructure, reinforcing the existing regime [17,37]. Furthermore, political and institutional lock-ins stemming from regulatory bias toward conventional industries and the lobbying of dominant actors have preserved the status quo and hindered the emergence of radical innovation [38]. The third level of analysis is the socio-technical landscape, which represents the broader external environment that influences socio-technical systems. Interactions at this macro-level create trends and pressures affecting innovation and transitions within regimes and niches. This landscape can cause sudden shocks and gradual changes in macroeconomic or macro-cultural aspects [39].

In the first phase, radical innovations of cleaner energy emerge in small niches at the periphery of the existing system. These innovations mature through a process of experimentation. These radical innovations establish themselves in one or more market niches in the second phase. As they become more stable, they converge into a dominant design and become institutionalized. However, niche innovations often struggle against an entrenched system locked in by a path that is dependent on fossil fuels [34].

In the third phase, the green innovations diffuse into the mainstream market and compete directly with the existing fossil fuel system. Landscape pressures challenge the regime, creating cracks and destabilization. When regime actors can no longer ignore or suppress these niche innovations, windows of opportunity open. In the fourth phase, if the green innovations are mature enough, they can break through and replace or transform the regime, leading to a systemic socio-technical transition [34].

The diffusion of niche innovations, including biofuel blends, EVs, and compressed natural gas (CNG), is gaining traction in emerging markets such as Indonesia, Thailand, and India, primarily driven by state policies and subsidy schemes [40–42]. However, they encounter challenges from well-established regimes focused on private vehicle ownership, road-centered infrastructure, and entrenched oil interests. Meanwhile, broader landscape pressures, primarily climate action and urban congestion, create a window of opportunities to scale these niche innovations and challenge existing regimes [40,41].

3. Materials and Methods

This study utilizes a qualitative method that employs HI and the MLP as analytical frameworks to examine Indonesia's energy transition toward sustainability in the transportation sector. These frameworks allow for a multi-dimensional analysis of institutional inertia, policy evolution, and socio-technical dynamics across different levels of actors over historical periods [43]. The HI approach identifies critical junctures by selecting factors determining crucial decision-making processes in Indonesia's energy transition policy, the net-zero transition [44].

Meanwhile, the MLP offers a robust analytical framework for examining interactions across three levels: landscape, regime, and niche [34]. Through MLP utilization, this study investigates how innovations emerge at the niche level and subsequently compete with existing transport energy regimes in Indonesia. Furthermore, the MLP framework evaluates the pressures from the socio-technical landscape, enabling eco-friendly fuel innovations to ultimately take advantage of "windows of opportunity" to transform the current energy regime [45].

Moreover, the integration between HI and the MLP provides an understanding of the long-term investment decisions that underpin energy sustainability policies, which are heavily influenced by institutional arrangements within a country [44]. Furthermore, HI and the MLP broaden the perspective to examine domestic institutional affairs through the lens of international relations [46]. This analysis covers the period from 1970 to 2025, with a focus on the 2015-to-2025 period and considering both the external factors of global socio-technical landscape dynamics, as well as the internal influence of the socio-technical landscape and the diffusion of niche innovations.

3.1. Data Collection

This study uses both primary data and secondary data. The primary data were collected to cross-check secondary data and explore institutional developments. This study utilized semi-structured interviews to gather primary data from twenty-eight actors representing various stakeholders associated with the energy transportation sector. The selection of these actors uses purposive sampling to capture representative key persons, not dominated by certain groups, and covers a broader range of institutions related to the transportation energy sector. These stakeholders consist of the government, energy companies, automotive manufacturers, transport providers, global institutions, energy associations, members of the public involved with transportation, NGOs, the standardization bureau, newspapers, and academics. Semi-structured interviews were conducted to explore several themes. These included the roles and involvement of stakeholders in developing institutions for energy transition, the impact of landscape-level factors on national energy policies, and the interactions between niche actors and existing regimes in the collaborative innovation of emerging AVF technologies. File S1 in Supplementary Materials displays the semi-structured interview format used in this study.

This study utilized stratified snowball sampling to ensure adequate representation of diverse stakeholders involved in Indonesia's transportation energy transition. Potential participants were approached through email, telephone, and LinkedIn. They were given one week to respond to the invitation to participate or to decline. Response times ranged from two days to three weeks after receiving the invitation. During the COVID-19 pandemic, interviews were conducted online using the MTeams, Zoom, or WhatsApp platforms. Online interviews were conducted from August 2020 to May 2023.

This study used secondary data collected from various sources, especially the Indonesian energy sector, for transportation regulations from 1970 until 2025, with a focus on the 2015-to-2025 period, international rules, global agendas, and technological advancement related to the concept of the sustainability of energy supply for transportation. Data were also collected from documents of international organizations on their websites, scientific publications, historical books, news articles on media websites, newspapers that contain historical developments, and multi-level actors' interactions in institutional arrangements in the energy sector relating to Indonesia's transportation sector. This study also gathered secondary data from quantitative analyses of survey results published in academic journals, institutional press releases, and newspapers. The quantitative analysis focuses specifically on public preferences regarding the adoption of EVs, including both the vehicles themselves and the supporting infrastructure, as part of Indonesia's emerging low-carbon technology initiative.

3.2. Analysis

The data analysis in this study has emphasized examining the HI characteristics and MLP framework of transport energy policy and explaining path dependencies, key events at critical points, and contingencies in energy sector institutional formulation and implementation. We also analyze the types of changes that occur and determine sources of the dominant change factors, for example, whether they originate from an exogenous global agenda or endogenous national environmental policies.

Interviews explored various statements from the participants regarding the institutional dynamics of energy for transportation, primarily focusing on Indonesia's mass public transportation, biofuels, and electrification as case studies. The documentation process for stakeholder responses was recorded using a digital voice recorder, and the transcript was processed using NVivo. NVivo enabled effective tracking of themes across interview transcripts and indicated when data saturation was achieved, signifying that no new themes emerged. This study follows four steps secondary data analysis methods: defining research topics and questions; identifying sources of data/information; gathering existing data; and structuring analysis and presentation of findings [47,48].

This study uses secondary data analysis methods to define research problems with the aim of new research related to energy transportation institution sustainability in Indonesia, which is different from previous research [49]. Secondary data research can utilize existing data to generate new ideas [50]. Secondary data analysis is a systematic research strategy utilizing the abundance of existing quantitative and qualitative data [49,51]. The secondary data analysis method also explores data from several different points of view [52].

This study examines the TCO to outline vehicle ownership costs across various technologies. The TCO comprises the purchase price, operational costs, maintenance, taxes, insurance, and resale value. It is vital for comparing ICEVs, HEVs, BEVs, and FCVs in consumer decisions and sustainable policies. Previous research shows that the TCO better reflects the competitiveness of new technologies, such as EVs, and is influenced by battery price declines, usage patterns, and energy price variations [53–55].

TCO calculation in this study uses the net present value (NPV) approach, based on the following formula:

$$TCO = \sum_{t=0}^{n} \frac{C_t}{(1+r)^t} - \frac{RV_n}{(1+r)^n} \quad (1)$$

In this formula, C_t represents the total cost in year t, which includes fuel or energy costs, maintenance, taxes, insurance, administrative expenses, and battery replacement if applicable. The variable r denotes the discount rate, n is the analysis period (10 years), and RV_n indicates the resale value at the end of year n.

To facilitate comparison between vehicles with different annual mileages, the TCO results are normalized accordingly.

$$TCO_{/\mathrm{km}} = \frac{TCO}{Annual\ Mileage \times n} \quad (2)$$

We used data triangulation to cross-validate our interview findings by comparing them with secondary data sources. This approach helps ensure accuracy and minimizes biases from both the researcher and participants. Previous studies have shown that using data triangulation enhances the reliability and credibility of qualitative research [56].

Qualitative analysis was carried out using content analysis on historical records related to international and national regulations, policies, global agreements, and events on the energy transition that affect transportation energy sustainable development in Indonesia. Critical junctures were identified, such as radical innovation of the sustainability concept in the context of global institutions and adoption into national policies or the constitution. Content analysis was also used to identify policies of institutions and global actors in the form of campaigns, prohibitions, boycotts, and technological advancements related to transportation energy to observe their interactions with the institutional development of the transition toward sustainable transportation energy. Socio-technical evolutions at multiple levels, including the micro-niche, meso-regime, and macro-landscape levels, such as shifts in energy, were also examined, and a time series of related key events was constructed. The validity of the analysis of the secondary data was verified using triangulation with the results of the primary data analysis of interview transcripts.

This research is not without limitations. The authors were unable to interview certain high-level government officials due to access barriers, which limited their understanding of the bargaining process at the elite level [57]. Responses from stakeholders like SOEs and large multinational corporations may align with official discourses, which can introduce bias. Documentary sources show mixed trends: corporate reports often present sustainability strategies in ideal perspectives, while NGOs focus on critical viewpoints. This study required careful triangulation of various data sources [58,59]. This research only covers developments up to early 2025, so findings may quickly become outdated due to policy changes or advancements in second-generation biofuels, EVs, and FCVs [36,60,61]. One of the authors, currently working in Indonesia's energy sector, might also introduce bias. To address this, the authors practiced reflexivity and discussed the findings periodically with all team members to ensure objectivity [62,63]. While these limitations are important, they do not invalidate the findings. Nevertheless, they should be taken into consideration when interpreting the study's conclusions.

4. Results

After Indonesia gained independence in 1945, Soekarno, the first president, implemented a nationalist strategy by nationalizing Dutch colonial institutions to manage natural resources. During this period, coal served as the primary energy source for transportation [64]. The energy transition of railways occurred with a switch to diesel power throughout the 1940s and 1950s, particularly in the US and Europe. Dieselization in Asia started in the 1950s, initiated by Japan. Diesel locomotives have been Indonesia's primary train movers since 1952. [65,66]. In response to the growing demand for petroleum, the government established a management institution in 1950. Indonesia joined the Organization of the Petroleum Exporting Countries (OPEC) in 1962 as a petroleum-producing nation. In 1968, Soeharto became Indonesia's second president, facing economic challenges from the failure of Soekarno's Guided Economy and longstanding defense and territorial issues. Figure 1 illustrates a schematic representation of the historical institutional evolution of energy transition within Indonesia's transportation sector. Landscape dynamics are represented in orange boxes, while the diffusion of niche innovations is illustrated in green boxes. Institutional dynamics are shown in blue boxes. Dashed line arrows indicate the creation of windows of opportunity.

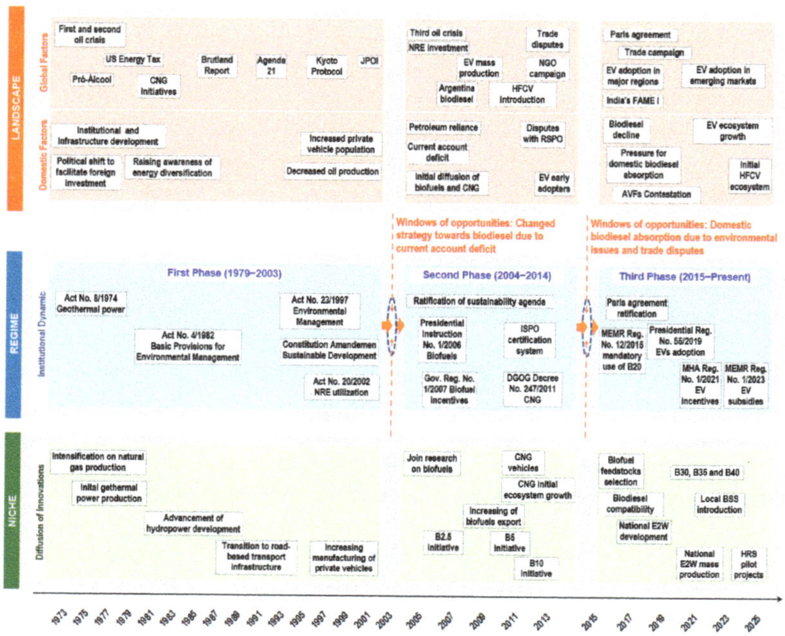

Figure 1. The national and global influences on Indonesian transportation energy sustainability institutional arrangements. Source: Authors.

4.1. Socio-Technical Landscape Dynamics

In the 1970s, Soeharto shifted away from Soekarno's nationalistic approaches by opening the economy to foreign investment. The government emphasized exploiting natural resources, particularly oil production and exports, to support economic recovery and infrastructure development [67]. Meanwhile, in the global sphere, the Organization of Arab Petroleum Exporting Countries imposed an oil embargo in 1973, which led to a massive oil crisis. A second crisis occurred in 1979 due to reduced oil production during the Iranian Revolution. These oil crises prompted many countries, including Indonesia, to seek alternative energy sources [68].

Several countries have developed biofuels as alternative vehicle fuels (AVFs). While biofuels have been used since the early 1900s, significant expansion began with Brazil's Pró-Álcool Program in 1975, which focused on sugarcane-based ethanol. This initiative, led by the Institute of Alcohol and Sugar (IAA), saw the launch of alcohol-fueled cars in 1978 [45,69]. The Energy Tax Act of 1978 initiated the development of ethanol in the US by requiring vehicle fuel to contain 10% ethanol (E10). Following the 2000s, implementing the Renewable Fuel Standard (RFS), subsidies, and the increased use of corn as feedstock significantly boosted ethanol production [70].

Since the 1980s, fuel diversification has involved using CNG as a low-emission AVF for public transportation, trucks, and taxis. Countries such as Iran, Pakistan, Argentina, Brazil, and China have widely adopted CNG vehicles. Strong government policies have supported the expansion of CNG refueling stations in these countries [71].

Concerns about environmental preservation began to gain global attention in the 1980s. The World Conservation Strategy has promoted "sustainable development" to protect the Earth's natural resources. This essential vision aims to stem practices that harm the environment, leading to significant effects on a global scale. Ongoing discussions on Global Sustainable Development have resulted in several key agreements, including the 1987 Brundtland Report, the 1992 Agenda 21, the 1997 Kyoto Protocol, and the 2002 Johannesburg Plan of Implementation (JPOI) [72].

From 1980 to the late 1990s, infrastructure development significantly improved Indonesia's economy and citizens' welfare. However, this development prioritized road infrastructure, without adequate planning for mass public transport, increasing private vehicle usage. This situation deteriorated further as declining oil production and increasing fuel consumption compelled the government to allocate more funds for subsidies [73].

The third oil crisis occurred between 2003 and 2008 due to a surge in global demand, the war in Iraq, and limited reserve production capacity. This oil crisis surged oil prices, causing the cost of fuel subsidies to rise from USD 4.4 billion to USD 7.4 billion [74]. The 2000s oil crisis has raised concerns about energy security, prompting European and Asian countries to include energy policies in their economic strategies. Many developed nations focused on reducing their reliance on oil and investing more in the innovation of renewable energy sources.

The diversification of AVF has included the development of biodiesel, which was institutionalized by the Argentine government through Biofuel Law 26.093 in 2006. This regulation mandates a 5% biodiesel blend (B5) in diesel fuel, with plans for future increases. Argentina's biodiesel production has rapidly grown, especially following the EU market's opening under the Renewable Energy Directive in 2009 [45].

The diffusion of biofuels has expanded to include countries in Southeast Asia. The Philippines has promoted domestic production to reduce crude oil imports by implementing a sugarcane-based E20 strategy. As the world's largest producer of coconuts, the Philippine government has also introduced B5 biodiesel derived from palm oil. In contrast, Thailand has diversified its feedstock using sugarcane and cassava to produce ethanol. The Thai government has widely adopted E20 and E85 ethanol blends for gasoline-fueled vehicles and B20, a palm oil-based blend, for heavy-duty vehicles [31,75].

Simultaneously with the re-emergence of biofuels following concerns about climate change and energy security, the world has seen the rebirth of EVs. Significant advancements in lithium-ion battery technology and government subsidies have driven China's transition to EVs since the 2000s [76,77]. Several developed countries, such as Germany and the United States, have pursued electrification for the energy transition. In 2004, Germany updated its vehicle tax system to reduce taxes on low-emission vehicles, including hybrid electric vehicles (HEVs), based on CO_2 emissions. The U.S. introduced the Energy Policy Act of 2005, which provides advanced vehicle credits for HEVs. The global trend toward EVs was propagated by the mass production of the Tesla Roadster in 2008 and the Nissan Leaf in 2010. With strong regulatory support and subsidies, the adoption of EVs rapidly spread across the EU, US, Japan, and South Korea during the 2010s [78].

Meanwhile, some countries in Latin America, Asia, and Africa have seen limited transition to EVs, as many emerging economies still face significant challenges, particularly high upfront purchase costs, and insufficient charging infrastructure. As an emerging economy, India launched the FAME I program in 2015 to offer subsidies for HEVs and battery electric vehicles (BEVs) based on battery capacity and vehicle efficiency. This program was replaced by FAME II in 2019, which focused primarily on transitioning to battery electric vehicles (BEVs). Thailand has led Southeast Asia in promoting HEVs. From 2007 to 2017, the country offered a lower tax of 10–15% on HEVs based on CO_2 emissions, while ICEVs faced higher rates of 20–30%. In 2015, the Eco Car Program Phase II included HEVs as low-emission vehicles. Starting in 2017, the focus shifted to BEVs and plug-in hybrid electric vehicles (PHEVs), with EV Roadmap 2030 aiming for 30% of car production to be electric [78].

Another AVF that has recently emerged is the hydrogen fuel cell vehicle (FCV). FCVs have evolved from laboratory concepts to commercial markets over the past sixty years. Initially used in NASA's 1960s space program, FCVs saw automotive applications in the 1990s with demonstration buses and prototype cars [79,80]. The 2010s marked the commercial launch of limited-production models like the Toyota Mirai and Honda Clarity [61,80]. However, global adoption of FCVs remains low and has not met deployment targets in early-adopter markets such as Japan, South Korea, and California [81]. The mass production of FCVs faces key barriers, including high costs, limited refueling infrastructure, and policies that favor EVs [60,61].

The recent literature emphasizes the shift in FCV innovations from passenger cars to heavy-duty transportation, where their range and rapid refueling offer significant advantages [60]. As battery and fuel cell costs decrease, their use in freight applications is expected to grow beyond 2030 [82]. Hydrogen refueling stations (HRSs) require intensive investments and are reliant on usage rates, worsening the chicken-and-egg dilemma [81]. Innovations are also focused on reducing reliance on platinum and enhancing durability with non-precious catalysts [61,80].

Climate change is a critical global issue that the Kyoto Protocol was not able to address effectively, primarily because it did not fully include developing nations due to the principle of differentiated responsibilities, resulting in the rejection of the protocol by the United States of America in the Byrd–Hagel Resolution. In response to this limitation and ongoing negotiations at the UNFCCC, the Paris Agreement was adopted at COP21 in Paris in 2015 with a goal to keep global temperature increases below 2 °C, with a target of limiting the rise to 1.5 °C above pre-industrial levels. It focuses on Nationally Determined Contributions (NDCs) and aims for carbon neutrality by the mid-century [2].

A reliance on fossil fuel-based routes has marked the landscape of Indonesia's energy transportation sector since 2000. In 2024, the transport sector contributed 36.11% to Indonesia's final energy consumption, making it the second-largest sector after industry at 45.94% [83]. The crude oil and oil products were the final energy consumption sources, totaling 507.76 million BOE out of 1292.34 million BOE, or 39.29% of the total consumption [83]. This fossil fuel path dependence has been triggered by the development of highway capacity and mass public vehicles, which have struggled to keep pace with the rapid growth of private vehicle ownership. At the same time, the global pressure for decarbonization highlights the necessity for low-emission transportation systems. However, these systems still face significant challenges related to infrastructure, technology, and societal habits.

4.2. Socio-Technical Regime Evolvement

Transportation is an energy-intensive sector in emerging economies, including Indonesia. Niche innovations and landscape dynamics have shaped the institutional dynamics of the energy transition. The evolution of the energy regime has gone through several policies and institutional changes, which are divided into three phases, as follows:

1. First phase: Development of energy institutions and infrastructure that fostered dependence on fuel oil (1970–2003).
2. Second phase: Encouragement of energy diversification through cleaner fuels, especially natural gas and biofuels (2004–2014).

3. Third phase: Development of affordable and clean energy, focusing on biofuels and electrification (2015 to present).

4.2.1. First Phase: Development of Energy Institutions and Infrastructure That Fostered Dependence on Fuel Oil (1970–2003)

During the first phase, the government focused on boosting economic growth and infrastructure development by leveraging crude oil production and export. The Soeharto administration enacted Law No. 8 of 1971, which designated Pertamina as the sole SOE responsible for managing the upstream and downstream oil and gas sectors [84].

The government has been exploring alternative energy sources to fossil oil, particularly natural gas. This exploration has drawn foreign investment through a production-sharing contract with Pertamina. In 1971, Mobil Oil Indonesia discovered natural gas fields in Arun and Lho Shukun, located in North Aceh. In 1972, Huffco Inc. discovered a second natural gas field, Badak, in Bontang, East Kalimantan [85].

The first and second oil crises were critical junctures that prompted the government to diversify energy sources beyond petroleum. During this phase, the diversification of renewable energy sources focused mainly on hydropower and geothermal energy in the electricity generation sector. Geothermal energy development has been formalized through several key policies: Presidential Decree No. 64/1971, Act No. 8/1974, and Presidential Decree No. 16/1974. These policies designated Pertamina to survey, explore, and exploit geothermal resources for electricity production. The operation of Kamojang Geothermal Power Plant Unit 1, with a capacity of 250 kW, in 1983 marked the beginning of the geothermal era in Indonesia. The government deregulated the sector to encourage geothermal development further and invited foreign investment through Presidential Decree No. 49/1991 [86].

During the Soeharto regime, especially in the 1980s and 1990s, there was a significant shift in the development of public transportation infrastructure from rail-based systems to road-focused construction. This transition to road-based transportation coincided with the growth of private car manufacturing. Policy changes prioritizing road transportation and private vehicles have led to "institutional lock-in" to petroleum dependency. As a result, the government's shift in focus promoted the use of private vehicles, as highlighted by Respondent 10:

> *"The development during the Soeharto administration [was intensive]. As we began building infrastructure, we were introduced to personal vehicle technology... We recognized the importance of mobility, ...but we became accustomed to relying on private cars. At that time, many of our graduates were influenced by the US, where the use of private vehicles was predominant." (Respondent 10, Academic).*

Our findings align with secondary sources, indicating a drastic increase in private vehicles compared to public transportation, such as buses. The number of motorcycles saw the most dramatic surge, from 2.67 million units in 1980 to a staggering 120.10 million units in 2018. There was a significant increase in passenger cars, from 639,464 units in 1980 to 16.44 million units in 2018. In contrast, buses experienced the slowest growth, from 86,284 units in 1980 to 2.54 million units in 2018 [87]. The unbalanced growth in private vehicle ownership, inadequate public mass transportation, and the growth in road capacity have created a pressing issue of congestion in urban areas [73]. Congestion greatly reduces environmental quality due to air and noise pollution. The congestion also reduces the efficiency of energy use in the transportation sector.

The substantial increases in fuel consumption, which began in the late 1980s, are a cause for concern. In 1987, fuel consumption was 0.493 MBPD, and it increased at an average rate of 4.53% to 1.72 MBPD in 2013. Along with the decline in Indonesian crude oil production, the government has been forced to import fuel oil to meet domestic needs. In 2004, Indonesia became a net importer of oil as consumption surpassed production [73].

To mitigate the adverse environmental effects of transport energy consumption, the government has taken a proactive approach. It passed Act No. 4/1982 on Basic Provisions for Environmental Management and enforced Road Traffic and Transportation Act No. 14/1992 in the transportation sector to curb air pollution and noise from motor vehicles, thereby safeguarding environmental sustainability. The government has committed to addressing climate change by ratifying international agreements, including the UNFCCC (Act No. 6 of 1994) and the Kyoto Protocol (Act No. 17 of 2004). The government also enacted Act No. 23 of 1997 and amended the Indonesian Constitution, Article 33, paragraph 4, in 2002 regarding environmental management. In line with this, the government introduced Act No. 20/2002, focusing on NRE source utilization [4].

4.2.2. Second Phase: Encouragement of Energy Diversification Through Cleaner Fuels, Especially Natural Gas and Biofuels (2004–2014)

The rise in global oil prices, along with the burden of fuel subsidies, led to a current account deficit in 2004. "Indonesia's current account deficit due to imports... Most of this deficit was attributed to fossil fuel imports, particularly petrol" (Respondent 2, Energy Company). The statement confirms OECD/IEA (88), which emphasizes that increased oil imports have burdened the budget, triggering a current account deficit, thus pressuring the government to balance energy supply with GHG emission reductions [88]. At this pivotal stage, the government has continuously attempted to balance the three critical challenges of the energy trilemma, ensuring that energy security is the main priority, which involves diversifying accessible and affordable alternative fuels. In addition, the government has implemented institutional arrangements to promote environmental sustainability. The government has also addressed key issues related to energy sustainability in the transportation sector.

Events leading to the current account deficit have become critical junctures, encouraging the government to institutionalize the energy transition. This institutional change has developed through layering, which involves the diffusion of low-emission vehicle fuels, including biofuels, CNG, and EVs. In 2005, the Indonesian government took significant steps to develop eco-innovations in biofuels to reduce reliance on fuel oil. The focus was on biodiesel made from palm oil and ethanol. These biofuels not only reduce dependency on traditional fuels but also open up new market opportunities for palm oil products [4]. Furthermore, the government established a regulatory framework for promoting biofuels by implementing Presidential Instruction No. 1/2006, followed by Presidential Decree No. 10/2006. Additionally, Government Regulation No. 1/2007 granted fiscal incentives for biofuel development [73].

The Roundtable on Sustainable Palm Oil (RSPO), established in April 2004, has impacted the global palm oil certification system by promoting sustainability. However, its implementation has raised concerns among stakeholders, particularly farmers, about limited engagement with local producers and the high certification costs affecting smallholders [4]. In response, the Ministry of Agriculture introduced the Indonesian Sustainable Palm Oil (ISPO) certification through MOA Regulation No. 19/2011. This initiative aims to reduce costs and eliminate membership fees, positioning ISPO as a viable alternative to the RSPO certification [4].

Indonesia pledged to reduce greenhouse gas (GHG) emissions at the 2009 COP conference in Copenhagen, resulting in policies focusing on energy efficiency, cleaner fuels, renewable energy, clean technologies in power plants and transportation, and sustainable mass transit. To support this, the government issued Presidential Regulation No. 61/2011 to launch the National Action Plan to Reduce GHG Emissions (RAN-GRK), targeting reductions in agribusiness, forestry and peatlands, energy, transportation, and waste management [73].

The Indonesian government has also implemented a decarbonization program that utilizes CNG for domestic transportation, as outlined in the Director General of Oil and Gas (DGOG) Decree No. 247/2011. This policy establishes the specifications for CNG fuel used in this sector. Additionally, the Minister of Energy and Mineral Resources (MEMR) has appointed two SOEs—PT Pertamina, under Ministerial Decree No. 2435/2014, and PT Perusahaan Gas Negara, under Ministerial Decree No. 2436/2014—to oversee the supply and distribution of gas fuel for road transportation [73].

Indonesian biodiesel diffusion and its sustainability have faced increased scrutiny from environmental NGOs. In March 2010, Greenpeace's report "Caught Red-Handed" revealed that Sinar Mas, a major Indonesian palm oil producer and Nestlé supplier, was expanding its cultivation into orangutan habitats in rainforests and peatlands. In October 2013, Greenpeace issued "License to Kill," exposing Wilmar International for sourcing palm oil from plantations in Tesso Nilo National Park, threatening endangered species like the Sumatran tiger [4]. In the same year, the EU imposed anti-dumping measures on Indonesian biodiesel [12], leading the Indonesian government to bring the dispute to the World Trade Organization in 2014. This commitment is further underscored by the increase in Indonesian biofuel production capacity from 1800 million liters in 2011 to 3000 million liters in 2014 [89].

4.2.3. Third Phase: Development of Affordable and Clean Energy, Focusing on Biofuels and Electrification (2015 to Present)

The dynamics of the Indonesian biodiesel diffusion, driven by environmental concerns and trade disputes, have led to a significant decrease in Indonesia's biofuel production to 1180 million liters in 2015 [89]. The impact at the farmer level has been more noticeable because it has led to the neglect of palm oil harvests. To address this, the MEMR issued Ministerial Regulation No. 12/2015, which mandates biofuel use and accelerated adoption in the transportation, industrial, commercial, and power generation sectors [73].

Concurrently, the government needs to decrease its dependence on petroleum imports, which has significantly strained the national budget. The changes in the socio-technical landscape have created critical junctures that have prompted the government to institute a policy mandating 20% biofuel blending (B20) in diesel fuel by 2016 [73]. To improve the competitiveness of biodiesel, the Palm Oil Plantation Fund Management Agency (BPDPKS) was established through Presidential Regulation No. 61/2015. This agency manages funds from palm oil export levies to support biodiesel sustainability in areas like human resources, research, plantation promotion, plantation rejuvenation, and infrastructure [90]. This initiative has encouraged energy companies and automotive manufacturers to take advantage of "windows of opportunity" and adopt biodiesel more widely as an AVF in Indonesia.

In 2016, the Indonesian government enacted Law Number 16 of 2016, which ratified the Paris Agreement. The ratification of the Paris Agreement has drawn significant attention from various groups in Indonesia regarding the global EV trend [54]. Promoting EVs through mainstream and social media has attracted interest from early adopters. Over time, these early adopters have shared their knowledge about the latest advancements in EV technology and their personal experiences in brand enthusiast groups on social media. The presence of these groups has educated the public about the benefits of EVs, contributing to an increase in the population of these eco-friendly vehicles.

At the end of the 2010s, under President Joko Widodo, Indonesia shifted toward closer ties with China. This change has led to critical junctures in the development and adoption of EVs. The socio-technical landscape has adjusted to focus on technology transfer and investments from foreign manufacturers such as Hyundai and Wuling.

Key policies include Presidential Regulation No. 55/2019, which introduced BEVs [91] and several fiscal incentives, such as tax reductions (Ministry of Home Affairs Regulation No. 1/2021) and title transfer fee exemptions (Government Regulation No. 74/2021). However, significant growth in EV adoption began following the introduction of limited subsidies under Minister of Industry (MoI) Regulation No. 6/2023 [91]. The government has started to provide regulatory support for the adoption of FCVs as potential future AVFs through Minister of Industry Regulation No. 29 of 2023 and the National Hydrogen and Ammonia Roadmap (RHAN) [92].

4.3. *Niche Innovations in Low-Carbon Mobilities*

4.3.1. Advancing Biofuels for Alternative Vehicle Fuels

The diffusion of biofuels in Indonesia began in 2006 with the formation of the National Biofuel Team (Timnas BBM). This organization comprises several key stakeholders: government agencies, energy companies, research institutions, the Indonesian Palm Oil Association (GAPKI), the Association of Biofuel Producers (APROBI), automotive manufacturers, and NGOs. In addition, research institutions and universities play a significant role in developing and diffusing biofuels through technology transfer and eco-innovation [93].

The Indonesian government initiated a mandatory program to incorporate biofuels into fuel oil, starting with a 2.5% biodiesel blend (B2.5) in 2006. The government's strategic focus on the EU market, following the adoption of the Renewable Energy Directive in 2009, underscores the country's dedication to expanding its biofuel exports. Indonesian biofuel companies have responded to take advantage of this "window of opportunity" by actively pursuing international certification for their products bound for the EU, a testament to Indonesia's potential in the global biofuel market. This figure gradually escalated to B5 by 2010 and B10 in 2013, demonstrating Indonesia's commitment to meeting the increasing global demand for biofuels [73].

The government has been instrumental in promoting biofuel production and enforcing regulations to fulfill domestic and international demands. In 2005, Indonesia's biofuel exports amounted to 6.82 million liters, dramatically growing to 607 million liters by 2008. In 2011, Indonesia began implementing its sustainability commitment, resulting in an increase in biofuel production capacity from 1800 million liters to 3000 million liters by 2014 [89].

In 2015, biofuel production declined due to sustainability issues and trade disputes. Bappenas utilized the Analytical Hierarchy Process (AHP) to identify suitable raw materials, evaluating potential biofuel plants based on factors such as feedstock availability and productivity. The plants were categorized into biodiesel producers (oil palm, coconut, and jatropha) and bioethanol producers (sugarcane, cassava, and corn) [93].

Automotive manufacturers were facing a challenging decision: whether to adopt biofuels or to shift toward EVs. To maintain their market dominance, the automotive industry leaders preferred to focus on increasing R&D investments to ensure that their existing engines were compatible with biodiesel rather than shifting their transition to EVs. Furthermore, automotive market leaders formed coalitions with palm oil companies to advocate for policies prioritizing biodiesel in the trajectories of AVF diffusion. This preference is reflected in the comments of the following respondents:

> *"Between 2015 and 2017, [the coalition of automotive industry leaders] united to resist the adoption of EVs. Consequently, to stay in business, they shifted their focus to biofuels" (Respondent 18, NGO).*

The policy mandating B20, supported by incentives, successfully restored biodiesel production levels in 2015 [94]. Biodiesel production gradually increased again in 2016, reaching 2700 million liters, and continued to rise to 3872 million liters by 2023 [89]. Following a steady increase from B30 in 2020 to B35 in 2023, the government plans to raise the biodiesel content to B40 in 2025. According to government reports, the progress of biodiesel diffusion in Indonesia has contributed to enhancing energy resilience, saving foreign exchange by reducing diesel imports, creating jobs, and lowering greenhouse gas emissions, as stated in the following:

> *"The mandatory biofuels program will help reduce petrol imports and save foreign exchange... The B40 program has brought important benefits for society and the environment. It has raised the value of CPO, ...created over 14,000 off-farm jobs, and provided about 1.95 million off-farm jobs. It has also helped cut GHG emissions by 41.46 million tons of CO_2 equivalent annually."* [95].

The advancement of biodiesel has led to a path dependence in the development of AVFs in Indonesia. The advancement of biodiesel has driven "institutional lock-in" in Indonesian biofuel development, resulting in path dependence for the diffusion of AVFs in Indonesia. In contrast, the adoption of ethanol as another biofuel—derived from sugarcane and jatropha—has not progressed rapidly. Since 2023, E5 fuel has been market-tested on a limited scale on the island of Java. The diffusion of ethanol as an AVF still faces several challenges in Indonesia, particularly due to limited local feedstock and a lack of incentive support. These issues make it difficult for ethanol to compete with gasoline, as stated by the Director of Biofuel at the MEMR, quoted by a secondary source.

> *"In its implementation, there are still many challenges, including the limited production of bioethanol and the absence of incentives to address the price difference between the Market Index Price of gasoline and bioethanol."* [96].

4.3.2. Natural Gas Utilization

The substantial growth in natural gas production has prompted the government to promote CNG technology as an AVF. In 2011, a key milestone was achieved by establishing four natural gas refueling stations in Palembang. In 2012, more CNG refueling stations were developed in East Java Province. During this time, the MEMR also initiated two pilot workshops for natural gas vehicles. By 2013, the Greater Jakarta Province had four CNG refueling stations, four mobile refueling units (MRUs), and a gas distribution pipeline in place. Four more CNG refueling stations were also constructed in East Kalimantan Province [97].

Several companies have contracted with Pertamina to manage the natural gas supply. The Palembang region's supply is sourced from Pertamina EP, while East Java receives its natural gas from Pertamina WMO and Santos. In the Greater Jakarta region, natural gas is provided by Pertamina EP, Pertamina ONWJ, Medco E&P, and Pertamina-Talisman Jambi Merang [97].

The government fostered CNG adoption by distributing converter kits for public vehicles, including buses, taxis, and motorized rickshaws. In 2011, 500 kits were distributed in Greater Jakarta and Palembang; the government increased the distribution to 1000 in 2012 and 2000 in 2013 [97]. CNG consumption grew from 1159 TJ in 2011 to 2349 TJ in 2012 but dropped to 983 TJ in 2022 due to the pandemic [98]. The use of natural gas for transportation in Indonesia is still very limited. In 2022, the industrial sector used 69.06% of natural gas, while transportation used only 0.14%. As of 2024, there are only 72 gas refueling stations and MRUs [98].

Despite the benefits of CNG, such as low emissions and competitive costs compared to petrol, private vehicle users remain hesitant to adopt it. This lack of interest among consumers has led to a shortage of automotive manufacturers producing CNG vehicles. Furthermore, the focus on biodiesel has impeded the adoption of CNG as an AVF. Additionally, the growing trend of EVs in Indonesia has further diverted attention from CNG among private vehicle users. Key barriers to CNG adoption include unclear regulations, inadequate infrastructure, and significant price differences with petrol, as stated below:

> *"The development of cars with CNG engines is currently hindered by three main obstacles: unclear government regulations, a lack of infrastructure, and the price difference between gasoline and CNG."* [99].

4.3.3. Electric Vehicles Diffusion

Global electrification prompted some Indonesian consumers to seek cleaner mobility in the early 2010s. Since 2012, the government has backed EV research. Several electric four-wheelers (E4Ws), such as electric car and bus prototypes, were developed in 2013 through a collaboration between universities and SOEs, but progress has stalled due to unclear regulations and insufficient funding [100]. Despite limited government support, early adopters of EVs have continued to grow by forming groups of brand enthusiasts. Sharing experiences and operational challenges related to EVs on social media has indirectly helped promote EVs to the public [54].

Evolving market demand and technological advancements have prompted industries to create new business opportunities through the adoption of EVs. Collaborative innovation has emerged among automotive manufacturers, energy companies, green suppliers, and research institutions to develop electric two-wheeler prototypes (E2Ws). Between 2017 and 2019, the development of E2Ws saw notable innovations led by companies like Viar and Gesits, which then shifted toward mass production. The enactment of Presidential Regulation No. 55/2019 has allowed both companies to market their E2W products to niche markets successfully. Support from both the central and regional governments has facilitated the transition to EVs within government and SOE fleets. Additionally, electrification has been implemented in buses for BRT TransJakarta, taxis, ride-hailing services, and logistics [101].

Since 2023, a range of fiscal and non-fiscal policies and minimal subsidies with minimal local content requirements have contributed to a significantly increasing number of local manufacturers entering the E2W market. The subsidy policy was provided for low-income communities. However, the initial stage's lack of coordination among government agencies temporarily delayed public interest in purchasing EVs. While subsidies increased EV purchases, stakeholders sought to expand the program, drawing inspiration from models in developing countries like India, which relate battery capacity to reduced GHG emissions [101].

The increasing number of EVs has led PLN, the state-owned electricity company, to expand its EV charging infrastructure. Similarly, oil and gas companies like Pertamina and Shell have established charging and battery swap networks at their fuel stations. The state has offered incentives for local and foreign private entities to invest in automotive manufacturing and the development of charging infrastructure. Some companies are working to facilitate the transition to EV battery charging by establishing a battery swap system (BSS) network. Gogoro has introduced its BSS system from Taiwan. Swap Energi and Volta have also developed innovative business models that embrace local cultural practices in energy purchasing, similar to conventional fuel refilling methods [101].

EVs are gaining popularity, especially as more people recognize their financial benefits. Low energy costs, affordable electricity predominantly sourced from coal, and reduced maintenance expenses have begun to change consumer perceptions about EVs [41]. The growing number of EVs has significantly supported the ongoing energy transition. In 2017, only 32 E2Ws were registered in the country. However, niche innovative developments and landscape pressures have led to regime policy changes that encourage EV adoption. As a result, by 2024, the number of registered E2Ws had surged to 160,578, alongside 34,504 units of E4Ws [102]. Figure 2 shows the growth of the diffusion of EVs in Indonesia from 2017 to 2024.

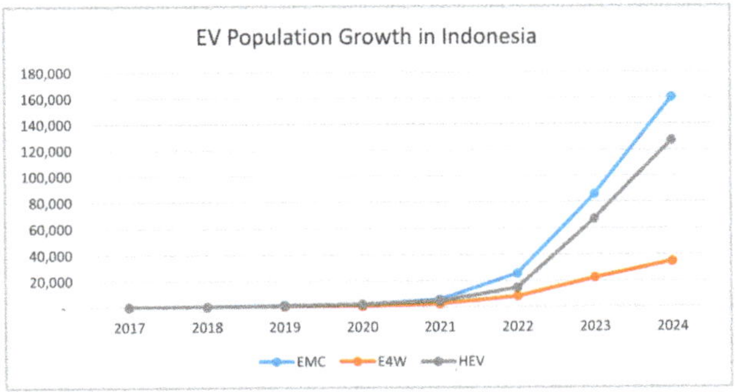

Figure 2. The growth of the diffusion of EVs in Indonesia. Source: Data from [102].

The adoption of EVs offers significant advantages, including reduced emissions and decreased reliance on fuel imports. Additionally, EVs facilitate the integration of renewable energy sources into the final energy mix, particularly within the transportation sector. However, Indonesia's persistent reliance on coal for electricity generation has led various stakeholders to advocate for a "movement aimed at discontinuing investments in coal-fired and other environmentally detrimental power plants" (Respondent 18, NGO).

The adoption of EVs will continue if stakeholders can overcome substantial challenges, primarily the high upfront purchase price, insufficient charging infrastructure, and lack of public awareness about sustainability [54]. To address these issues, automotive leaders have introduced HEVs as a socio-technical transition. In 2024, Indonesia had 127,243 HEVs, nearly reaching the number of E2Ws [102]. However, regulatory support for HEVs in Indonesia is still limited. "To reduce emissions and fuel consumption, the government should provide proportional incentives for hybrid vehicles, instead of offering the same incentives as for BEVs" [103].

4.3.4. Hydrogen Fuel Cell Vehicle Innovation

In Indonesia, HFCVs are still in the early stages of development, but pilot projects indicate a gradual transition from innovation discourses to demonstration phases. These hydrogen initiatives include the mobility and energy sectors. Regulatory support for hydrogen utilization in Indonesia's transportation sector was evidenced by Minister of Industry Regulation No. 29 of 2023 and the National Hydrogen and Ammonia Roadmap (RHAN) [92]. PLN launched a pilot HRS in Senayan, Jakarta, in 2024. This station was designed to serve buses starting in August 2024 and passenger vehicles from October 2024, with commercialization planned for 2027 [104]. In February 2025, Toyota Indonesia inaugurated the country's first hydrogen refueling station at its xEV Centre in Karawang. This facility is equipped with both 350-bar and 700-bar dispensers, catering to demonstration vehicles such as the Toyota Mirai, as well as forklifts [105].

Pertamina Geothermal Energy (PGE) is advancing into green hydrogen production utilizing geothermal power. The Ulubelu Green Hydrogen Pilot Project in Lampung, which launched in September 2025, aims to produce approximately 100 kg of hydrogen per day through anion exchange membrane (AEM) electrolysis [106]. Additionally, PGE has entered into a Joint Study Agreement with Pertagas to explore the integration of geothermal electricity in the production of hydrogen and ammonia. This initiative will leverage Pertagas' existing infrastructure for potential distribution [107]. Furthermore, Hyundai Motor Group has proposed a waste-to-hydrogen ecosystem in West Java. This initiative aims to integrate hydrogen production from waste with on-site refueling stations, with support from Pertamina and local governments, and is expected to be operational by 2027 [108].

Recent research demonstrates that health-conscious and AI-driven energy management enhances FCVs. Jia et al. [109] developed a thermal management strategy that integrates cabin climate control with fuel cell and battery systems, optimizing coolant flow and compressor operation. The results showed an 8–9% increase in system efficiency and an over 12% reduction in fuel cell ageing, confirming that this approach lowers hydrogen consumption [109]. Additionally, studies by Huang et al. [110] and Sellali et al. [111] support the notion that health-conscious control frameworks extend stack life while ensuring passenger comfort, thereby promoting sustainable FCV operation [110,111].

Building upon these frameworks, Jia et al. [112] developed an advanced energy management system utilizing an enhanced deep deterministic policy gradient (DDPG) algorithm with driving-intention speed prediction and health-aware optimization. By integrating LSTM predictions with deep reinforcement learning, the model improved the hydrogen economy by 6% and reduced degradation by 15–16% [112]. Research by Zhang et al. [113] further supports the benefits of predictive and degradation-aware algorithms. Overall, these studies demonstrate that AI-driven, health-focused control strategies are crucial for next-generation fuel cell vehicles, enhancing efficiency, prolonging component life, and improving hydrogen utilization for low-carbon mobility [113].

4.4. TCO Comparison

This study compares the TCO of various fuel types for passenger cars available in the Indonesian market. We selected several ICEV variants, including the Daihatsu Ayla (A-segment hatchback gasoline), Mitsubishi Xpander (low MPV gasoline), Toyota Innova Reborn (mid-MPV diesel), and Toyota Innova Zenix (mid-MPV gasoline). Additionally, we included hybrid cars such as the Suzuki Ertiga Hybrid (a mild hybrid MPV) and the Honda HR-V e:HEV (a full hybrid C-SUV). For the BEV segment, we examined the BYD Atto 1 (A-segment hatchback) and the Hyundai Kona Electric (Crossover). Lastly, the FCV segment is represented by the imported Toyota Mirai. Table 1 presents the assumptions utilized in the TCO calculation. These assumptions include the exchange rate (1 USD $\approx$ IDR 16,600), energy tariffs, annual mileage (15,000 km/year), a discount rate of 10%, vehicle tax (approximately 1.50% of the book value), insurance rates (1.50%), administrative costs, and the residual value at year 10 for each powertrain type. Additionally, it includes EV charging tariffs (public and premium) and an estimate of retail hydrogen fuel prices in USD. Meanwhile, Table 2 summarizes the TCO calculations for 10 years of usage for various types of passenger cars available on the Indonesian market. File S2 in Supplementary Materials presents a summary of the TCO components for several types of vehicles examined in this study.

Table 1. The assumptions utilized in the TCO calculation.

Parameter	Value	Parameter	Value (USD)
Exchange rate USD to IDR	16,600	Maintenance cost per km for gasoline ICE	0.04
Annual mileage (km)	15,000	Maintenance cost per km for HEVs	0.04
Discount rate	10%	Maintenance cost per km for diesel ICE	0.05
Motor vehicle tax rate annual	1.50%	Maintenance cost per km for BEVs	0.02
Insurance rate annual	1.50%	Maintenance cost per km for FCVs	0.05
Resale year 5 (pct of OTR)	55.00%	Non-subsidized gasoline (per liter)	0.78
Resale year 10 (pct of OTR)	30.00%	Non-subsidized diesel (per liter)	0.84
BEV battery replacement year	8	Home charging (per kwh)	0.10
HEV battery replacement year	7	Public charging (per kwh)	0.21
Admin insurance per year (USD)	10	Fast charging (per kwh)	0.24
Battery pack cost per kWh (USD)	115	Hydrogen liquid (per kg est)	14.50
HEV aux batt replacement (USD)	900		

Table 2. Ten-year TCO calculation for several different fuel passenger cars.

Vehicle	Type	OPEX Cost (USD/km)	Upfront Purchase (USD)	TCO NPV 10 y (USD)	TCO per km 10 y (USD per km)
Daihatsu Ayla 1.2	ICE Petrol	0.09	9789	14,682	0.098
BYD Atto 1	BEV	0.04	12,952	15,860	0.106
Toyota Avanza	CNG	0.08	16,438	19,524	0.130
Suzuki Ertiga Hybrid	Hybrid Mild	0.09	17,280	21,949	0.146
Mitsubishi Xpander	ICE Petrol	0.12	19,053	24,530	0.164
Toyota Innova Reborn	ICE Diesel	0.14	26,114	31,989	0.213
Honda HR-V e:HEV	Hybrid	0.09	26,744	31,035	0.207
Toyota Innova Zenix	ICE Petrol	0.13	27,672	32,883	0.219
BYD Atto 3	BEV	0.05	29,819	32,311	0.215
Hyundai Kona Electric	BEV	0.05	37,795	39,574	0.264
Toyota Mirai (import est.)	FCEV	0.26	93,012	103,719	0.691

The TCO differs among vehicle types. The BYD Atto 1, a compact BEV priced at about USD 13,000, offers a ten-year TCO of approximately USD 31,000, including an eight-year battery replacement. Hybrids like the Honda HR-V e:HEV (USD 27,000) and Suzuki Ertiga Hybrid (USD 17,200) are also competitive. The HR-V has a ten-year TCO of USD 46,000 (USD 0.31/km), while the Ertiga's TCO is about USD 39,000 (USD 0.27/km) for ten years. Both hybrids show strong long-term cost stability, closing in on small BEVs.

Conventional ICEs, such as the Daihatsu Ayla, at almost USD 9700, face long-term disadvantages due to rising fuel costs, totaling around USD 14,600 over ten years (USD 0.098/km). Larger models such as the Mitsubishi Xpander (USD 19,000), Toyota Innova Reborn Diesel (USD 26,000), and Innova Zenix gasoline (USD 27,600) have ten-year total ownership costs of approximately USD 24,500, USD 31,900, and USD 32,800, respectively, all of which are higher per kilometer than BEVs, hybrids, and CNG. Mid-range BEVs, like the BYD Atto 3 (USD 32,300) and the Hyundai Kona Electric (USD 39,500), are more affordable than larger ICEVs. However, with battery costs estimated at USD 7000 to USD 7400, their ten-year TCO is USD 32,300 to USD 39,500. The Toyota Avanza CNG (USD 16,400) has a ten-year TCO of around USD 19,500 (USD 0.13 per kilometer). The Toyota Mirai, starting at USD 93,000, has a ten-year TCO of USD 103,700 (USD 0.69 per kilometer).

The BYD Atto 1, as a small BEV, has the lowest cost efficiency per kilometer over the first five years of utilization. However, over ten years, the cost efficiency per kilometer changes as follows: gasoline low-cost green cars (Ayla), small BEVs (Atto 1), hybrids (Ertiga, HR-V), ICE MPVs/SUVs (Xpander, Innova), medium BEVs (Atto 3, Kona), and FCEVs (Mirai). Figure 3 displays the TCO per component for various vehicles with different fuel types.

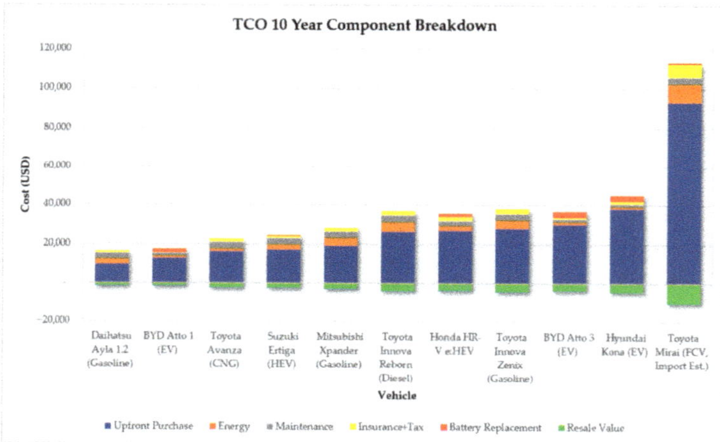

Figure 3. TCO per component for various vehicles with different fuel types.

The 2025 TCO analysis reveals an exciting insight: Indonesia's adoption of low-carbon technology is significantly driven by strong institutional factors rather than market forces alone. The government's strategic fiscal planning, thoughtful policy implementation, and focused actions play a crucial role in determining the success of these technologies.

Our study also examines several secondary sources that gather data through surveys of the Indonesian public on the adoption of EV technology in Indonesia. Indira Ayu et al. (2025) [114] surveyed 1360 respondents across six provinces in Java and conducted expert interviews with stakeholders from government agencies, OEMs, and charging infrastructure operators. Using a Mixed Multinomial Logit (MMNL) model, they identified purchase price, operating cost, range, and charging duration as the most significant factors influencing the adoption of BEVs. Consumers showed a preference for BEVs priced below USD 25,000 with operational costs under USD 0.05 per kilometer. Stakeholder perspectives highlighted critical barriers, including policy inconsistency, reliance on battery supply, and limited charging infrastructure, all of which impede market growth [114].

According to complementary findings from PwC (2025) [115], 60% of Indonesian consumers plan to purchase an EV within the next five years. Currently, 7% of respondents already own an EV, while 78% are potential adopters, and 15% remain skeptical about the transition. The primary motivations for considering an EV include fuel cost savings—up to 60% compared to gasoline—and environmental benefits. However, the main concerns among consumers are related to charging time, range limitations, and battery lifespan. The survey highlights a growing interest in EVs among the urban middle class, but gaps hinder actual sales conversion in infrastructure, as does uncertain policy direction [115].

5. Discussion

The findings indicate that Indonesia's transition in transportation and energy is a historically layered process, characterized by institutional continuity and selective adaptation rather than radical transformation. When analyzed through the frameworks of HI and MLP, this transition can be interpreted as a multi-scalar interplay between landscape pressures, regime resilience, and niche experimentation, mediated by policy feedback, layering, and conversion mechanisms.

5.1. Synthesizing Historical Institutionalism and Multi-Level Perspective in the Context of Indonesia's Political Economy of Transport

Indonesia's transportation energy sector has undergone phases of inertia and transition, often influenced by political dynamics. Throughout the Soeharto era, which began in the 1970s, the country relied heavily on fossil fuels, using oil production and exports to finance national development. However, during this socio-technical period, there was a significant shift in resource management policies, including efforts to invite foreign investment.

Initially, the economy relied on fossil fuels to drive development and ensure national energy security. However, external factors such as the 1970s energy crisis and fluctuations in petroleum prices have significantly impacted the sector. These landscape dynamics highlighted the critical junctures that prompted the government to diversify its energy and fuel sources. As a result of these institutional changes, there has been an increase in exploration and production activities focused on cleaner and more environmentally friendly sources, such as natural gas and geothermal energy. Indonesia's substantial fossil fuel reserves have impacted the country's energy diversification strategy, especially in promoting renewable energy from an energy security standpoint. However, the policy to shift toward renewable sources has mainly concentrated on the electricity generation sector. Consequently, renewable electrical energy has not yet been used to fuel vehicles or trains in the first phase.

The socio-technical landscape has evolved due to various global sustainability initiatives established in the 1980s and 1990s. Key global sustainability agendas included the Brundtland Report in 1987, Agenda 21 in 1992, and the Kyoto Protocol in 1997. The adoption of these initiatives led to significant institutional changes at the meso-level regime, marked by the enactment of Act No. 4/1982 on Basic Provisions for Environmental Management, Act No. 23/1997 on Environmental Management, and the fourth amendment to the 1945 Constitution. Our findings indicate that landscape dynamics pressured the socio-technical regime to formulate national policies that aligned energy production advancement with environmental preservation. Our study emphasizes the literature regarding the impact of the socio-technical landscape on institutional regime changes in energy transitions [34].

Our study reveals that urban-centered development has resulted in economic growth and a rise in population. However, this urbanization has not been well managed. Additionally, there has been a shift from trains to road transportation, leading to an increase in private vehicles without proper road planning or improvements to public transport. Evolutions in the landscape and regime policies have entrenched petroleum dependence.

Incorrectly planned policies have led to significant traffic congestion. Despite commitments to global sustainability initiatives, the government has not effectively implemented fuel efficiency and emission restrictions.

According to the HI framework, this regime orientation describes enduring path-dependence, where policy choices focused on fuel price stabilization and state dominance in energy have yielded increasing returns that constrain future options [116]. Fuel subsidies, once viewed as redistributive instruments, have evolved into political institutions that shape state–citizen relations and fiscal priorities. When global oil prices spiked in the 2000s, the government faced a dual fiscal and legitimacy crisis, prompting a reactive institutional layering approach through gradual diversification policies without dismantling entrenched subsidy mechanisms [30,43,117,118]. Our study also revealed that these events had become critical junctures that pressured the regime to alter its policies to reduce fuel imports, thus creating "windows of opportunity" for enhancing the development of alternative vehicle fuels.

The MLP has elucidated how these macro-institutional structures interact with niche-level technological innovations. Oil price volatility and global climate commitments have changed the landscape, increasing pressure on regimes to implement diversification programs. However, state actors within these regimes have selectively adapted through institutional conversion—repurposing policy instruments to maintain rent distribution while incorporating new technologies.

The evolution of low-carbon mobility, including CNG, biofuels, EVs, and FCVs, illustrates this incremental institutional adaptation. Presidential Instruction No. 1/2006 represents a significant milestone in establishing biofuels institutionalization in Indonesia. The subsequent implementation of Government Regulation No. 1/2007, which introduces fiscal incentives, has effectively attracted increased private investment in biofuel development, including sugarcane-based ethanol and palm oil-based biodiesel. Our study highlighted the vital role of the National Biofuels Team in facilitating technology transfer from other countries that have successfully developed biofuels and promoting niche innovations from local companies. Thus, biofuel companies effectively have taken advantage of the "window of opportunity" offered by opening the biofuel market in the EU and the US. Our findings align with recent research highlighting the state's role in fostering and facilitating niche innovations by private and SOEs during the energy transition [119].

The transition to CNG in countries like Iran, Pakistan, Argentina, Brazil, and China has prompted governments to encourage its use in public transportation. This shift was formalized with the issuance of DGOG Decree No. 247/2011, which was supported by the establishment of two SOEs to manage the supply and distribution of CNG. However, despite Indonesia's abundant natural gas production, the CNG conversion program within the transportation sector has faced challenges and has been unsuccessful. We found that landscape influences and government support have not significantly affected the adoption of CNG at the micro-niche level. Moreover, while CNG has primarily been utilized in public vehicles, fostering niche innovations for private vehicles has encountered obstacles. These obstacles include unclear government regulations, inadequate infrastructure, and the price disparity between gasoline and CNG. The diffusion of CNG has reflected early diversification efforts but suffered policy drift due to inter-ministerial fragmentation and insufficient investment. Our findings are consistent with the literature showing that newcomers to radical innovation who encounter high levels of uncertainty and require significant infrastructure will enter and exit the socio-technical system [120]. The emergence of environmental issue reports, trade disputes, and the Paris Agreement between 2011 and 2015 has significantly changed the landscape of Indonesia's biodiesel exports. These developments have led to a decline in biodiesel production and neglected harvests at the farmer level. Our study highlights that these global and domestic shifts have created an exogenous shock, pushing the state to modify the institutional framework encompassing biodiesel. Conversely, this exogenous shock has opened "windows of opportunity" to promote energy transition through AVFs and reduce fuel imports.

Our study also reveals a contention between two groups of stakeholders: one coalition group advocates for the absorption of domestic biodiesel, while the other coalition prefers a transition to EVs. The results of the AHP conducted by Bappenas have become instrumental in determining the outcome of this contestation, ultimately establishing biodiesel as the state's primary preference. This decision has led to a path dependence in the development of AVFs in Indonesia. Indonesia's biofuel development trajectory differs from that of Thailand and the Philippines, which focus on ethanol rather than biodiesel. Significant institutional changes have occurred with the implementation of MEMR Regulation No. 12/2015, which mandates the use of B20 biodiesel in various sectors, including transportation. The establishment of BPDPKS in 2015 stabilized the biodiesel market through export levies, embedding a policy feedback loop that favors the interests of palm oil [121,122]. These developments represent critical junctures so that energy companies can leverage the "windows of opportunity" to foster the expansion of Indonesian biodiesel. Our study contributes to the body of literature on the contention among stakeholders shaping energy transition projections through layering on the existing system, particularly in emerging economies [123].

The technological disruption brought about by the proliferation of EVs in Asia since 2017 cannot be overlooked. Early adopters have indirectly socialized the concept of EVs, leading to collaborative innovations at the micro-niche level. The diffusion of EV trends and a shift in political orientation have altered the socio-technical landscape and prompted government support for EV adoption. Presidential Regulation No. 55 of 2019 and MHA Regulation No. 1 of 2021 marked the initial institutional support for EVs in Indonesia, representing a more recent form of institutional conversion—redirecting fiscal instruments from industrial promotion to decarbonization objectives. This regulatory framework has facilitated the development of local products, starting with the E2W segment. Niche innovations in the battery swapping business model have emerged as a transition from the existing ICEV energy purchasing system. The increased regime support has become more noticeable with the implementation of a limited subsidy scheme since 2023. Gradually, the public has begun to recognize the benefits of transitioning to EVs, particularly from an economic perspective.

Meanwhile, MoI Regulation No. 29 of 2023 and the National Hydrogen and Ammonia Roadmap (RHAN) have institutionalized the utilization of hydrogen as an AVF. Furthermore, FCV pilot projects, such as those involving the construction of hydrogen refueling stations by PLN, Hyundai, and Toyota in recent years, have demonstrated initial displacement dynamics, although institutional support remains minimal.

The findings from the quantitative TCO analysis and the qualitative consumer perception survey present a consistent yet nuanced view of Indonesia's EV transition. The TCO analysis indicates that BEVs have achieved a level of economic competitiveness, with operational energy costs around USD 0.106 per kWh when charged at home. This cost is significantly lower than that of CNG, petrol, or diesel vehicles over a ten-year ownership period. Although BEVs have higher initial costs and incur mid-life battery replacement expenses, small-to-medium BEVs and hybrid models, like the Honda HR-V e:HEV (approximately USD 0.03 per km), demonstrate long-term cost-efficiency and resilience against fuel price volatility. In contrast, ICEVs face rising fuel costs, while infrastructural and refueling challenges hinder the adoption of CNG and FCVs.

The findings from our techno-economic analysis are supported by consumer insights from PwC Indonesia's 2023 nationwide survey conducted in eight major cities: Jakarta, Bekasi, Tangerang, Bogor, Surabaya, Semarang, Medan, and Depok [115]. The survey revealed that 78% of respondents are interested in purchasing an electric car, and 74% are interested in an electric motorcycle. However, public adoption is limited by concerns about affordability. Specifically, 87% of respondents feel that battery prices are prohibitively high, and 83% are worried about the cost of expensive spare parts. Additional recurring expenses, such as annual maintenance costs (USD 200–300) and mid-life battery replacement costs, are also significant deterrents [115].

This perception aligns with the TCO outcomes, which indicate that while BEVs have lower operating costs, the overall lifetime cost advantage only becomes clear after several years of use. This advantage is dependent on stable electricity tariffs and government fiscal incentives. PwC further emphasizes that tax exemptions ranging from USD 1000 to USD 1500 per vehicle could significantly reduce this psychological cost barrier, underscoring the importance of state-led incentives in promoting the adoption of EVs [115].

Triangulating these datasets indicates that Indonesia's transition to EVs hinges on more than just affordable technology; it also depends on consumer confidence and reliable infrastructure. TCO analysis shows that BEVs and hybrids are increasingly economically viable. However, survey results highlight the need for improved support in areas such as battery warranty transparency, charging accessibility, and post-purchase guarantees to convert interest into actual adoption. To address these challenges, sustained fiscal incentives, improved public charging infrastructure, and consistent policies are crucial for reducing the gap between cost-competitiveness and perceived consumer risk in Indonesia's EV market [114,115].

Our study found that Indonesia followed a different path in its electrification transition by facilitating the transition from HEVs to BEVs less. This approach contrasts with other developing countries such as India and Thailand, which provide substantial incentive support for HEVs to reduce GHG emissions. Our findings enrich the existing literature on the various trajectories of developed and developing countries in their energy transition toward sustainability in the transportation sector [124].

Hydrogen-as-an-AVF initiatives include the Indonesian mobility and energy sectors. However, they remain pilot-scale and heavily dependent on foreign partnerships and SOEs initiatives, contrasting sharply with the more established ecosystems in Japan, South Korea, or Germany. Collaborative networks have been established between state-owned energy companies and foreign manufacturers and licensors to develop a hydrogen ecosystem. Additionally, a pilot project has demonstrated the potential to utilize geothermal green energy sources for producing green hydrogen. This finding aligns with recent studies that emphasize the necessity for significant investment, particularly in infrastructure development, to enhance the adoption of HFCVs, especially considering the low uptake of HFCVs in Indonesia as a developing country [61,80].

5.2. *Comparative Insights from Global and Regional Transitions*

Our study utilizes a comparative analysis that positions Indonesia's hybrid trajectory within a broader framework of transition pathways among developing countries. Brazil's Pró-Álcool and RenovaBio programs serve as key examples of a coherent institutional structure where state intervention effectively aligns agricultural policy, energy security, and industrial innovation. The Brazilian government has implemented stable fiscal incentives, performance-based mandates, and carbon intensity assessments to ensure the competitiveness of ethanol without relying solely on subsidies [125].

Meanwhile, Argentina prioritizes soybean-based biodiesel, supported by the National Biodiesel Promotion Law (No. 26093/2006). However, the country faces challenges due to volatility stemming from weak institutional consistency and heavy reliance on exports. Frequent policy reversals, such as the suspension of the biodiesel mandate in 2017, underscore the impact of a lack of robust policy feedback on sustainability [126]. Similarly, Indonesia's biofuel governance relies on commodity rents; however, greater bureaucratic centralization through BPDPKS has set its biodiesel diffusion apart from that of Argentina [127].

In the diffusion of e-mobility, China's EV strategy has demonstrated the effectiveness of coordination among different levels of government institutions in achieving large-scale progress. By integrating EV development into its national industrial planning, including the Five-Year Plans and the new energy vehicle (NEV) credit system, China captured nearly two-thirds of the global EV market share by 2024. This strategy shows how a centralized institutional design can realign existing frameworks [60,128]. On the other hand, India's FAME-II policy combines fiscal incentives with domestic manufacturing requirements, reflecting Indonesia's ambitions in EV manufacturing. However, India has benefited from stronger inter-ministerial coordination [60].

Regional electrification in Southeast Asia is highlighted by Thailand's "30@30" policy, which aims for 30% of vehicle production to be electric by 2030. This initiative is supported by the Board of Investment (BOI) through tax breaks and straightforward integration with the expansion of renewable energy [60,61]. Vietnam has demonstrated effective public–private collaboration through VinFast, while Malaysia is advancing its hybridization efforts with the Proton program. Singapore is notable for its comprehensive EV approach, which combines carbon taxes, incentives, and the electrification of public transportation under a unified governance framework [60].

For comparison, Indonesia's transportation transition has been shaped by its hybrid institutionalism, where partial conversion occurs within a deeply entrenched structure of rent-seeking. The HI approach demonstrates that incremental change happens because self-interest is tied to the interest of large palm oil corporations, fossil fuel suppliers, and the fiscal dependence of SOEs, all of which limit radical transformation. The MLP contextualizes Indonesia's situation as a path of reconfiguration, where niche technologies—especially biofuels and EVs—coexist rather than completely replacing incumbents. Therefore, while landscape pressures, such as declining battery costs and global hydrogen initiatives, have impacted the regime, it is the adaptation of institutions that has led to different outcomes [32,92].

6. Conclusions

This study provides a comprehensive understanding of Indonesia's transport-energy transition by synthesizing the HI and MLP frameworks, demonstrating that the evolution of sustainability transitions in developing countries is inseparable from their political economy and institutional legacies. The findings reaffirm that Indonesia's pathway toward sustainable mobility is not driven by spontaneous technological competition, but rather by the historical sequencing of institutions, power-laden governance structures, and policy feedback mechanisms that shape how innovations emerge, compete, and stabilize within the socio-technical regime.

From a historical perspective, Indonesia's transport-energy trajectory reveals the deep imprint of state developmentalism and rent-based governance that emerged during the Soeharto era (1966–1998). This developmental model institutionalized the provision of cheap energy as a social contract, embedding fuel subsidies as a central mechanism of state legitimacy. Over time, these subsidies evolved into self-reinforcing institutions, creating fiscal path dependencies and constraining subsequent reform attempts. The transition from oil exporter to net importer in 2004 intensified the structural tension between fiscal sustainability and political stability. By the 2010s, fuel subsidies had become a significant burden on the national budget, limiting investment in alternative energy and infrastructure diversification. This historical legacy represents the archetype of increasing returns [19], where the cost of exiting established institutions grows over time, perpetuating fossil fuel-dependence.

Within the MLP framework, these institutional path-dependencies mediate how global pressures and technological niches—such as biofuels, CNG, EVs, and FCVs—are domestically embedded, producing a hybrid transition regime where existing and new institutions coexist. Empirical findings from semi-structured interviews indicate that policy change primarily occurs through layering and conversion, rather than radical displacement. Biofuel mandatory programs, backed by palm oil fund management, along with EV incentives, illustrate how new policies can reinforce existing fiscal frameworks. In contrast, the stagnation of CNG and FCVs reflects a drift owing to inconsistent support. In contrast, CNG has stagnated due to institutional drift, as political interest has decreased, and policy coherence has dissolved. Meanwhile, emerging FCV pilots hint at displacement dynamics, representing the early stages of a potential challenge to the existing regime. However, these niche innovations have not yet led to systemic reconfiguration of the institutional arrangement. Quantitative TCO analysis complements these insights, revealing that cost-competitiveness among technologies depends on the intensity of institutional support—through fiscal incentives, industrial mandates, and energy pricing—rather than inherent market efficiency. In summary, economic viability is shaped by institutional factors rather than being solely determined by technological advancements.

Compared to other developing economies, Indonesia's transition to sustainable energy is notably different. For instance, Brazil has successfully institutionalized its ethanol transition through stable fiscal incentives. In China, centralized coordination has enabled the widespread adoption of EVs. India's FAME programs have demonstrated a gradual yet consistent shift toward hybrid technologies. Meanwhile, Thailand and Malaysia have achieved moderate progress through industrial localization. In contrast, Indonesia's transition has been slower and more politicized, hindered by fragmented governance and fiscal dependencies that prioritize short-term stability over long-term structural decarbonization efforts.

This study makes theoretical contributions by integrating the HI–MLP framework in three ways. First, it shows that institutional mechanisms—such as layering, conversion, feedback, and drift—mediate the effects of multi-level pressures on policy outcomes. Global technological trends and growing environmental norms impact Indonesia's regime only through domestic policies.

Second, it describes hybrid transition regimes, where existing and new institutions coexist, as exemplified by Indonesia's biodiesel sector, which is supported by subsidies. The hybrid transition regimes challenge the straightforward regime vs. niche divide in early MLP theory, suggesting a continuum influenced by the domestic political economy. Finally, methodologically, this study combines qualitative institutional analysis with quantitative TCO modeling, bridging interpretive and economic approaches, grounding transition theory in measurable outcomes.

Policy implications emphasize the need to engage and consider the aspirations of multilayered stakeholders in formulating sustainability policies, particularly SOEs, the private sector, academics, NGOs, and the broader society. To effectively enhance the growth of the EV market, policymakers should focus on the TCO and the learning curve. China's experience demonstrates that time-limited and declining consumer incentives, along with a robust charging network, can help EVs move beyond the early adoption phase. In

Thailand, incorporating export credit can help mitigate risks during the initial production phase. In Indonesia, reallocating fossil fuel subsidies to provide TCO-closing incentives—such as purchase rebates or tax credits for local content—can stimulate demand for EVs. Fiscal incentives should be targeted at making EVs affordable to enable mass adoption. This socio-technical transition must be inclusive, focusing on electric two-wheelers and three-wheelers in rural areas, supported by term subsidies, microcredit, and programs like EVs to villages. This initiative combines off-peak electricity rate discounts, micro-depot charging, and the public procurement of basic service fleets. Charging infrastructure needs to be developed along the rural–urban corridor using a matching grant model for local operators. This approach is similar to China's NEVs to the countryside campaign and India's FAME II, both of which have successfully increased penetration in non-metro areas. According to IEA data, the market share of EVs rises when charging infrastructure and incentives are developed simultaneously [60].

Biofuel policies should support smallholder farmers with minimum price contracts and sustainability premiums to ensure their long-term viability. Aligning ISPO with global standards is essential for market access [4,129]. Indonesia needs to develop second-generation biofuels from lignocellulose and agricultural waste while enforcing environmental conservation rules and offering performance incentives. Encouraging the use of used cooking oil (UCO) through collection systems and quality standards, like in the EU, is also important [130,131]. The government also needs to improve fuel quality standards and ensure vehicle compatibility, with support from manufacturers, similar to flex-fuel vehicles in Brazil [132].

The government needs to adopt more proactive policies that support emerging innovations, such as FCVs. Indonesia should establish carbon intensity standards for hydrogen, prioritize its use in fleets and logistics before introducing it to retail markets, and develop multi-energy hubs within Pertamina's and multinational corporations' gas station networks. Additionally, a regulatory sandbox for energy management in FCVs should be introduced. To further enhance research collaboration, it is important to address energy management concerns in FCVs, particularly by integrating driving intention, speed prediction, and health-aware control technologies. Providing financial support through green industrial credit based on technological readiness will ensure that Indonesia keeps pace with global energy innovation trajectories [133,134].

The limitation in this study occurs because of the limited number of interviews, as it is challenging to gain qualitative insight from all of the experiences of stakeholders involved in the sustainability of Indonesian transportation energy. Research encompasses many formations of institutional stability and institutional change over time and under the duality of external and exogenous influences on interactions between actors, including global, national, and local interactions. Conducting comparative studies in various countries would provide insights into the development of historical institutional theory, particularly in dealing with emerging technology and projecting the implementation of the technology trajectory in the future.

Supplementary Materials: The following supporting information can be downloaded at: https://www.mdpi.com/article/10.3390/en18215723/s1.

Author Contributions: Conceptualization. J.C.L. and F.F.; methodology. J.C.L. and F.F.; validation. J.C.L., M.M.R. and F.F.; formal analysis. J.C.L. and F.F.; investigation. J.C.L. and F.F.; resources. J.C.L. and F.F.; data curation. M.M.R.; writing—original draft preparation. J.C.L. and F.F.; writing—review and editing. J.C.L., M.M.R. and F.F.; visualization. J.C.L. and F.F.; supervision. J.C.L.; project administration J.C.L. and F.F. All authors have read and agreed to the published version of the manuscript.

Funding: The contribution of J.C.L. was funded by the project "Creating resilient sustainable micro-grids through hybrid renewable energy systems" (EPSRC. EP/R030243/1).

Data Availability Statement: The data presented in this study are available upon request from the corresponding author because the data are part of an ongoing study.

Acknowledgments: Ferry Fathoni is fully supported by scholarships from PT Pertamina (Persero).

Conflicts of Interest: Ferry Fathoni was employed by the PT Pertamina (Persero). The remaining authors declare that the research was conducted in the absence of any commercial or financial relationships that could be construed as a potential conflict of interest.

References

1. Green, J.F.; Reyes, R.S. The history of net zero: Can we move from concepts to practice? *Clim. Policy* **2023**, *23*, 901–915. [CrossRef]
2. Fankhauser, S.; Smith, S.M.; Allen, M.; Axelsson, K.; Hale, T.; Hepburn, C.; Kendall, J.M.; Khosla, R.; Lezaun, J.; Mitchell-Larson, E.; et al. The meaning of net zero and how to get it right. *Nat. Clim. Change* **2022**, *12*, 15–21. [CrossRef]
3. Schrijver, N. Managing the global commons: Common good or common sink? *Third World Q.* **2016**, *37*, 1252–1267. [CrossRef]
4. Astari, A.J.; Lovett, J.C.; Wasesa, M. Sustainable pathways in Indonesia's palm oil industry through historical institutionalism. *World Dev. Sustain.* **2025**, *6*, 100200. [CrossRef]
5. North, D.C. Institutional change: A framework of analysis. In *Institutional Change: Theory and Empirical Findings*; Routledge: New York, NY, USA, 1993; pp. 35–46.
6. Thelen, K. Historical institutionalism in comparative politics. *Annu. Rev. Political Sci.* **1999**, *2*, 369–404. [CrossRef]
7. Williamson, O.E. Nobel Prize Lecture Presentation: Transaction Cost Economics, the Natural Progression Stockholm: Nobel Prize. 2009. Available online: https://www.nobelprize.org/prizes/economic-sciences/2009/williamson/lecture/ (accessed on 26 March 2024).
8. March, J.G.; Olsen, J.P. *Elaborating the "New Institutionalism"*; Rhodes, R.A.W., Binder, S.A., Rockman, B.A., Eds.; Oxford University Press: Oxford, UK, 2006; pp. 3–20.
9. Immergut, E.M. Historical-institutionalism in political science and the problem of change. In *Understanding Change*; Palgrave Macmillan: London, UK, 2006; pp. 237–259.
10. Pierson, P.; Skocpol, T. Historical institutionalism in contemporary political science. *Political Sci. State Discip.* **2002**, *3*, 1–32.
11. Thelen, K.; Steinmo, S. Historical institutionalism in comparative politics. In *Structuring Politics: Historical Institutionalism in Comparative Analysis*; Steinmo, S., Thelen, K., Longstreth, F., Eds.; Cambridge University Press: Cambridge, UK, 1992; pp. 1–15.
12. Official Journal of the European Union (European Union). *Council Implementing Regulation (EU) No 1194/2013 of 19 November 2013 Imposing a Definitive Anti-Dumping Duty and Collecting Definitively the Provisional Duty Imposed on Imports of Biodiesel Originating in Argentina and Indonesia*; European Union: Brussels, Belgium, 2013.
13. Lovett, J.C.; Hofman, P.S.; Morsink, K.; Clancy, J. Technology transfer and global markets. In *Low-Carbon Technology Transfer from Rhetoric to Reality*; Ockwell, D., Mallett, A., Eds.; Routledge: Oxford, UK, 2012; pp. 322–339.
14. Obidzinski, K.; Andriani, R.; Komarudin, H.; Andrianto, A. Environmental and social impacts of oil palm plantations and their implications for biofuel production in Indonesia. *Ecol. Soc.* **2012**, *17*, 1–21. [CrossRef]
15. al Irsyad, M.I.; Firmansyah, A.I.; Harisetyawan, V.T.F.; Supriatna, N.K.; Gunawan, Y.; Jupesta, J.; Yaumidin, U.K.; Purwanto, J.; Inayah, I. Comparative total cost assessments of electric and conventional vehicles in ASEAN: Commercial vehicles and motorcycle conversion. *Energy Sustain. Dev.* **2025**, *85*, 101599. [CrossRef]
16. Yuniaristanto; Afraah, S.M.; Sutopo, W.; Hisjam, M. (Eds.) *The Sustainability-Oriented Total Cost of Ownership Model of the Electric Motorcycle*; Springer Nature: Singapore, 2023; pp. 365–373.
17. Eitan, A.; Hekkert, M.P. Locked in transition? Towards a conceptualization of path-dependence lock-ins in the renewable energy landscape. *Energy Res. Soc. Sci.* **2023**, *106*, 103316. [CrossRef]
18. Hall, P.A.; Taylor, R.C.R. Political science and the three new institutionalisms. *Political Stud.* **1996**, *XLIV*, 936–957. [CrossRef]
19. Pierson, P. Not just what, but when: Timing and sequence in political processes. *Stud. Am. Political Dev.* **2000**, *14*, 72–92. [CrossRef]
20. Eckersley, P. Cities and climate change: How historical legacies shape policy-making in English and German municipalities. *Politics* **2017**, *37*, 151–166. [CrossRef]
21. Streeck, W.; Thelen, K.A. Beyond continuity: Institutional change in advanced political economies. In *Beyond Continuity: Change in Advanced Political Economies*; Streeck, W., Thelen, K.A., Eds.; Oxford University Press: Oxford, UK, 2005; pp. 1–39.
22. Van der Heijden, J. Institutional layering: A review of the use of the concept. *Politics* **2011**, *31*, 9–18. [CrossRef]
23. Fioretos, K.O.; Falleti, T.G.; Sheingate, A.D. *The Oxford Handbook of Historical Institutionalism*; Oxford University Press: Oxford, UK, 2016.

24. Capoccia, G. When do institutions "bite"? Historical institutionalism and the politics of institutional change. *Comp. Political Stud.* **2016**, *49*, 1095–1127. [CrossRef]
25. Thoenig, J.-C. Institutional theories and public institutions: Traditions and appropriateness. In *Handbook of Public Administration*; Peters, G.B., Pierre, J., Eds.; SAGE Publishing: London, UK, 2003; pp. 127–137.
26. Collier, D.; Collier, R.B.; Hershberg, E. Shaping the political arena: Critical junctures, the labor movement, and regime dynamics in Latin America. *Can. J. Lat. Am. Caribb. Stud.* **2004**, *29*, 327.
27. Mahoney, J.; Thelen, K.A. *Explaining Institutional Change: Ambiguity, Agency, and Power*; Mahoney, J., Thelen, K.A., Eds.; Cambridge University Press: Cambridge, UK, 2010.
28. Greif, A.; Laitin, D.D. A Theory of Endogenous Institutional Change. *Am. Political Sci. Rev.* **2004**, *98*, 633–652. [CrossRef]
29. Lockwood, M.; Kuzemko, C.; Mitchell, C.; Hoggett, R. Historical institutionalism and the politics of sustainable energy transitions: A research agenda. *Environ. Plan. C Politics Space* **2017**, *35*, 312–333. [CrossRef]
30. Ichsan, M.; Pye, O.; Saputra, W.; Bakhtiar, I. Institutional change and the framing of palm oil-based biofuel development in Indonesia: In the national interest? *South East Asia Res.* **2025**, *33*, 183–204. [CrossRef]
31. OECD. *A Growth-Friendly and Inclusive Green Transition Strategy for Thailand*; OECD: Paris, France, 2024. [CrossRef]
32. IEA. *World Energy Outlook 2024*; IEA: Paris, France, 2024. Available online: https://iea.blob.core.windows.net/assets/6a25abba-1973-4580-b6e3-ba014a81b458/WorldEnergyOutlook2024.pdf (accessed on 20 November 2024).
33. Carvalho, A.D.P.; Cunha, S.K.d.; Lima, L.F.d.; Carstens, D.D. The role and contributions of sociological institutional theory to the socio-technical approach to innovation theory. *RAI Revista de Administração e Inovação.* **2017**, *14*, 250–259. [CrossRef]
34. Geels, F.W. The Multi-Level Perspective on Sustainability Transitions: Background, overview, and current research topics. *Camb. Open Engag.* **2024**. [CrossRef]
35. Hansmeier, H.; Schiller, K.; Rogge, K.S. Towards methodological diversity in sustainability transitions research? Comparing recent developments (2016–2019) with the past (before 2016). *Environ. Innov. Soc. Transit.* **2021**, *38*, 169–174. [CrossRef]
36. Klitkou, A.; Bolwig, S.; Hansen, T.; Wessberg, N. The role of lock-in mechanisms in transition processes: The case of energy for road transport. *Environ. Innov. Soc. Transit.* **2015**, *16*, 22–37. [CrossRef]
37. Męczyński, M.; Ciesiółka, P. Toward a Green Transformation Based on Eco-Innovations of Enterprises in Old Industrial Areas: The Case of Silesia Province. In *The Changing Economic Geography of Companies and Regions in Times of Risk, Uncertainty, and Crisis*; Routledge: Milton Park, UK, 2024; pp. 105–127.
38. Helmrich, A.; Chester, M.; Miller, T.R.; Allenby, B. Lock-in: Origination and significance within infrastructure systems. *Environ. Res. Infrastruct. Sustain.* **2023**, *3*, 32001. [CrossRef]
39. Geels, F.W. Micro-foundations of the multi-level perspective on socio-technical transitions: Developing a multi-dimensional model of agency through crossovers between social constructivism, evolutionary economics and neo-institutional theory. *Technol. Forecast. Soc. Change* **2020**, *152*, 119894. [CrossRef]
40. Hidayat, R.; Cowie, J. A framework to explore policy to support the adoption of electric vehicles in developing nations: A case study of Indonesia. *Transp. Res. Procedia* **2023**, *70*, 364–371. [CrossRef]
41. Novizayanti, D.; Prasetio, E.A.; Siallagan, M.; Santosa, S.P. Agent-based modeling framework for electric vehicle adoption transition in Indonesia. *World Electr. Veh. J.* **2021**, *12*, 73. [CrossRef]
42. Mohamad, M.; Songthaveephol, V. Clash of titans: The challenges of socio-technical transitions in the electrical vehicle technologies—The case study of Thai automotive industry. *Technol. Forecast. Soc. Change* **2020**, *153*, 119772. [CrossRef]
43. Cameron, R.; Geels, F.W. Conditions for politically accelerated transitions: Historical institutionalism, the multi-level perspective, and two historical case studies in transport and agriculture. *Technol. Forecast. Soc. Change* **2019**, *140*, 221–240. [CrossRef]
44. Ydersbond, I.M. (Ed.) Aiming to be environmental leaders, but struggling to go forward: Sweden and Norway on energy system transformation. In Proceedings of the Renewable Energy Research Conference, RERC 2014, Oslo, Norway, 16–18 June 2014; pp. 16–23.
45. Nikas, A.; Koasidis, K.; Köberle, A.C.; Kourtesi, G.; Doukas, H. A comparative study of biodiesel in Brazil and Argentina: An integrated systems of innovation perspective. *Renew. Sustain. Energy Rev.* **2022**, *156*, 112022. [CrossRef]
46. Farrell, H.; Newman, A.L. Making global markets: Historical institutionalism in international political economy. *Rev. Int. Political Econ.* **2010**, *17*, 609–638. [CrossRef]
47. Smith, A.K.; Ayanian, J.Z.; Covinsky, K.E.; Landon, B.E.; McCarthy, E.P.; Wee, C.C.; Steinman, M.A. Conducting high-value secondary dataset analysis: An introductory guide and resources. *J. Gen. Intern. Med.* **2011**, *26*, 920–929. [CrossRef]
48. Wallace Foundation. *Workbook B: Conducting Secondary Research*; Wallace Foundation: New York, NY, USA, 2015.
49. Heaton, J. *Reworking Qualitative Data*; SAGE: London, UK, 2004.
50. Fielding, N. Getting the most from archived qualitative data: Epistemological, practical and professional obstacles. *Int. J. Soc. Res. Methodol.* **2004**, *7*, 97–104. [CrossRef]
51. Johnston, M.P. Secondary data analysis: A method of which the time has come. *Qual. Quant. Methods Libr.* **2017**, *3*, 619–626.

52. Hinds, P.S.; Vogel, R.J.; Clarke-Steffen, L. The possibilities and pitfalls of doing a secondary analysis of a qualitative data set. *Qual. Health Res.* **1997**, *7*, 408–424. [CrossRef]
53. Woody, M.; Adderly, S.A.; Bohra, R.; Keoleian, G.A. Electric and gasoline vehicle total cost of ownership across US cities. *J. Ind. Ecol.* **2024**, *28*, 194–215. [CrossRef]
54. Sasongko, T.W.; Ciptomulyono, U.; Wirjodirdjo, B.; Prastawa, A. Identification of electric vehicle adoption and production factors based on an ecosystem perspective in Indonesia. *Cogent Bus. Manag.* **2024**, *11*, 2332497. [CrossRef]
55. Wróblewski, P.; Drożdż, W.; Lewicki, W.; Dowejko, J. Total Cost of Ownership and Its Potential Consequences for the Development of the Hydrogen Fuel Cell Powered Vehicle Market in Poland. *Energies* **2021**, *14*, 2131. [CrossRef]
56. Creswell, J.W.; Poth, C.N. *Qualitative Inquiry and Research Design: Choosing Among Five Approaches*, 4th ed.; Sage Publications: Los Angeles, CA, USA, 2018.
57. Creswell, J.W. *Research Design: Qualitative, Quantitative, and Mixed Methods Approaches*, 4th ed.; Sage Publications, Inc.: Los Angeles, CA, USA, 2014.
58. Yin, R.K. *Case Study Research and Applications: Design and Methods*, 6th ed.; SAGE Publications, Inc.: Thousand Oaks, CA, USA, 2018.
59. Bowen, G.A. Document Analysis as a Qualitative Research Method. *Qual. Res. J.* **2009**, *9*, 27–40. [CrossRef]
60. IEA. *Global EV Outlook 2025*; IEA: Paris, France, 2025. Available online: https://iea.blob.core.windows.net/assets/7ea38b60-3033-42a6-9589-71134f4229f4/GlobalEVOutlook2025.pdf (accessed on 14 August 2025).
61. Reddy, V.J.; Hariram, N.P.; Maity, R.; Ghazali, M.F.; Kumarasamy, S. Sustainable Vehicles for Decarbonizing the Transport Sector: A Comparison of Biofuel, Electric, Fuel Cell and Solar-Powered Vehicles. *World Electr. Veh. J.* **2024**, *15*, 93. [CrossRef]
62. Uwe, F. *The Sage Handbook of Qualitative Research Quality*; SAGE Publications: London, UK, 2024.
63. Wright, R.; Chapla, K.; Booth, A.; Nelson, K.E.; Peeler, A.; Swain, C.; Won, S.; Wu, D.S. Enhancing Rigor, Quality, and Patient Engagement in Qualitative Research: A Step-By-Step Guide to Applying Reflexive Thematic Analysis to the Experience-Based Co-Design Methodology. *Qual. Health Res.* **2024**, *16*, 10497323241291798. [CrossRef]
64. Kontan. Understanding the History and Revitalisation of Indonesia's Railway System Jakarta: Kontan. 2024. Available online: https://jelajahekonomi.kontan.co.id/ekonomi-infrastruktur/news/mengenal-sejarah-hingga-pembenahan-wajah-baru-kereta-api-di-indonesia (accessed on 23 April 2025).
65. KAI. Raliways History Jakarta: KAI. 2017. Available online: https://www.kai.id/corporate/about_kai/#tabs-1-item-0 (accessed on 17 July 2020).
66. Kompas. The Last Steam Train Jakarta: Kompas. 2010. Available online: https://travel.kompas.com/read/2010/06/19/17304062/Kereta.Uap.Terakhir (accessed on 12 March 2025).
67. Lindblad, J.T. The Indonesian Economy in the Early Independence Period. *Itinerario* **2010**, *34*, 7–8. [CrossRef]
68. Corsello, F.; Gomellini, M.; Pellegrino, D. Inflation and energy price shocks: Lessons from the 1970s. *Bank Italy Occas. Pap.* **2023**. [CrossRef]
69. Grandis, A.; Fortirer, J.; Pagliuso, D.; Buckeridge, M. Scientific Research on Bioethanol in Brazil: History and Prospects for Sustainable Biofuel. *Sustainability* **2024**, *16*, 4167. [CrossRef]
70. Sajid, Z.; da Silva, M.A.B.; Danial, S.N. Historical Analysis of the Role of Governance Systems in the Sustainable Development of Biofuels in Brazil and the United States of America (USA). *Sustainability* **2021**, *13*, 6881. [CrossRef]
71. Shamsapour, N.; Hajinezhad, A.; Noorollahi, Y. Developing a system dynamics approach for CNG vehicles for low-carbon urban transport: A case study. *Int. J. Low-Carbon Technol.* **2020**, *16*, 577–591. [CrossRef]
72. Heinrichs, H.; Martens, P.; Michelsen, G.; Wiek, A. *Sustainability Science*; Springer: Berlin/Heidelberg, Germany, 2016; pp. 1–4.
73. BPPT. Indonesia Energy Outlook 2018: Sustainable Energy for Land Transportation Jakarta: PPIPE BPPT. 2018. Available online: https://doi.org/10.5281/zenodo.8216479 (accessed on 6 June 2019).
74. U.S. Energy Information Administration. *Indonesian Petroleum Infographic*; U.S. Energy Information Administration: Washington, DC, USA, 2020. Available online: https://www.eia.gov/international/data/country/IDN/infographic/petroleum-infographic?pd=44&p=0000000000000000000000000000000000000000000000000000024&u=0&f=A&v=mapbubble&a=-&i=none&vo=value&t=C&g=none&l=249{-}{-}104&s=315532800000&e=1483228800000 (accessed on 15 July 2020).
75. ERIA. *Review of Biofuels Sustainability Assessment and Sustainability Indicators in East Asia Summit Countries*; ERIA: Jakarta, Indonesia, 2021. Available online: https://www.eria.org/uploads/media/Research-Project-Report/2021-18-Analysis-of-Future-Mobility-Fuel-Scenarios-/8_ch.3-Review-Biofuel-Sustainability-Assessment.pdf (accessed on 6 May 2025).
76. Kalthaus, M.; Sun, J. Determinants of electric vehicle diffusion in China. *Environ. Resour. Econ.* **2021**, *80*, 473–510. [CrossRef]
77. Zhuge, C.; Dong, C.; Wei, B.; Shao, C. Exploring the role of technology innovations in the diffusion of electric vehicle with an agent-based spatial integrated model. *Resour. Conserv. Recycl.* **2021**, *174*, 105806. [CrossRef]
78. IEA. *Global EV Outlook 2024*; IEA: Paris, France, 2024. Available online: https://iea.blob.core.windows.net/assets/a9e3544b-0b12-4e15-b407-65f5c8ce1b5f/GlobalEVOutlook2024.pdf (accessed on 20 November 2024).

79. Ehsani, M.; Gao, Y.; Longo, S.; Ebrahimi, K. *Modern Electric, Hybrid Electric, And Fuel Cell Vehicles*, 3rd ed.; CRC Press: Boca Raton, FL, USA, 2018.
80. Fang, T.; Vairin, C.; von Jouanne, A.; Agamloh, E.; Yokochi, A. Review of Fuel-Cell Electric Vehicles. *Energies* **2024**, *17*, 2160. [CrossRef]
81. Trencher, G.; Wesseling, J. Roadblocks to fuel-cell electric vehicle diffusion: Evidence from Germany, Japan and California. *Transp. Res. Part D Transp. Environ.* **2022**, *112*, 103458. [CrossRef]
82. Link, S.; Stephan, A.; Speth, D.; Plötz, P. Rapidly declining costs of truck batteries and fuel cells enable large-scale road freight electrification. *Nat. Energy* **2024**, *9*, 1032–1039. [CrossRef]
83. MEMR. *Handbook of Energy and Economic Statistics of Indonesia 2024*; MEMR: Jakarta, Indonesia, 2025. Available online: https://www.esdm.go.id/assets/media/content/content-handbook-of-energy-and-economic-statistics-of-indonesia-2024.pdf (accessed on 13 June 2025).
84. Pertamina. Milestone PT Pertamina (Persero). 2020. Available online: https://www.pertamina.com/en/milestone-pt-pertamina-persero (accessed on 7 July 2020).
85. Sunarya, W.; Taufik, G.A. *Introduction to Indonesian Oil and Gas Law*; Wibowo & Rekan Law Firm: Jakarta, Indonesia, 2018.
86. Pertamina Geothermal Energy. PT Pertamina Geothermal Energy Profile. 2019. Available online: https://www.pge.pertamina.com/en/energizing-green-future (accessed on 7 July 2020).
87. BPS. *Development of the Number of Motor Vehicles by Type (Unit), 2015–2022*; BPS: Jakarta, Indonesia, 2024. Available online: https://www.bps.go.id/assets/statistics-table/2/NTcjMg==/perkembangan-jumlah-kendaraan-bermotor-menurut-jenis{-}{-}unit-.html (accessed on 29 July 2025).
88. OECD/IEA. *Energy Policies Beyond IEA Countries: Indonesia 2015*; OECD/IEA: Paris, France, 2015. Available online: https://www.oecd.org/content/dam/oecd/en/publications/reports/2015/02/energy-policies-beyond-iea-countries-indonesia-2015_g1g4f6ea/9789264065277-en.pdf (accessed on 12 April 2025).
89. OECD-FAO. OECD-FAO Agricultural Outlook 2017–2026 Paris: OECD-FAO. 2024. Available online: https://stats.oecd.org/Index.aspx?DataSetCode=HIGH_AGLINK_2017# (accessed on 5 March 2024).
90. CIFOR. Bioenergy Development in Indonesia: Opportunities and Challenges for Biodiesel Industry Policy: CIFOR. 2018. Available online: https://www.cifor.org/publications/pdf_files/WPapers/WP242Dharmawan.pdf (accessed on 21 May 2020).
91. DEN. *Indonesia Energy Outlook 2024*; DEN: Jakarta, Indonesia, 2024. Available online: https://www.esdm.go.id/assets/media/content/content-laporan-kinerja-sekjen-den-tahun-2024.pdf (accessed on 4 March 2025).
92. GAIKINDO. *Hydrogen Car Prices in Indonesia Remain High*; GAIKINDO: Jakarta, Indonesia, 2024. Available online: https://www.gaikindo.or.id/gaikindo-mobil-hidrogen-di-ri-harganya-masih-tinggi/ (accessed on 6 October 2025).
93. APROBI. Finding the Ideal Biofuel Diversification in Indonesia Jakarta: APROBI. 2022. Available online: https://www.aprobi.or.id/mencari-diversifikasi-bahan-bakar-nabati-yang-ideal-di-indonesia/ (accessed on 29 July 2025).
94. Indonesia Investments. Palm Oil 2017. Available online: https://www.indonesia-investments.com/id/bisnis/komoditas/minyak-sawit/item166 (accessed on 14 July 2020).
95. DDTCNews. Mandatory Biodiesel 40 Percent to Take Effect in January 2025, Could Save Foreign Exchange Jakarta: DDTCNews. 2025. Available online: https://news.ddtc.co.id/berita/nasional/1808045/mandatory-biodiesel-40-persen-berlaku-januari-2025-bisa-hemat-devisa (accessed on 17 March 2025).
96. Kontan. Market Test for 5% Bioethanol Content Fuel (E5) Will Be Conducted Mid-Year Jakarta: Kontan. 2023. Available online: https://newssetup.kontan.co.id/news/uji-pasar-bbm-kandungan-bioethanol-5-e5-bakal-di-lakukan-pada-tengah-tahun-ini (accessed on 17 May 2025).
97. MEMR. *Fuel to Gas Diversification in the Transport Sector*; MEMR: Jakarta, Indonesia, 2013. Available online: https://www.migas.esdm.go.id/post/Diversifikasi-BBM-ke-Bahan-Bakar-Gas-Pada-Sektor-Transportasi (accessed on 12 June 2023).
98. IEA. *Final Consumption of Gas by Sector in Indonesia*; IEA: Paris, France, 2023.
99. Republika. Three Challenges Facing by CNG Cars in Indonesia Jakarta: Republika. 2013. Available online: https://republika.co.id/berita/otomotif/mobil/13/12/19/my1x6d-tiga-kendala-mobil-cng-di-indonesia (accessed on 8 May 2025).
100. CNBC Indonesia. The Sad Story of Electric Cars Made in Indonesia: A Dream Unfulfilled! Jakarta: CNBC Indonesia. 2023. Available online: https://www.cnbcindonesia.com/news/20230206065851-4-411170/kisah-sedih-mobil-listrik-buatan-ri-mimpi-tak-kesampaian (accessed on 20 March 2025).
101. BCG. Electrifying Indonesia's Two-Wheeler Industry Jakarta: BCG AEML. 2022. Available online: https://www.bcg.com/publications/2022/electrifying-indonesias-two-wheeler-industry (accessed on 2 February 2024).
102. Tempo. Outlook 2025: Awaiting the Development of Electric Vehicle Infrastructure Jakarta: Tempo. 2025. Available online: https://www.tempo.co/ekonomi/outlook-2025-menanti-perkembangan-infrastruktur-kendaraan-listrik-1196867 (accessed on 14 February 2025).

103. Kompas. The Government Has Not yet Provided Incentives for Hybrid Cars, Says An Observer Jakarta: Kompas. 2024. Available online: https://otomotif.kompas.com/read/2024/07/30/150100515/pemerintah-belum-beri-insentif-mobil-hybrid-ini-kata-pengamat (accessed on 17 May 2025).
104. PLN. 2024 Annual Report Jakarta: PLN. 2025. Available online: https://web.pln.co.id/statics/uploads/2025/07/AR-PLN-2024-1507.pdf (accessed on 28 July 2025).
105. CNN Indonesia. Toyota Indonesia Builds Hydrogen Charging Station in Karawang Jakarta: CNN Indonesia. 2025. Available online: https://www.cnnindonesia.com/otomotif/20250211125054-603-1196944/toyota-indonesia-bangun-stasiun-pengisian-hidrogen-di-karawang (accessed on 22 September 2025).
106. PGE. Pertamina Unveils Modern Electrolysis Technology for Green Hydrogen Production in Ulubelu. Jakarta: PGE2025. 2025. Available online: https://www.pge.pertamina.com/en/press-release/pertamina-unveils-modern-electrolysis-technology-for-green-hydrogen-production-in-ulubelu (accessed on 22 September 2025).
107. PGE. Supporting Decarbonization, PGE and Pertagas Strengthen Synergy for Clean Energy through Green Hydrogen Study. Jakarta: PGE. 2025. Available online: https://www.pge.pertamina.com/en/press-release/supporting-decarbonization-pge-and-pertagas-strengthen-synergy-for-clean-energy-through-green-hydrogen-study (accessed on 22 September 2025).
108. *Hyundai Motor Group Presents Action Plans for Waste-To-Hydrogen Ecosystem in Indonesia*; Hyundai: Jakarta, Indonesia, 2025.
109. Jia, C.; Liu, W.; He, H.; Chau, K.T. Health-conscious energy management for fuel cell vehicles: An integrated thermal management strategy for cabin and energy source systems. *Energy* **2025**, *333*, 137330. [CrossRef]
110. Huang, X.; Zhang, J.; Ou, K.; Huang, Y.; Kang, Z.; Mao, X.; Zhou, Y.; Xuan, D. Deep reinforcement learning-based health-conscious energy management for fuel cell hybrid electric vehicles in model predictive control framework. *Energy* **2024**, *304*, 131769. [CrossRef]
111. Sellali, M.; Ravey, A.; Betka, A.; Kouzou, A.; Benbouzid, M.; Djerdir, A.; Kennel, R.; Abdelrahem, M. Multi-Objective Optimization-Based Health-Conscious Predictive Energy Management Strategy for Fuel Cell Hybrid Electric Vehicles. *Energies* **2022**, *15*, 1318. [CrossRef]
112. Jia, C.; Liu, W.; He, H.; Chau, K.T. Superior energy management for fuel cell vehicles guided by improved DDPG algorithm: Integrating driving intention speed prediction and health-aware control. *Appl. Energy* **2025**, *394*, 126195. [CrossRef]
113. Zhang, Y.; Zhang, C.; Fan, R.; Deng, C.; Wan, S.; Chaoui, H. Energy management strategy for fuel cell vehicles via soft actor-critic-based deep reinforcement learning considering powertrain thermal and durability characteristics. *Energy Convers. Manag.* **2023**, *283*, 116921. [CrossRef]
114. Indira Ayu, A.; Bambang, P.; Santoso, M.S.; Mohamed Yusuf Faridian, W.; Liane, O.; Prawira Fajarindra, B. Battery Electric Vehicle Adoption in Indonesia: Insights from Consumer Preferences and Stakeholder Perspectives. *J. Civ. Eng. Forum* **2025**, *11*, 334–350. [CrossRef]
115. PwC. PwC: More Indonesians Interested in Buying Electric Vehicles Jakarta: PwC. 2025. Available online: https://www.pwc.com/id/en/media-centre/pwc-in-news/2023/indonesian/pwc-makin-banyak-orang-indonesia-minat-beli-kendaraan-listrik.html#:~:text=Angkanya%20pun%20lebih%20tinggi%20dibandingkan,72%25%20responden%20merupakan%20generasi%20milenial (accessed on 6 October 2025).
116. Pierson, P. Increasing returns, path dependence, and the study of politics. *Am. Political Sci. Rev.* **2000**, *94*, 251–268. [CrossRef]
117. Geels, F.W.; Turnheim, B. *The Great Reconfiguration: A Socio-Technical Analysis of Low-Carbon Transitions in UK Electricity, Heat, and Mobility Systems*; Cambridge University Press: Cambridge, UK, 2022.
118. IEA. *World Energy Outlook 2023*; IEA: Paris, France, 2023. Available online: https://iea.blob.core.windows.net/assets/86ede39e-4436-42d7-ba2a-edf61467e070/WorldEnergyOutlook2023.pdf (accessed on 12 February 2024).
119. Sembiring, M. Critical Geopolitics of Global Environmental Norms: Exploring Indonesia's Response to Climate Mitigation Agenda. *Geopolitics* **2025**, *30*, 1–24. [CrossRef]
120. Geels, F.W. Disruption and low-carbon system transformation: Progress and new challenges in socio-technical transitions research and the Multi-Level Perspective. *Energy Res. Soc. Sci.* **2018**, *37*, 224–231. [CrossRef]
121. Astari, A.J.; Lovett, J.C. Does the rise of transnational governance 'hollow-out'the state? Discourse analysis of the mandatory Indonesian sustainable palm oil policy. *World Dev.* **2019**, *117*, 1–12. [CrossRef]
122. Choiruzzad, S.A.B. Save Palm Oil, Save the Nation: Palm Oil Companies and the Shaping of Indonesia's National Interest. *Asian Politics Policy* **2019**, *11*, 8–26. [CrossRef]
123. Wibisono, H.; Lovett, J.C.; Anindito, D.B. The contestation of ideas behind Indonesia's rural electrification policies: The influence of global and national institutional dynamics. *Dev. Policy Rev.* **2023**, *41*, e12650. [CrossRef]
124. Qadir, S.A.; Ahmad, F.; Al-Wahedi, A.M.A.B.; Iqbal, A.; Ali, A. Navigating the complex realities of electric vehicle adoption: A comprehensive study of government strategies, policies, and incentives. *Energy Strategy Rev.* **2024**, *53*, 101379. [CrossRef]
125. IEA Bioenergy. Annual Report 2023 Paris: IEA Bioenergy. 2024. Available online: https://www.ieabioenergy.com/wp-content/uploads/2024/06/Annual-Report-2023.pdf (accessed on 20 July 2025).

126. Giraudo, M.E.; Grugel, J. Imaginaries of Soy and the Costs of Commodity-led Development: Reflections from Argentina. *Dev. Change* **2022**, *53*, 796–826. [CrossRef]
127. Farobie, O.; Hartulistiyoso, E. Palm oil biodiesel as a renewable energy resource in Indonesia: Current status and challenges. *BioEnergy Res.* **2022**, *15*, 93–111. [CrossRef]
128. Xu, W.; Mo, W. Institutional unlocking or technological unlocking? The logic of carbon unlocking in the new energy vehicle industry in China. *Energy Policy* **2024**, *195*, 114369. [CrossRef]
129. de Vos, R.E.; Suwarno, A.; Slingerland, M.; van der Meer, P.J.; Lucey, J.M. Pre-certification conditions of independent oil palm smallholders in Indonesia. Assessing prospects for RSPO certification. *Land Use Policy* **2023**, *130*, 106660. [CrossRef]
130. Panoutsou, C.; Germer, S.; Karka, P.; Papadokostantakis, S.; Kroyan, Y.; Wojcieszyk, M.; Maniatis, K.; Marchand, P.; Landalv, I. Advanced biofuels to decarbonise European transport by 2030: Markets, challenges, and policies that impact their successful market uptake. *Energy Strategy Rev.* **2021**, *34*, 100633. [CrossRef]
131. Suhara, A.; Karyadi; Herawan, S.G.; Tirta, A.; Idris, M.; Roslan, M.F.; Putra, N.R.; Hananto, A.L.; Veza, I. Biodiesel Sustainability: Review of Progress and Challenges of Biodiesel as Sustainable Biofuel. *Clean Technol.* **2024**, *6*, 886–906. [CrossRef]
132. Mosquéra, L.R.; de Oliveira, M.N.; Martins, P.H.d.S.; Bispo, G.D.; Borges, R.V.; Serrano, A.L.M.; Pompermayer, F.M.; Neumann, C.; Gonçalves, V.P.; Bork, C.A.S. Biofuel Dynamics in Brazil: Ethanol–Gasoline Price Threshold Analysis for Consumer Preference. *Energies* **2024**, *17*, 5265. [CrossRef]
133. Agrawal, R.; Agrawal, S.; Samadhiya, A.; Kumar, A.; Luthra, S.; Jain, V. Adoption of green finance and green innovation for achieving circularity: An exploratory review and future directions. *Geosci. Front.* **2024**, *15*, 101669. [CrossRef]
134. Sadiq, M.; Le-Dinh, T.; Tran, T.K.; Chien, F.; Phan, T.T.H.; Quang Huy, P. The role of green finance, eco-innovation, and creativity in the sustainable development goals of ASEAN countries. *Econ. Res. Ekon. Istraz.* **2023**, *36*, 2175010. [CrossRef]

MDPI AG
Grosspeteranlage 5
4052 Basel
Switzerland
Tel.: +41 61 683 77 34

Energies Editorial Office
E-mail: energies@mdpi.com
www.mdpi.com/journal/energies

www.ingramcontent.com/pod-product-compliance
Lightning Source LLC
Chambersburg PA
CBHW061442210926
51850CB00008B/1583
* 9 7 8 3 7 2 5 8 6 2 3 4 4 *